ELEMENTS OF
PETROLEUM
PROCESSING

ELEMENTS OF
PETROLEUM
PROCESSING

D. S. J. Jones
Consultant Chemical Engineer, Calgary, Canada

JOHN WILEY & SONS

Chichester · New York · Brisbane · Toronto · Singapore

Copyright ©1995 by John Wiley & Sons Ltd,
Baffins Lane, Chichester,
West Sussex PO19 1UD, England
Telephone National (01243) 779777
International +44 1243 779777

Other Wiley Editorial Offices

John Wiley & Sons, Inc., 605 Third Avenue,
New York, NY 10158-0012, USA

Jacaranda Wiley Ltd, 33 Park Road, Milton,
Queensland 4064, Australia

John Wiley & Sons (Canada) Ltd, 22 Worcester Road, Rexdale,
Ontario M9W 1L1, Canada

John Wiley & Sons (SEA) Pte Ltd, 37 Jalan Pemimpin #05-04,
Block B, Union Industrial Building, Singapore 2057

Library of Congress Cataloging-in-Publication Data

Jones, D. S. (D. Stan)
    Elements of petroleum processing / D. S. Jones.
        p.   cm.
    Includes index.
    ISBN 0-471-95254-0
    1. Petroleum—Refining.   I. Title.
TP690.J615   1995   95-19286
665.5′3—dc20   CIP

British Library Cataloguing in Publication Data

A catalogue record for this book is available from the British Library

ISBN 0 471 95254 0

Typeset in 10/12pt Times by Dobbie Typesetting Limited, Tavistock, Devon
Printed and bound in Great Britain by Bookcraft (Bath) Ltd, Midsomer-Norton, Avon

CONTENTS

LIST OF FIGURES

PREFACE

Much of this book was assembled and compiled by myself into a reference manual for an African oil refining company. The work was in response to a brief by which the client company required a document to serve as a book of reference encompassing basic process calculation techniques and rules of thumb relating to the refining processes. The idea was to improve the knowledge and expertise of the refinery's technical and engineering staff in the process performance evaluation and operation of the refinery. Although the refining company had up-to-date computerized process simulation packages its more junior staff had some difficulty inputting these programs due to their limited refinery process experience. The manual was also meant to improve this situation.

The completion and the successful use of the manual over the past few years has prompted the writing of *Elements of Petroleum Processing*. Some changes and additions have been made to the original work to provide a textbook which is oriented not only to refinery operation but also, in many instances, to basic design practices and techniques. The entire work concerns only the energy-related Refinery and does not include those data, correlations, etc. met with in lube oil and petrochemical oil refineries.

This book is divided into three parts, Part 1 includes a brief description of the chemical make-up of crude oil, the various types of crude and their products. It continues with the rules of thumb and techniques used to define product quality and basic correlation. The part also includes a brief description of the more common processes found in an energy-related refinery and makes reference also to the more common non energy refineries. Finally, it provides a simple example of how Part 1 is used to define and establish a refinery process configuration. Part 2 is the first of two parts that introduce the reader to detailed calculation techniques and data concerning the refinery processes. It encompasses those techniques used in the design and operation of the primary crude atmospheric and vacuum distillation plants. It follows with a similar treatment relating to the fractionation of the light fractions of the crude oil. Each chapter of this part includes a description of the process unit in some detail and provides related calculations in stepwise form. A worked example of the calculation topic completes each chapter. Part 3 covers a selection of those units often included in refinery configurations which upgrade low-grade products. This is accomplished by converting them to a different chemical composition. Each selected process is the subject of the individual chapters contained in this part. The format of these chapters follows closely that of Part 2. However, in some cases in this part there is a more detailed discussion covering the reaction chemistry and design techniques for the processes described.

Throughout this work the data and techniques used have been well proven over years of use in design and operation before the days of modern high-speed computer programmes covering the same subject. Much of the contents are taken from personal files and experience over a period of 42 years in or associated with the oil industry. This experience spans the years spent in refinery operations, process development and process engineering with a major international engineering and construction company and, more recently, as a consultant engineer.

Stan Jones
Calgary, Alberta

Part 1

AN INTRODUCTION TO CRUDE OIL
AND ITS PROCESSING

1 CRUDE OIL AND ITS PRODUCTS

Oil refining and the development of the internal combustion engine have continued since the latter part of the nineteenth century. Each have influenced the other in the course of this development, but it is the oil industry and its effect on almost every part of modern living that has been the most spectacular. This part of the book offers an introduction to this industry in the form of outlining the nature of crude oil, its products and the manner in which it is processed to arrive at those products that have become a part of today's living. The purpose of these introductory chapters is also to describe and define the basic measures and parameters used in the industry. These set the stage for the more detailed examination of those refinery processes provided in subsequent chapters.

1.1 Composition

Crude oil is a mixture of literally hundreds of hydrocarbon compounds ranging in size from the smallest, methane, with only one carbon atom, to large compounds containing 200 or more. A major portion of these compounds are paraffins or isomers of paraffins. These are straight-chain hydrocarbon compounds such as butane, i.e.:

Normal butane
(denoted simply as nC_4)

ISO butane
(denoted simply as iC_4)

The remaining hydrocarbon compounds are either cyclic paraffins called naphthenes or aromatics. Below are examples of those compounds:

Cyclohexane (naphthene)

Benzene (aromatics)

These families of hydrocarbons are called homologues, and because of the large quantity of these compounds which exist in crude oil, only the simplest of compounds in the homologues can be isolated to some degree of purity on a commercial scale. Generally, in refining projects, isolation of comparatively pure products are restricted to those compounds lighter than C_7 s. Most compounds, however, have been isolated and identified but only under strict and delicate laboratory conditions.

In the industry the products are restricted to groups of these hydrocarbons boiling between selective boiling ranges. Thus products are usually denoted by boiling points rather than by the compounds they contain.

Not all compounds contained in crude oil are pure hydrocarbons. The crude does contain certain inorganic impurities such as sulphur, nitrogen and metals. By far the most common of these impurities are the organic sulphur compounds called mercaptans. These compounds are similar to hydrocarbons but with the addition of one or more sulphur atoms, e.g.:

Ethyl mercaptan

More complicated sulphur compounds also exist such as thiophenes in the high carbon number range homologues and disulphides in the middle distillate range. Because of the close relationship that exists between these sulphur compounds and neighbouring hydrocarbons in such characteristics as vapour pressures, the compounds cannot be isolated by distillation processes on a commercial scale.

Organic chloride compounds are not usually removed from the crude as products but the corrosive effect attributed to these compounds on parts of refinery plants is always a source of concern. A considerable amount of investigation has been, and is still being, carried out on prevention of corrosion by HCl generated in the process from these organic chlorides.

The metals contained in crude are usually nickel, sodium and vanadium. Because these are not very volatile they are found in the heavier products of crude such as fuel oil. They only become a nuisance in certain cases when they can affect further processing of the oil, or if they exist in such large quantities in fuel oil as to limit the saleable value of the product. Their removal can be effected in such cases by an extraction process where they are removed as a part of a bituminous extract called asphaltenes.

Before leaving the subject of crude oil composition it is necessary to mention yet another series of organic compounds which, although not always present in the original crude, are often formed during the refining processes. These are unsaturated or double-bond compounds. As mentioned earlier, most compounds contain carbon and hydrogen and are so made up that each carbon atom contains its full quota of four atoms, in the form of either hydrogen or carbon. However, some of these molecules break down under conditions of high temperature and permanently lose one or more of those attached atoms. Such compounds are unstable and readily combine with themselves or other similar compounds to form polymers. They are called unsaturates while the stable hydrocarbons such as paraffins, cycloparaffins and aromatics are called saturated hydrocarbons. An example of such a compound is ethylene, which has the following formula:

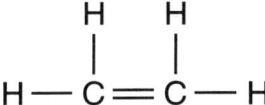

Note the double-bond linking the two carbon atoms.

1.2 Types of Crude Oil

Crude oil found in various parts of the world can vary considerably in characteristics. This can readily be seen in the difference in specific gravity that exists between the crudes. For instance, Zelten crude (Libyan) has an API gravity of 39.0 while Bachequero crude has a gravity of 16.0. This difference is due to the fact that although each crude oil contains basically the same hydrocarbon compounds, the proportion of these hydrocarbon vary considerably from one crude to another. Thus some crudes are relatively rich in paraffins, and this is reflected by the waxy nature of the crude (most Middle East crudes fall into this group). Others contain more cycloparaffins and aromatics (such as Nigerian and some American West Coast crudes).

Although it is theoretically possible to produce any type of refined product from any crude, it is not usually economically feasible to do so. For instance, better yields of reformer stock for aromatics production is obtained from Nigerian crude than from Kuwait, while considerably more residium for fuel oil is obtained from Kuwait than Nigeria. To satisfy the demand of these two products, refineries often blend two such devious crudes, changing the proportion of the blend to satisfy the particular need. If product demands are seasonal, as in the case of many gas oils, imports of specifically selected crudes are scheduled to optimize the production of such cuts for the season.

Selectivity of crude type is comparatively easy to effect and is common practice in the refining industry where most oil companies have their own oil wells in various parts of the world. However, occasionally the demand for a particular product or the elimination of a low-value one becomes so persistent that major conversions of a chemical nature have to be adopted. Such processes as thermal and catalytic cracking and, more recently, hydrocracking are used to effect this. These, as processes, are discussed in detail in later chapters of this book.

1.3 Products from Crude Oil

In general, the products that are normally obtained from crude oil can be grouped as follows:

(1) *Volatile products* (these are the lightest products)
● Propane LPG (liquefied petroleum gas)
● Butane LPG
● Light naphtha (C_5s and nC_6)
(2) *Light distillates*
● Gasolines
● Heavy naphtha (petrochem naphtha and base for paints)
● Kerosene and jet fuels
(3) *Middle distillates*
● Automotive diesel
● Heating oils
● Gas oils
(4) *Fuels oils*
● Marine diesel
● Bunker fuels (for ships)
(5) *Lubricating oils*
● Motor
● Spindle
● Machine oils
(6) *Waxes*
● Food and paper coating grade
● Pharmaceutical grade
(7) *Bitumen*
● Asphalt
● Coke

Products in these groups are produced from distillation processes and treated to meet certain specifications. These specifications are the result of a compromise between desirable performance characteristics in the product and the ability to make such products from the crude and the processing facilities at hand. Wide ranges in physical properties are often tolerated in these specifications in order to cater for crude oil source, sales area and the product's ultimate use.

2 THE CRUDE ASSAY

The crude assay is a compilation of laboratory and pilot plant data that defines the properties of the specific crude oil. At a minimum the assay should contain a true boiling point curve and a specific gravity curve for the crude oil. However, most crude oil suppliers extend the scope of the assay to include sulphur contents, viscosity, pour points and many other properties.

Figures A2.1 to A2.5 in appendix 2 are typical curves taken from a Kuwait crude assay. The character of the crude has changed slightly since these curves were developed, so more recent data must now be used for definitive design work. These curves are:

Figure A2.1 is a plot of the true boiling point of the crude against total distillate recovery (the TBP curve).

Figure A2.2 is a plot of specific gravity against mid-boiling points of the crude oil fractions.

Figure A2.3 shows the sulphur content of the crude oil fractions against their mid-boiling point.

Figure A2.4 is a similar plot of the freezing, pour, and cloud points of the crude fractions against their mid-boiling point.

Figure A2.5 is a plot of the viscosity of the crude fractions against their mid-boiling points. These are plotted at 70 °F, 100 °F, 120 °F and 210 °F.

2.1 Assay Data and Laboratory Tests

To understand fully the significance of these assay curves and data it is necessary to define and explain some of the basic laboratory tests and data used to compile the assay. These are as follows:

● *TBP curve (true boiling point curve)*   This is a plot of the boiling points of almost pure components, contained in the crude oil or fractions of the crude oil. It is accomplished under laboratory conditions using a complex batch-distillation apparatus of a hundred or more equilibrium stages and with a large volume of reflux. A typical TBP curve for a whole crude is shown as Figure A2.1. Initially the temperature goes up in well-defined steps representing single components. However, as the temperature rises, the number of close boiling point components increases and the steps become increasingly less distinct until they merge into a smooth curve. This test is rarely used on a routine basis.

- *ASTM distillation*   This is carried out in a simple apparatus designed to boil the test liquid and to condense the vapours as they are generated. Vapour temperatures are noted as the distillation proceeds and are plotted against the distillate recovered. Because only one equilibrium stage is used and no reflux is returned, the separation of components is poor and mixtures are distilled. The initial boiling point is higher than that for the TBP and the final boiling point is lower. ASTM distillations are employed as routine tests to measure quality of refinery products.
- *Equilibrium flash vaporization* (EFV)   When a mixture is heated without allowing the vapour to separate from the remaining liquid the vapour assists in causing the high boiling parts of the mixture to vaporize. In almost all commercial plant operations involving change of liquid vapour phase continuous flash vaporization occurs. The EFV curve of an oil is determined in a laboratory using an apparatus which confines liquid and vapour together until the required degree of vaporization is achieved. The percentage vaporized is plotted against temperature for several runs to give the EFV curve. Separation is poorer for this type of distillation than for an ASTM, therefore the initial point will be higher than the ASTM but final boiling point will be lower. Empirical correlations exist to convert from TBP ASTM and EFV.
- *API gravity*   This is an expression of the density of an oil. Unless otherwise stated, the API gravity refers to density at 60 °F. Its relationship with specific gravity is given by the expression degrees API = (141.5 − 131/SP GR).
- *Reid vapour pressure* (RVP)   This is a laboratory test used to determine the vapour pressure of natural gasoline at 100 °F. The true vapour pressure is about 5–9% higher than the figure obtained by this test.
- *Flash points*   The flash point of an oil is the temperature at which the vapour above the oil will momentarily flash or explode. This temperature is determined by laboratory testing using an apparatus consisting of a closed cup containing the oil, heating and stirring equipment, and a special adjustable flame. The type of apparatus used for middle distillate oils and fuels is called the Pensky Marten (PM) apparatus while the apparatus used for kerosene and light distillates is called the Abbel. Full details of test method and description of the equipment is given in the ASTM standards, Part 7 Petroleum Products and Lubricants. This test serves to indicate the temperature below which an oil can be handled safely. There are many empirical methods for determining flash point from the ASTM distillation curve. One such correlation is given by the expression.

Flash point °F = 0.77 (ASTM 5% °F–150 °F).
- *Octane number*   This is a measure of a gasoline's resistance to knock or detonation in a cylinder of a petroleum engine. The higher this resistance, the higher is the efficiency of the fuel to produce work. There is also a relationship between the anti-knock characteristic of a gasoline (octane number) and the compression ratio

of the engine in which it is to be used. The higher the octane number, the higher the compression ratio that can be used. By definition, the octane number of a gasoline is that percentage of Iso-octane in a blend of Iso-octane and normal heptane which exactly matches the knocking behaviours of the gasoline. Thus a 90-octane gasoline matches the knock characteristic of a blend containing 90% Iso-octane and 10% normal heptane.

The knock characteristics are determined in the laboratory using a standard single-cyclinder test engine which is equipped with a super-sensitive knock meter. The reference fuels (Iso_8 and $N\ C_7$ blend) are run and compared with a second run using the gasoline. Full details of this test is given in the ASTM standards, Part 7 Petroleum Products and Lubricants.

Two octane numbers are usually determined. The first is the research octane number (ON (res)) and the second is the motor octane number (ON (mm)). The same basic equipment is used for both but the engine speed for the motor method is higher than that for research. The actual octane number obtained in a commercial engine would be somewhere between these two. The significance of these two octane tests is to evaluate the sensitivity of the gasoline to the severity of operating conditions in the engine. Invariably the research octane number is higher than the motor octane, the difference between them being quoted as 'the sensitivity of the gasoline'.

● *Viscosity*   The viscosity of an oil is a measure of its resistance to internal flow and is an indication of its lubricating qualities. In the oil industry it is usual to quote viscosities in centistokes (which is the unit for kinematic viscosity), seconds Saybolt Universal, seconds Saybolt Furol or seconds Redwood. These units have been correlated and such correlations can be found in most data books. In the laboratory, test data on viscosity are usually determined at temperatures of 100 °F, 130 °F or 210 °F. In the case of fuel oils temperatures of 122 °F and 210 °F are used.

● *Cloud and pour points*   These are tests to indicate the relative waxiness of an oil. They do not, however, measure the actual wax content of the oil. The tests merely consist of reducing the temperature of the oil under controlled conditions. The temperature at which the oil becomes cloudy or hazy is taken as its cloud point, while the temperature at which the oil ceases to flow is taken as its pour point.

● *Sulphur content*   This is self-explanatory, and is usually the measure of total sulphur in the oil.

There are, of course, many other laboratory tests and data that define the qualities and characteristics of crude oil and its products. The ones listed above are among the most common, and other tests and definitions can be found in the appropriate volumes of ASTM Tests and the API manuals.

3 PSEUDO-COMPONENTS AND PREDICTING PRODUCT PROPERTIES

It is not possible to separate the components of crude oil into commercial processes but groups of these components are produced to meet the requirements of the various refinery products. These are called 'cuts' and each identified as cut ranges. Each of these 'cuts' can be further defined by dividing them up into pseudo- (not real) components and relating their properties (such as SG, viscosities, etc.) to comparable real and identified components. Some of the technology used and the method of developing pseudo-components are discussed below.

3.1 Terminology

- *Cut point*   A cut point is defined as that temperature on a whole TBP curve that represents the limits of a fraction to be produced. Consider a typical TBP curve shown in Figure 3.1.
  A fraction with a cut-off point of 100 °F represents a yield of 20% of the whole as that fraction. A boiling range is, more often than not, quoted as the yield between two cut points. Thus a boiling range (TBP) of 100–200 °F represent a fraction with a yield of 30% − 20% = 10% vol. on whole feed.
- *End points*   A cut point is an ideal temperature used to predict a yield of a fraction. When that fraction is actually produced commercially, however, its final TBP boiling point will be considerably higher than the quoted cut point. This is due to the inability of the process to separate perfectly the components of one fraction from those in the adjacent one. Its actual final boiling point is called the end point. Thus the relationship of the product TBP curves to the whole crude TBP appears as in Figure 3.1.
- *Mid-boiling point components*   In compiling the assay, narrow boiling fractions are distilled from the crude and are analysed to determine their properties.

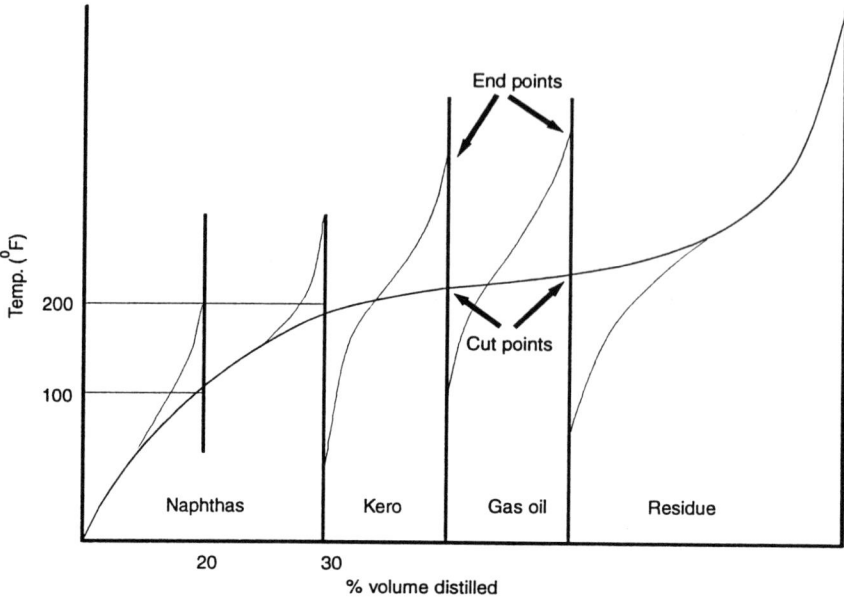

Figure 3.1. Definition of cut points and end points

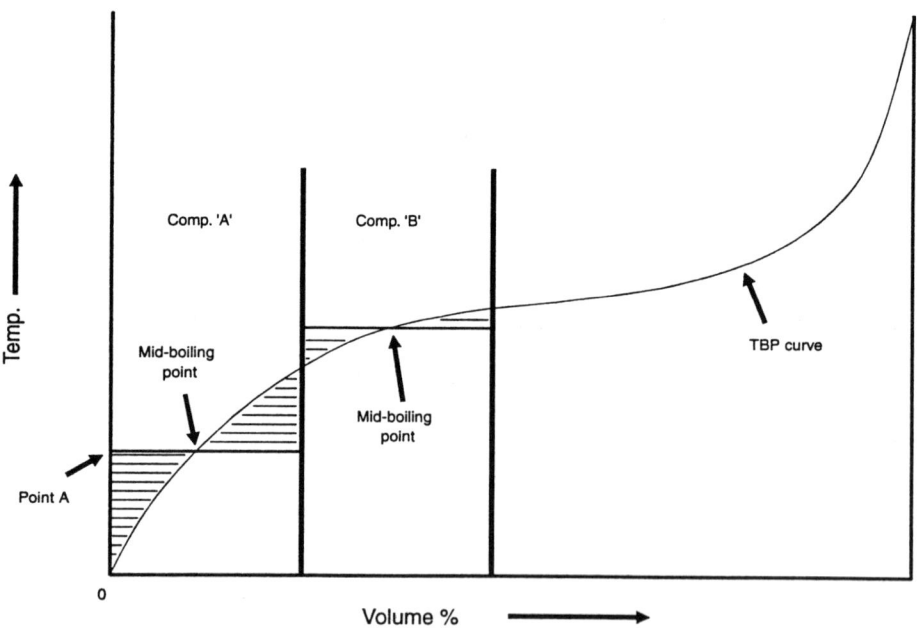

Figure 3.2. Definition of mid-boiling points

These are then plotted against the mid-boiling point of these fractions to produce a smooth correlation curve. To apply these curves for a particular calculation it is necessary to divide the TBP curve of the crude or fraction of the crude into mid-boiling point components. To do this consider Figure 3.2. For the first component take an arbitrary temperature point A. Draw a horizontal line through this point from the 0% volume. Extend this line until the area between the line and the curve on both sides of the temperature point A are equal. The length of this horizontal line measures the yield of component A having a mid-boiling point A °F. Repeat for the next component and continue until the whole is divided into these mid-boiling point components.

● *Mid-volume percentage point components* Sometimes the assay has been so constructed as to correlate crude oil properties against components on a mid-volume percentage base instead of mid-boiling point. In using such data as this it becomes necessary to divide the TBP curve into mid-volume point components. This is easier than the mid-boiling point concept and requires only that the curve is divided into a number of volumetric sections. The mid-volume point of each section is the arithmetical mid-point in this case.

● *Predicting TBP and ASTM curves from assay data* The properties of products can be predicted by constructing mid-boiling point components from a TBP curve and assigning the properties to each of these components. These assigned properties may be obtained from the assay data, known components of similar boiling points or established relationships such as gravity, molecular weights and boiling points. However, before these mid-boiling point (pseudo-) components can be developed it is necessary to know the shape of the product TBP curve. The following is a method by which this can be achieved.

Good, Connel et al (1) accumulated data to relate the ASTM end point to a TBP cut point over the light and middle distillate range of crude. Their correlation curves are given in Figure A1.3 in Appendix 1, and are self-explanatory. Thrift (2) derived a probable shape of ASTM data. The probability graph that he developed is given as Figure A1.4 in Appendix 1. The product ASTM curve from a well-designed unit would be a straight line from 0% vol. to 100% vol. on this graph. Using these two graphs it is possible to predict the ASTM distillation curve of a product knowing only its TBP cut range. This is shown in the following example.

Example Calculation

Calculate the ASTM curve for kerosene, cut between 387 °F and 432 °F cut points on Kuwait crude.

SOLUTION

Yield of crude	= 3.9% vol.
Cut range	= 27.3%–31.2% vol.
90% of TBP cut = 30.81% vol.	= 430 °F
From Figure A1.3. curve B ASTM EP = 432-13 °F = 419 °F	
From Figure A1.3. curve G ASTM 90% = 430-24 = 406 °F	

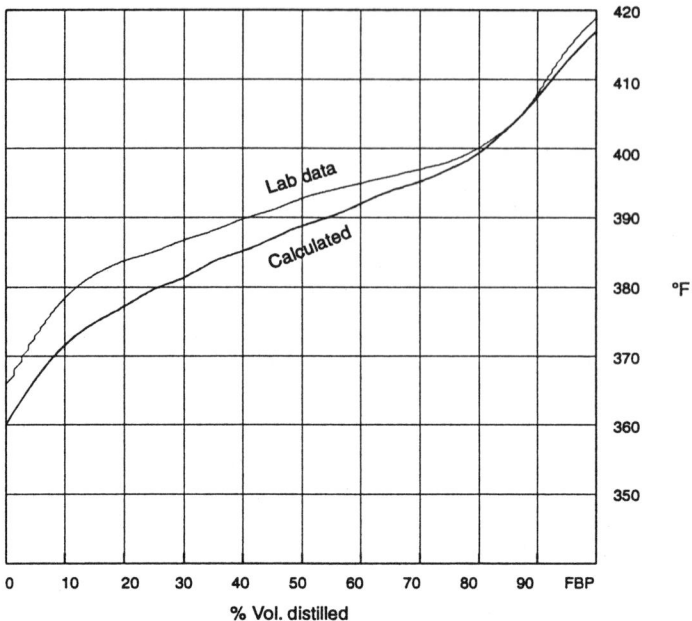

Figure 3.3. Comparison between calculated ASTM curve and laboratory data

These two points are plotted on Figure A1.4 and a straight line drawn through them to define the probable ASTM distillation of the cut. This is plotted linearly in Figure 3.3 and can be seen to compare well with laboratory results of the actual product from a crude distillation unit. The predicted ASTM curve may now be used to arrive at the TBP and EFV curves of the product by the methods described later in this book (see Section 6.3).

3.2 Predicting Product Properties

The following describes the prediction of product properties using pseudo-components (mid-boiling point) and assay data. A diesel cut with TBP cut points 432–595 °F on Kuwait crude will be used to illustrate these calculations. The actual TBP of this cut is predicted using the method already described. The curve is then divided into about six pseudo-mid-boiling point components as described earlier and is shown in Figure 3.4.

● *Predicting the gravity of the product* Using the mid-boiling point versus specific gravity curve from the assay given in Appendix 2 the SG for each component is obtained. The weight factor for each component is then obtained by multiplying the vol.% of that component by the specific gravity. The sum of the weight factors divided by the 100% volume total is the specific gravity of the gas oil cut.

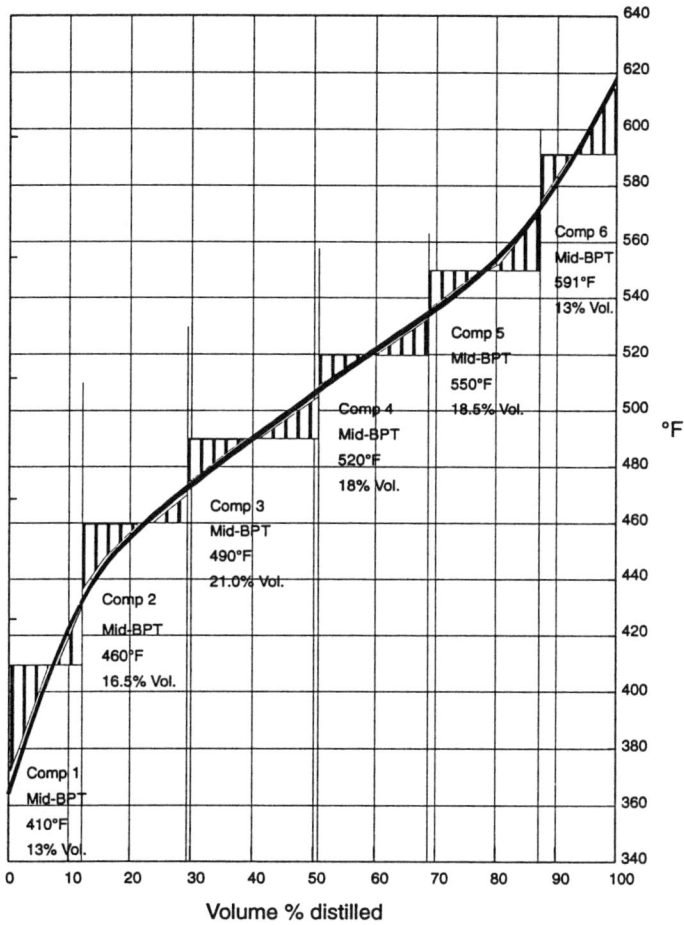

Figure 3.4. Pseudo-component breakdown by mid-boiling points

Component	Volume %	Mid-BPT (°F)	SG at 60 °F	Weight factor
1	13.0	410 (210 °C)	0.793	10.3
2	16.5	460 (238 °C)	0.801	13.2
3	21.0	490 (254 °C)	0.836	17.6
4	18.0	520 (271 °C)	0.844	15.2
5	18.5	550 (288 °C)	0.846	15.7
6	13.0	591 (311 °C)	0.850	11.1
Total	100.00	—	—	83.1

SG = 0.831 (actual plant data for this cut was 0.829)
°API = 38.8

● *The prediction of product sulphur content* The prediction of sulphur content is similar to the method used for specific gravity. First, the TBP curve for the product is determined and split into pseudo-mid-boiling point components. The weight factor is then determined for each component as before. (*Note:* sulphur content is always quoted as a percentage weight.) Using the relationship of percentage sulphur to mid-boiling point given in the assay (Appendix 2) the sulphur content of each component is read off. This is multiplied by the weight factor for each component to give a sulphur factor. The sum of the total sulphur factors divided by the total weight factor gives the wt% sulphur content of the fraction. For example, using the same gas oil cut as before, its sulphur content is determined as follows:

Component	Weight factor	Mid-BPT (°F)	Sulphur % wt	Sulphur factor
1	10.3	410 (210 °C)	0.20	2.06
2	13.2	460 (238 °C)	0.410	5.41
3	17.6	490 (254 °C)	0.840	14.78
4	15.2	520 (271 °C)	1.160	17.63
5	15.7	550 (288 °C)	1.350	21.20
6	11.1	591 (311 °C)	1.500	16.65
Total	83.1	—	—	77.73

Sulphur % weight = (77.7/83.1) × 100 = 0.935% wt (actual plant data gave 0.931% wt)

● *Viscosity prediction from the crude assay* Unlike sulphur contents and gravity, viscosities cannot be arithmetically related to components. To determine the viscosity of a blend of two or more components a blending index must be used. A graph of these indices are given in Maxwell's *Data Book on Hydrocarbons* (3), and part of this graph is reproduced as Figure 3.5. Using the blending index and having split the fraction into components as before, the viscosity of the fraction can be predicted as shown in the following example:

Component	Volume %	Mid-BPT (°F)	Viscosity CS at 100 °F	Blending index	Viscosity factor
	(A)			(B)	(A × B)
1	13.0	420 (210 °C)	1.49	63.5	825.5
2	16.5	460 (238 °C)	2.0	58.0	957.0
3	21.0	490 (254 °C)	2.4	55.0	1155.0
4	18.0	520 (271 °C)	2.9	52.5	945.0
5	18.5	550 (288 °C)	3.7	49.0	906.5
6	13.0	591 (311 °C)	4.8	46.0	598.0
Total	100.0	—	—	—	5387.0

Overall viscosity index $= \dfrac{5387.0}{100} = 53.87$

From Figure 3.5 viscosity = 2.65 cSt
(actual plant result was 2.7 cSt)

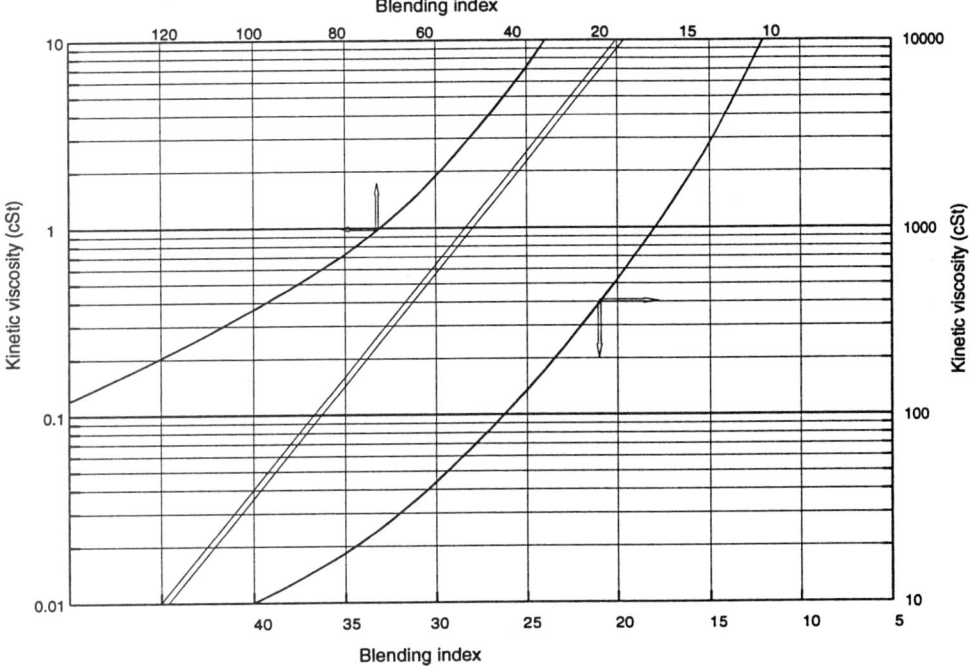

Figure 3.5. Viscosity blending chart (Maxwell, Data Book on Hydrocarbons, used by permission.)

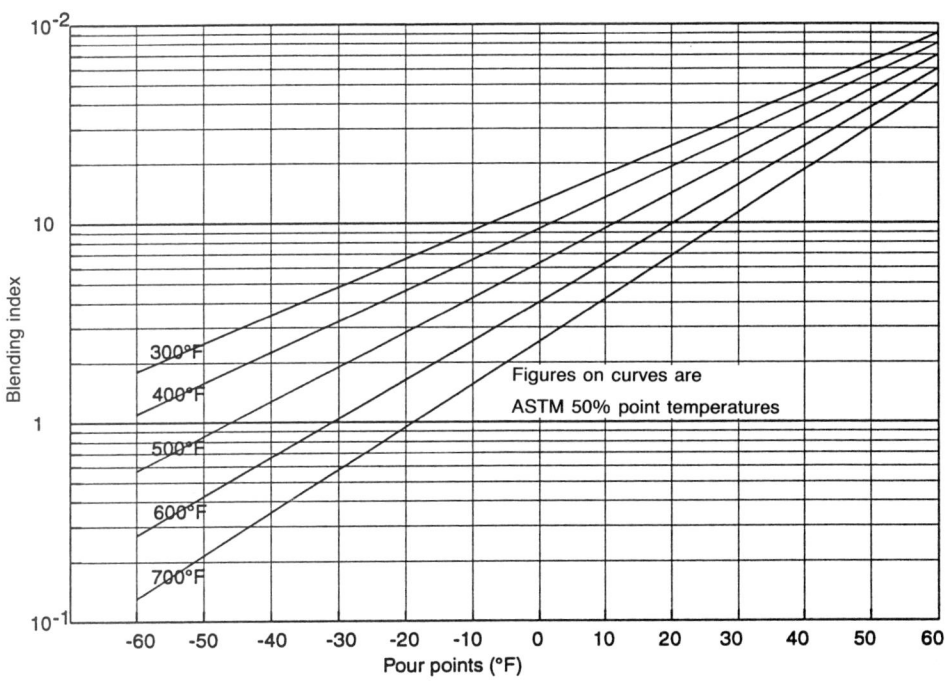

Figure 3.6. Pour point blending chart

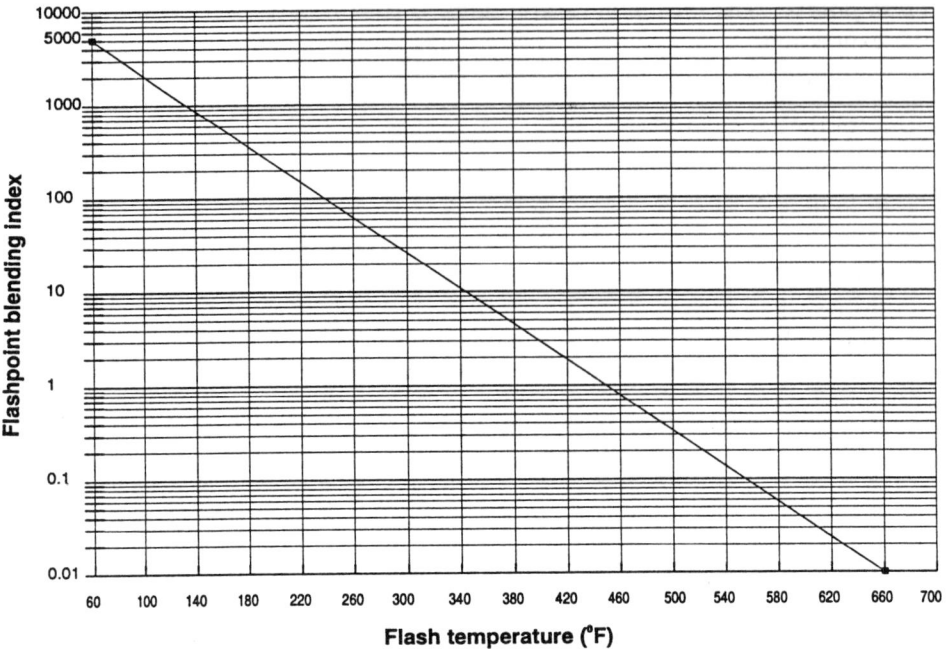

Figure 3.7. Flash point blending indexes can be combined linearly by volume (Reprinted by permission from Hydrocarbon Processing, June 1963, copyright Gulf Publishing Co all rights reserved.)

● *Cloud and pour points*   In predicting these properties it is not necessary to breakdown the product TBP as we have done for API, sulphur, etc. The accuracy of the tests and of blending indices does not warrant this. These properties are therefore read off directly from the mid-boiling point of the whole product. Considering the gas oil used in the previous example, its mid-boiling point is about 510 °F. From the crude assay its pour point is − 5 °F and cloud point is + 4 °F.

Determining pour point for a blend of two or more products is rather more difficult. In this case blending indices are used. A graph of these indices is given as Figure 3.6. This is self-explanatory and its application is explained by the following example: 2000 BPSD of a gas oil having a pour point of − 5 °F and ASTM distillation 50% point of 500 °F is to be blended with 4000 BPSD of a waxy distillate having a pour point of 30 °F and a 50% ASTM distillation point of 700 °F. The pour point of the final blend is predicted below:

Components	Composition BPSD	Composition Fraction	ASTM 50% 50%	ASTM 50% Factor	Pour Point Pour PT	Pour Point Index	Pour Point Factor
		(A)	(B)	(A×B)		(C)	(A×C)
Gas oil	2000	0.33	500	165	− 5	5.8	1.9
Waxy dist.	4000	0.67	700	469	+ 30	12.7	8.5
Total	6000	1.00	—	634	—	—	10.4

From the chart in Figure 3.6 the pour point corresponds to an index of 10.4 and ASTM 50% of 634 °F is 22 °F. This is the pour point of the blend.

● *Flash points*   The flash point of a product is related to its ASTM distillation by the expression:

Flash point = 0.77 (ASTM 5% in °F–150 °F).

Thus for the gas oil product in the Example (given below) the flash point is predicted as follows:

Flash point °F = 0.77 (420 – 150)
             = 208 °F

● *Blending products to meet a flash point specification*   As with pour points and viscosities, the flash point of a blend of two or more components is determined by using a flash blending index. Figure 3.7 gives these indices. Again the indices are blended linearly by volume as in the case of viscosities. Thus consider the following example. 2000 BPSD of kerosene with a flash point of 120 °F is to be blended with 8000 BPSD of fuel oil with a flash point of 250 °F. What will be the flash point of the final blend?

Solution

Component	Volume	Fraction	Flash (°F)	Flash index	Factor
		(A)		(B)	(A × B)
Kero	2000	0.2	120	310	62.0
Fuel	8000	0.8	250	5.5	4.4
Total	10 000	1.0	—	—	66.4

From Figure 3.7 the flash point corresponding to 66.4 index = *166 °F*
This is the final blend flash point

● *Predicting mole weights of products*   It is important to know the molecular weights of the hydrocarbon streams to be able to carry out many of the design or other calculations relating to refinery processes or operations. This is particularly so in distillation calculations and where chemical reactions occur. The term 'moles' refers to pound moles. A relationship exists between the mean average boiling point (MEABP), API and the molecular weight of petroleum fractions (4) and this is given in Figure 3.8 (see also Figure A1.5 in Appendix).

Example calculation

Referring again to the example gas oil fraction used earlier, the ASTM distillation curve corresponding to this fraction is as follows:

% vol.	°F
0	406
10	447
30	469
50	487
70	507
90	538
100	578

Slope of the curve is $\dfrac{t90\% - t10\%}{80}$

$$= \frac{538 - 447}{80} = 1.14\ °F/\%$$

Vol. average boiling point $= \dfrac{T_{10} + (2 \times T_{50}) + T_{90}}{4}$

$$= \frac{447 + 974 + 538}{4}$$

$$= 490$$

Mean average boiling point $= 490 + (-5\ °F)$. See top curve of Figure 3.8

$$= 485\ °F$$

°API is 38.8 as calculated earlier.

Mole weight of this fraction is *194* (From Figure 3.8)

A rigorous calculation method resulted in a mole weight of 201.

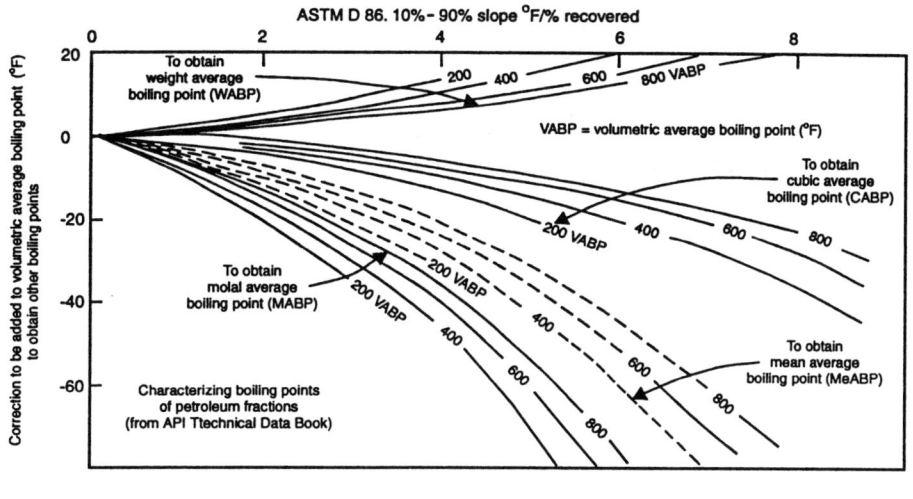

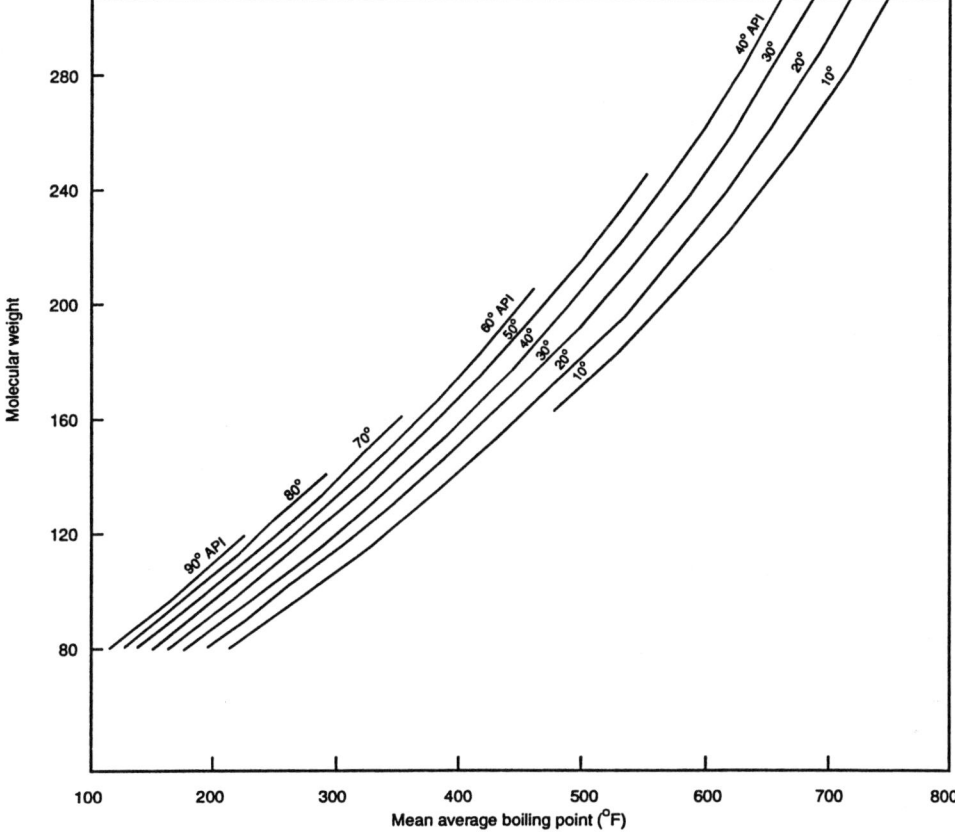

Figure 3.8. Correlations between MeABP, SG and mol. wt

4   BASIC PROCESSES

This chapter provides an introduction to some of the most common of the processes included in a fuel-oriented refinery. These processes are only discussed here in summary form but later in the book they are dealt with in some detail.

4.1   The Atmospheric Crude Distillation Unit

In refining the crude oil it is first broken up into those raw stocks that are the basis of the finished products. This break-up of the crude is achieved by separating the oil into a series of boiling point fractions which meet the distillation requirements and some of the properties of the finished products. This is accomplished in the crude distillation units. Normally there are two units that accomplish this splitting-up function:

● An atmospheric unit
● A vacuum unit

The crude oil first enters the atmospheric unit where it is desalted (dissolved brine is removed by washing) and heated to a predetermined temperature. This is accomplished by heat exchange with hot products and finally by a direct-fired heater. The hot and partially vaporized crude is 'flashed' in a trayed distillation tower.

Here the vapour rising in the tower is selectively condensed by cooled reflux streams and taken off at various parts of the tower as distillate side-streams. The light oils not condensed in the tower are taken off at the top of the tower to be condensed externally as the overhead product. The unvaporized portion of the feed leaves the bottom of the tower as the atmospheric residue.

The unit operates at a small positive pressure around 5–10 psig in the overhead drum, thus its title of 'atmospheric' crude unit. Typical product streams leaving the distillation tower are as follows:

Overhead distillate	*Full-range naphtha*	*Gas to 380 °F cut point*
1st side-stream	Kerosene	380–480 °F
2nd side-stream	Light gas oil	480–610 °F
3rd side-stream	Heavy gas oil	610–690 °F
Residue	Fuel oil	+ 690 °F

Full details of this unit are given in Chapter 6.

4.2   The Crude Vacuum Unit

Further break-up of the crude is often required to meet the refinery's product slate. This is usually required to produce low-cost feed to cracking units or the basic stocks for lubricating oil production. To achieve this, the residue from the atmospheric unit is distilled under sub-atmospheric conditions in the crude vacuum distillation unit. This unit operates similarly to the atmospheric unit in that the feed is heated by heat exchange with hot products and then in a fired heater before entering the distillation tower. In this case, however, the tower operates under reduced pressure (vacuum) conditions. These units operate at overhead pressures as low as 10 mm Hg absolute. Under these conditions the hot residue feed is partially vaporized on entering the tower. The hot vapours rise up the tower to be successively condensed by cooled internal reflux stream moving down the tower as was the case in the atmospheric distillation unit. The condensed distillate streams are taken off as side-stream distillates. There is no overhead distillate stream in this case.

The high-vacuum condition met within these units is produced by a series of steam ejectors attached to the unit's overhead system. Typical product streams from this unit are as follows:

Top side-stream	Light vacuum gas oil	690–750 °F
2nd side-stream	Heavy vacuum gas oil	750–985 °F
Residue	Bitumen	+ 985 °F

This unit is further described and discussed in Chapter 7.

4.3   The Light End Unit

The full-range naphtha distillate as the overhead product from the atmospheric crude unit is divided further into the basic components of the refinery's volatile and light oil products. This is done in the light end plant which contains four separate distillation units:

- The debutanizer
- The depropanizer
- The de-ethanizer
- The naphtha splitter

The full-range naphtha from the atmospheric crude overhead is routed first to the debutanizer unit. The stream is heated by heat exchange with hot products before entering the feed tray of the debutanizer column. This is a distillation column containing between 30 and 40 trays. Separation of butanes and lighter gas from the naphtha occurs in this tower by fractionation. The butanes and lighter are taken off as an overhead distillate while the naphtha is removed as the column's bottom product. The overhead distillate is then heated again by heat exchange with hot streams and fed into a depropanizer column. This also has about 30 to 40 distillation trays and separates a butane stream from the propane and lighter material stream by fractionation. The butanes leave as the column's bottom product to become the butane LPG product after further 'sweetening' treatment (sulphur removal). The column's overhead distillate is fed to a de-ethanizer column after preheating. Here the propane is separated from the lighter materials and leaves the column as the bottom product. This stream becomes part of the refinery's propane LPG product after some further sweetening treatment. There will be no overhead distillate product from this unit. The material lighter than propane leaves the overhead drum as a vapour containing mostly ethane.

The debutanized naphtha leaving the bottom of the debutanizer is subsequently fractionated in the naphtha splitter to give a light naphtha stream as the overhead distillate and a heavy naphtha as the column's bottom product. The light naphtha is essentially C5s and nC6s. This stream is normally sent to the refinery's gasoline pool as blending stock. The heavy naphtha stream contains the cycloparaffin components and the higher paraffin isomers necessary in making good catalytic reformer feed. This stream is therefore sent to the catalytic reformer after it has been hydrotreated for desulphurization.

4.4   The Catalytic Reformer Plant

The purpose of the catalytic reformer plant is to upgrade low-octane naphtha to the high-octane material required for motor gasoline fuel. It achieves this by reforming some of the hydrocarbons in the feed to hydrocarbons of high octane value. Notable among those reactions is the conversion of cycloparaffin content of the feed to aromatics. This reaction also gives up hydrogen molecules which are subsequently used in the refinery in other processes.

The feed from the bottom of the naphtha splitter is hydrotreated in the naphtha hydrotreater for the removal of sulphur and nitrogen. It leaves this unit to be pre-heated to the reforming reaction temperature by heat exchange with products and by a fired heater. The feed is mixed with a recycle hydrogen stream before entering the first of three reactors. The reforming reactions take place in these reactors and the reactor temperatures are sustained and controlled by intermediate fired heaters. The effluent leaves the last reactor to be cooled and partially condensed by heat exchange with cold feed and a condenser. This cooled effluent is routed to a flash drum from which a hydrogen-rich stream is removed as a gas while the reformate is removed as a liquid stream and sent to a stabilizer column. The bottoms from this column is debutanized reformate and is sent to the gasoline pool for blending to meet motor gasoline specifications. Part of the gas leaving the flash drum is recycled to the reactors as the unit's recycle stream. The remaining gas is normally sent to the naphtha hydrotreater for use in that process. Details of this unit are given in Chapter 9.

4.5   The Hydrotreating Units (Desulphurization)

Most streams from the crude distillation units contain sulphur and other impurities such as nitrogen, and metals in some form or other. By far the most common of these impurities is sulphur, and this is also the least tolerable of these impurities. Its presence certainly lowers the quality of the finished products and, in the processing of the crude, invariably affects the performance of the refining processes. Hydrotreating the raw distillate streams removes a significant amount of the sulphur impurity by reacting the sulphur molecule with hydrogen to form hydrogen sulphide (H_2S). This is then removed as a gas.

Two types of desulphurizing hydrotreaters are presented in this book (see Chapter 10):

● Naphtha hydrotreating—Once-through hydrogen
● Diesel hydrotreating—Recycle hydrogen

In naphtha hydrotreating the naphtha from the naphtha splitter is mixed with the hydrogen-rich gas from the cat reformer unit and preheated to about 700 °F by heat exchange and a fired heater. On leaving the fired heater the streams enters a reactor containing a desulphurizing catalyst (usually a CoMo on alumina base). The sulphur components of the feed combine with the hydrogen to form H_2S. The effluents from the reactor are cooled and partially condensed before being flashed in a separator drum. The gas phase from this drum is still high in hydrogen content and is routed to other downstream hydrogen-user processes. This stream contains most of the H_2S produced in the reactors. The remainder leaves the flash drum with the desulphurized naphtha liquid to be removed in the hydrotreater's stabilizer column.

Diesel hydrotreating has very much the same process configuration as the naphtha unit. The main difference is that this unit will almost invariably have a hydrogen-rich stream recycle. The recycle is provided by the flashed gas stream from the flash drum. This is returned to mix with the feed and a fresh hydrogen make-up stream before entering the preheater system. The recycle gas stream in these units is often treated for the removal of H_2S before returning to the reactors.

4.6   The Fluid Catalytic Cracking Unit (FCCU)

This cracking process is among the oldest in the oil industry. Although developed in the mid-1920s it first came into prominence during the Second World War as a source of high-octane fuel for aircraft. In the early 1950s its prominence as the major source of octane was somewhat overshadowed by the development of the catalytic reforming process with its production of hydrogen as well as high-octane material. The prominence of the FCCU was re-established in the 1960s by two developments in the process:

- The use of highly active and selective catalysts (zeolites)
- The establishment of riser cracking techniques

These developments enabled the process to produce higher yields of better-quality distillates from lower-quality feedstocks. At the same time, catalyst inventory and consumption costs were significantly reduced.

The process consists of a reactor vessel and a regenerator vessel interconnected by transfer lines to enable the flow of finely divided catalyst powder between them. The oil feed (typically, HVGO from the crude vacuum unit) is introduced to the very hot regenerated catalyst stream leaving the regenerator en route to the reactor. Cracking occurs in the riser inlet to the reactor due to the contact of the oil with the hot catalyst. The catalyst and oil are dispersed in the riser so that contact between them is very high, exposing a large portion of the oil to the hot catalyst. The cracking is completed in the catalyst fluid bed in the reactor vessel. The catalyst fluidity is maintained by steam injection at the bottom of the vessel. The cracked effluent leaves the top of the reactor vessel as a vapour to enter the recovery section of the plant. Here the distillate products of cracking are separated by fractionation and forwarded to storage or further treating. An oil slurry stream from this recovery plant is returned to the reactor as recycle.

The catalyst from the reactor is transferred to the regenerator on a continuous basis. In the regenerator the catalyst is contacted with an air stream which maintains the catalyst in a fluidized state. The carbon on the catalyst is burned off by contact with the air and converted into CO and CO_2. The reactions are highly exothermic, raising the temperature of the catalyst stream to well over 1000 °F and thus providing the heat source for the oil cracking mechanism.

Details of this process together with typical yield data are given in Chapter 11.

4.7 Hydrocracking

This process is fairly new to the industry becoming prominent in its use during the late 1960s. As the name suggests, the process cracks the oil feed in the presence of hydrogen. It is a high-pressure process operating normally around 2000 psig. This makes the unit rather costly and because of this has diminished its importance in the industry compared with the FCCU and thermal cracking. However, the process is very flexible. It can handle a wide spectrum of feeds including virgin gas oils, vacuum gas oils, thermal cracker gas oils, FCCU cycle oils and the like. Its products need very little downstream treating to meet finished specifications. The naphtha stream it produces is particularly high in naphthenes, making it a good reformer stock for atomatic production.

The process consists of one or two reactors, a preheat system, recycle gas section, and a recovery section. The oil feed (typically, the vacuum gas oils) is preheated by heat exchange with reactor effluent streams and by a fired heater. Make-up and recycle hydrogen streams are introduced into the oil stream before entering the reactor(s). (*Note:* in some configurations the gas streams are also preheated prior to joining the oil.) The first section of the reactor is often packed with a desulphurizing catalyst to protect the more sensitive cracking catalyst further down in the reactor from injurious sulphur, nitrogen and metal poisoning. Cracking occurs in the reactor(s) and the effluent leaves the reactor to be cooled and partially condensed by heat exchange. The stream enters the first of two flash drums. Here the drum pressure is almost that of the reactor. A gas stream rich in hydrogen is flashed off and recycled back to the reactors as recycle gas. The liquid phase from the flash drum is routed to a second separator which is maintained at a much lower pressure (around 150–100 psig). Because of this reduction in pressure a second gas stream is flashed off. This will have a much lower hydrogen content but will contain C_3s and C_4s. For this reason, the stream is often routed to an absorber column for LPG recovery. The liquid phase leaves from the bottom of the low-pressure absorber to enter the recovery side, where products are separated by fractionation and sent to storage. Further details of this process are given in Chapter 12.

4.8 Thermal Cracking Units

Thermal cracking processes are the true workhorses of the oil-refining industry. The processes are relatively cheap when compared with the fluid cracker and the hydro-cracker, but go a long way to achieving the oil-cracking objective of converting low-quality material into more valuable oil products. The process family of thermal crackers has three members:

- Thermal crackers
- Visbreakers
- Cokers

The term 'thermal cracking' is usually given to those processes that convert heavy oil (usually fuel oil or residues) into lighter product stock such as LPG, naphtha, and middle distillates by applying only heat to the feed over a prescribed element of time. When applied to a specific process the term refers to the processing of atmospheric residues (long residue) to give the lighter products. 'Visbreaking' refers to the processing of vacuum residues (short residues) to reduce the viscosity of the oil only and thus to meet the requirements of a more valuable fuel oil stock. 'Coking' refers to the most severe process in the thermal cracking family. Either long or short residues can be fed to this process whose objective is to produce the lighter distillate products and oil coke only. The coker process is extinctive—that is, it converts *all* the feed. In the other two processes there is usually some unconverted feed, although the thermal cracker can be designed to be 'extinctive' by recycling the unconverted oil.

The three thermal cracking processes have the same basic configuration. This consists of a cracking furnace, a 'soaking' vessel or coil, and a product-recovery fractionator(s). The feed is first preheated by heat exchange with hot product streams before entering the cracking furnace or heater. The cracking furnace raises the temperature of the oil to its predetermined cracking temperature. This is always in excess of 920 °F, and by careful design of the heater coils the oil is retained in the furnace at a prescribed cracking temperature for a predetermined period of time (the residence time). In some cases an additional coil section is added to the heater to allow the oil to 'soak' at a fixed temperature for a longer period of time. In other cases the oil leaves the furnace to enter a drum which retains the oil at its cracking temperature for a short time. In the coker process the oil leaves the furnace to enter one of a series of coker drums in which the oil is retained for a longer period of time at its coking temperature for the production of coke.

The cracked oil is quenched by a cold heavy oil product stream on leaving the soaking section to a temperature below its cracking temperature. It then enters a fractionator where the distillate products are separated and taken off similar to the crude distillation unit. In the case of the cokers the coke is removed from the drums by high-velocity water jets on a regular batch basis. The coking process summarized here refers to the more simple 'delayed coking' process. There are other coking processes which are more complicated, such as the fluid coker and the proprietary Flexi coker.

Further details on thermal cracking are provided in Chapter 13. This chapter also includes the treating of residues by hydrocracking and a method of calculating yields and properties of cracked stock.

4.9   Other Refinery Processes

The processes summarized above are the more common to be included in a fuel refinery's configuration. In addition to these there will also be gas-treating processes and often some additional gasoline octane-enhancement processes. These are described in more detail in Chapters 14 and 15, respectively.

Gas treating is always required to remove the H_2S impurity generated by hydro-treating or cracking from the refinery fuel gas or hydrogen recycle streams. The removal of H_2S for these purposes is accomplished by absorbing the sulphide into an amine or similar solution that readily absorbs H_2S. Stripping the rich amine removes the H_2S from the system to be further reacted with air to produce elemental sulphur in a sulphur plant.

The octane-enhancement process detailed in this book is the alkylation plant. These types of processes are usually proprietary and are provided to refiners under licence. The process described in Chapter 14 is the HF alkylation process, which utilizes hydrogen fluoride as the catalyst to convert unsaturated C_4s to high-octane isobutanes. The unit's recovery side is the aspect dealt with in some detail together with a descriptive item on the safe handling of HF.

4.10   The Non-energy Product Refineries

This book deals with the energy product refineries because this is the basic and largest of the refinery types. There are, in addition, those petroleum refineries or sections of a refinery that produce lubricating oils and others that produce only feedstock for converting to petrochemicals. This section briefly describes the type units that are common to these non-energy refineries.

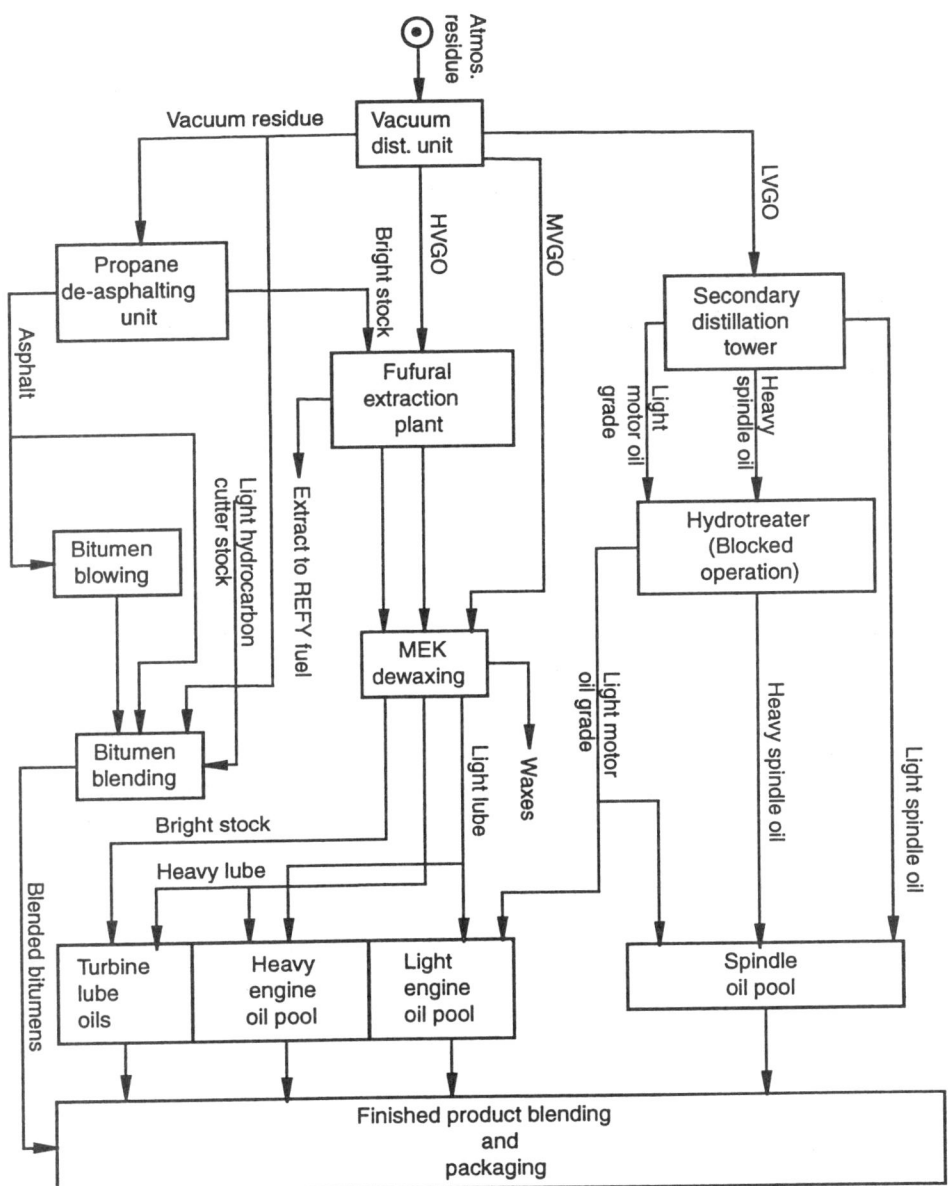

Figure 4.1. A lube oil refinery configuration

THE LUBE OIL REFINERY

The schematic flow diagram (Figure 4.1) shows a typical lube oil producing refinery configuration. Only about eight or nine base lube oil stocks are produced from refinery streams. The many hundreds of commercial grades of lubricating oils used in industry and transportation are blends of these base stocks with some small amounts of proprietary additives (mostly organic acid derivatives) included to meet their required specifications. There are also two quite important by-products to lube oil: bitumen and waxes. Most refineries include bitumen blending in their configuration but all except the older refineries no longer process the waxes. These are exported to manufacturers specializing in wax and grease production.

Lube oil production starts with the vacuum distillation of atmospheric residue. This feedstock is usually cut into three distillate streams, each having a boiling range which gives streams with viscosities meeting the finished blending product specifications. The lighter stream is taken off as the top side-stream and is further distilled again under vacuum to three light lube oil blending cuts. These are called spindle oils and when finished will form the basis of light lubes used for domestic purposes such as sewing machine, bicycle and other home lubricant requirements. Some of the heavier spindle oils are also used as blend stocks for light motor oils. These spindle oils require very little treatment for finishing, usually only mild hydrotreating suffices to meet colour requirement.

The second distillate side-stream is de-waxed and sent to the motor oil pool. It may also be blended with the heavier bottom side-stream as heavy engine oil stock. The bottom side-stream is one of the base blending stocks for heavy engine oils and the turbine oil stocks. To meet colour and other specifications these heavier oils must be treated for the removal of undesirable componets (such as heavy aromatics) by solvent extraction. This is accomplished prior to the stream being dewaxed and routed to storage. The heavy vacuum residue from the vacuum tower is routed to a propane de-asphalting unit. Here the very thick bituminous asphaltenes are removed by extraction with liquid propane. The raffinate from this extraction process is the heaviest lube oil blending stream commonly called Bright stock. This stream is also routed to the solvent extraction unit and the de-waxing process before storage.

Solvent extraction is accomplished in a trayed column by contacting the oil feed and solvent counter currently in the tower. The lighter raffinate stream leaves the top of the tower to be stripped free of the solvent in an associated stripper column before entering the de-waxing unit. The extracted components leave the bottom of the tower also to be stripped free of the solvent in an associated stripper column. The extract in this case may be routed to the propane de-asphalting unit or simply sent to the refinery fuel supply. The solvent in modern refineries is either fufural, phenol or a proprietary solvent based on either of these chemicals. In earlier plants oleum or liquid SO_2 was used for this purpose.

The oil streams routed to the de-waxing plant are contacted and mixed with a crystallizing agent such as methyl ethyl ketone (MEK) before entering a series of chiller tubes. Here the oil/MEK mix is reduced in temperature to a degree that the wax contained in the oil crystallizes out. The stream with the wax now in suspension enters a series of drum filters where the wax and oil are separated. Both streams are stripped free of the MEK in separate columns. The MEK is recycled while the de-waxed oil is sent to storage and blending. The wax may be retained as a solid in a suitably furnished warehouse or remelted and stored in special tanks with an inert gas cover.

The asphalt from the propane de-asphalting unit is stripped free of propane and any other light ends using inert gas as the stripping agent. It leaves the unit to proceed either directly to the bitumen pool or to be further treated by air blowing. The air-blowing process increases the hardness of the bitumen where this is required to meet certain specifications. It is accomplished either as a batch process or on a continuous basis. The hot stripped asphalt from the de-asphalting unit enters the air blower reactor under level control (if the process is continuous). Air is introduced via a small compressor to the bottom of the reactor vessel and allowed to bubble up through the hot oil phase. The air removes some of the heavy entrained oils in the asphalt and reacts mildly to partially oxidize the asphaltenes. The hot oil vapours and the unused air leave the top of the reactor to be burned in a suitably designed incinerator. The blown asphalt leaves the reactor as a side-stream to bitumen storage or blending.

The production of lube oils usually takes place in a section of an energy refinery. The various grades of the oils are also produced in a blocked operation using storage facilities between the units. This is feasible as the amounts of lube oils required to be produced are relatively small and normally do not justify separate treating facilities for each grade.

REFINING FOR PETROCHEMICALS

Feedstocks for the production of petrochemicals originate from refineries with processes similar to those described in this book for producing fuels. Indeed, there are only a few refineries worldwide that cater only for petrochemical requirements. Most petrochemical feedstocks are produced by changing operating parameters of the normal fuel refinery processes. In catering for the petrochemical needs much of the refinery product streams are tailored as follows:

● Aromatic streams—high in benzene, toluene, xylenes
● Olefin streams—high in ethylene, propylene and C_4s

Producing the aromatic feedstock

The production of aromatics feedstocks originates with the catalytic reforming of a refinery stream of a heavy naphtha range (say, 120–420 °F) and rich in naphthenes. A typical stream that meets this criterion would be a naphtha stream from a hydrocracker. Thus, in meeting petrochemical needs, a hydrocracker forming part of a fuel refinery configuration would be operated to maximize naphtha production. This would mean running the unit on a low space velocity with a higher oil recycle rate (that is, most recovered product heavier than the naphtha would be recycled back to the reactors).

Another source of high naphthene feed to the cat reformer would be hydrotreated FCCU naphtha. Of course, the hydrotreating of unsaturates has a high demand on the refineries hydrogen system, but this is balanced to some extent by the additional hydrogen produced in reforming the naphthenes. Should the refinery configuration include a thermal cracker and/or a steam cracker the hydrotreating of the naphtha cut from these units would also yield high naphthene cat reformer feedstock.

Catalytic reforming of the high naphthene content naphtha produces aromatics but there is also present some unreacted paraffins and some naphthenes. The

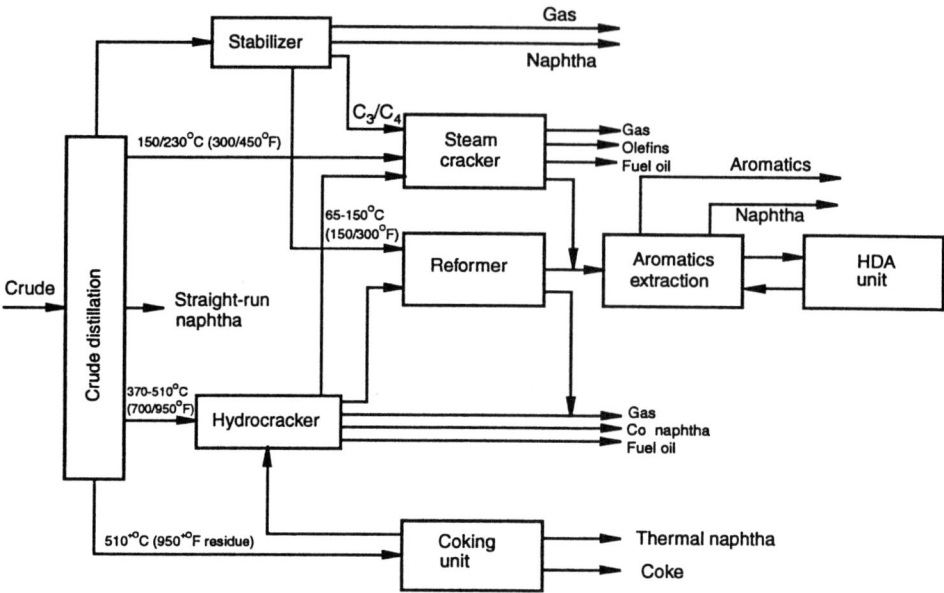

Figure 4.2. A petrochemical refinery configuration

downstream petrochemical units that separate and purify the aromatic reformate are expensive in both capital and operating costs. The specification for the BTX (benzene, toluene, xylene) feed therefore is very stringent and excludes non-aromatic components as much as possible. Another process may therefore be included in the refinery configuration to 'clean up' this aromatic feed stream before leaving the refinery. This is an aromatic extraction plant, i.e. a licensed process using a solvent to separate the paraffins and aromatics by counter-current extraction. The rich aromatic stream is then forwarded to the BTX plant where benzene, toluene, ethyl benzene and o-xylene are separated by fractionation while the *para*-xylene is usually separated by crystallization. The meta 'xylene' may also be recovered by super-distillation but more often than not it is converted into o-xylene in an isomerization unit.

Producing the olefin feedstock

The source of olefins in a refinery configuration is from the FCCU, a thermal cracker, or a steam cracker. The olefins produced as a gas are ethylene, propylene, and the C_4s such as butylenes, butadiene, etc. Liquid olefinic products from these units are normally hydrotreated to make reformer feedstock and thus the BTX feed. All products, of course, are treated for sulphur control and clean-up before leaving the refinery as petrochemical feedstock. The specifications for these products are stringent and usually the 'clean-up' plants are dedicated to the treatment of these products.

Olefins are used mostly in the production of polymers such as the vinyl polymers (vinyl chloride, vinyl acetate and the like), the polyethylene products, and the polypropylene products. The heavier C_4s are a major constituent in the production of synthetic rubbers.

Figure 4.2 shows a configuration for an actual petrochemical feedstock-producing refinery. This configuration differs a little from the general description given above insofar as there is no FCCU. Instead, all the olefins are produced in a steam cracker and coker. Hydrotreaters are not shown but they do exist. Note that the steam cracked naphtha in this case is sent directly to the aromatics and the HDA units. The latter unit accomplishes the hydrogenation of this stream and the extracted naphtha is then recycled to the reformer.

4.11   Oxygenated Gasolines

The concentration of vehicles on the roads in most of the cities in the modern world has increased dramatically over the last two decades. The emission of pollutants from these vehicles is adding significantly to the already-critical problem of atmospheric pollution. The problem is now so acute that governments of most First World countries are seeking legislation to curb and minimize this pollution and most countries will see the implementation of Clean Air Acts in the 1990s.

Petroleum-refining companies have been working diligently for many years to satisfy the requirements of Clean Air legislation. This began in the 1970s with the elimination of tetra-ethyl lead from most gasoline requirements. Processes such as isomerization and polymerization of refinery streams were developed together with a surge in the use of the alkylation process. However, the further decrease in pollutants now requires a move away from the traditional gasoline octane enhancers such as the aromatics and the RVP adjusters such as the C_4s.

Catalytic reforming produces gasoline streams to meet octane requirements mainly by converting cycloparaffins to light aromatics. Fluid catalytic cracking also produces gasoline blending stocks by cracking paraffins to light olefins and the products from these two processes still make up the bulk of a refinery's gasoline pool. Unfortunately, the aromatics are 'dirty compounds' in that they produce a sooty exhaust emission— unacceptable in meeting the Clean Air requirements. The light olefins are also unacceptable because of their 'evaporative emission' properties. Considerable work is begin done with alcohols to take the place of aromatics as octane enhancers and the C_4s used for meeting other gasoline specifications such as RVP. The work has had some success and some companies in North America use ethanol in the gasoline blend.

The greatest success in reducing aromatics in gasoline to date, however, has been in the production and blending of oxygenated compounds into the gasoline pool. A press release by a number of companies in 1989 is quoted as follows:

- Adding oxygenates reduces the amount of exhaust emissions (hydrocarbons and carbon monoxide) and the benefits have been quantified.
- Reducing aromatics and/or boiling range of gasoline can either reduce or increase exhaust emissions, depending upon vehicle type.

Oxygenates are ether compounds derived from their respective alcohols. There are three candidates of these ethers to meet the gasoline requirements:

- Methyl tertiary butyl ether (MTBE)
- Ethyl tertiary butyl ether (ETBE)
- Tertiary amyl methyl ether (TAME)

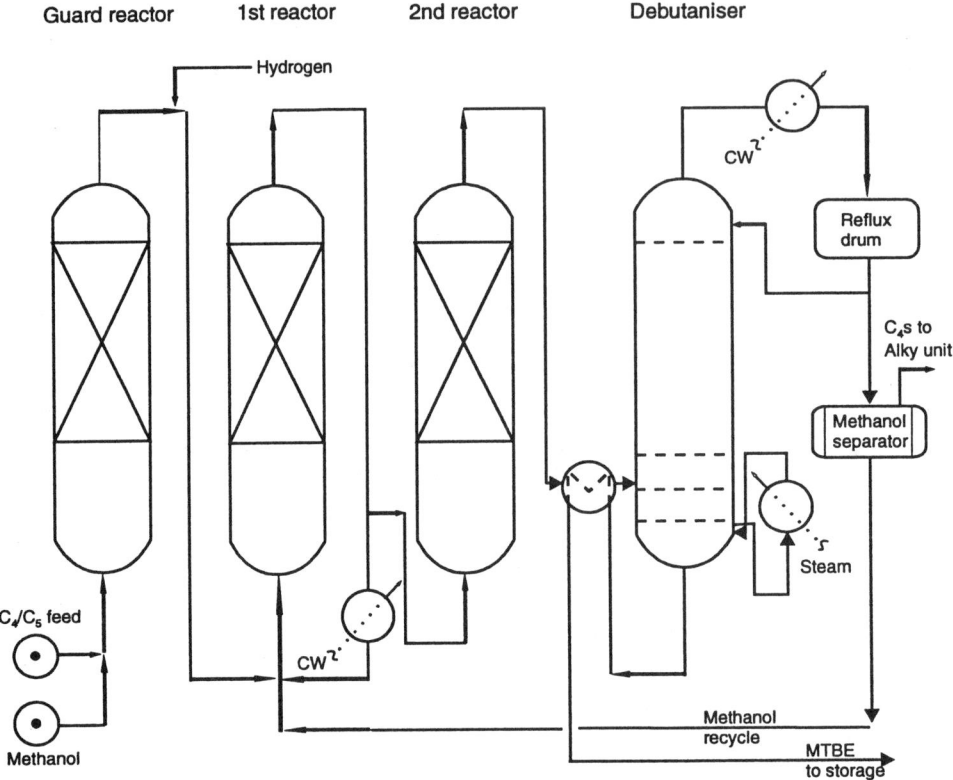

Figure 4.3. A typical flow diagram for the production of MTBE

Of these three candidates MTBE looks most likely to be the one that will be most satisfactory in meeting all the gasoline pool objectives. This compound has the blending quality of 109 octane, a RVP blending of 8–10 psi and a boiling point of 131 °F. There are a number of commercial units already in operation producing MTBE.

ETBE production is currently still in the pilot plant stage. Its octane-blending properties seem to be slightly better than MTBE and so does its RVP blending qualities. The ethanol feed stream, of course, is not so readily available as methanol. It is believed in North America that the first ETBE plant will be built in the Mid-West, where ethanol is more readily available.

TAME has an average octane number of 104 and a RVP blending of 3–5 psi. Except for the lower octane value, TAME has similar blending properties to ETBE. The ether in this case is formed by the reaction of 2-methyl-2-butene and 2-methyl-1-butene olefine feed and methanol. Commercial plants already operate in the UK and parts of Europe producing TAME. The compound is used in Europe as a gasoline product and not as a gasoline blend stock. The front end of a cracked gasoline stream is used in this manufacture to provide the olefin.

Considerable work in the research of ethers as fuel components is still in its early stages, but it seems that the oxygenated gasolines offer a solution to curbing emission pollutants. The physical characteristics of heavy hydrocarbons in gasoline suggests that the problem compounds are at the higher end of the gasoline boiling range. Further work currently being done with the higher ethers, alcohols, and esters may contain the solution of replacing the present troublesome heavy hydrocarbons to meet the Clean Air philosophy.

THE PRODUCTION OF THE ETHERS (MTBE)

There are several licensed processes for the production of methyl tertiary butyl ether by the etherification of a C_4/C_5 olefin stream and methanol. These processes have very similar configurations and are flexible enough to be converted quite simply to the production of the other ethers. Figure 4.3 is a typical flow diagram for this process.

The olefin feed from a FCCU or a steam cracker is combined with a methanol stream to enter a guard reactor which removes any impurities. A small hydrogen stream is added to the hydrocarbon from the guard reactor prior to entering the first ether reactor. This reactor contains a special resin catalyst and the reactor feed flows upward through this catalyst bed at moderate temperature and pressure and in the liquid phase. The reaction is exothermic and temperature control is maintained by externally cooling a recycle stream from the first of two reactor vessels.

The catalyst in this case performs three reactions simultaneously: etherification of branched olefins, selective hydrogenation of the unwanted diolefins, and hydro-isomerization of olefins by a double-bond switch. The reactor effluent leaves the top of the second reactor vessel to be heated in a feed heat exchanger with the debutanizer bottoms product. The overheads from the debutanizer is a C_4 and methanol stream. These streams are separated in a separation vessel, the methanol stream is recycled to the first reactor while the C_4s are returned to the FCCU light ends unit. The bottom product is C_5^+ enriched with MTBE (or TAME, depending on the olefin feed used).

5   REFINERY CONFIGURATION DEVELOPMENT

Much work has to be completed before a refining company can commit the large expenditure required to upgrade an existing refinery or to build a new facility. Certainly, during the early phases of the study work many different processing configurations will be developed and evaluated. This will be necessary to determine which processing route will be the best to meet the refinery's commercial objectives. All these configuration studies will be initiated by predicting the yields and quality of the products obtainable from the selected crudes the refinery intends to process. Calculations similar to those given in this chapter will be used to develop these preliminary data. The evaluation studies will continue to include capital and operating costs of the various units considered in these early configurations and then examine their economic impact to arrive at a selected processing route.

In this chapter a very simple refinery configuration is developed from knowledge of the correlations and processes given in this book. The product specifications for this exercise is given at the end of the book as Appendix 3.

Consider the following scenario. A small oil company desires to build a refinery to produce the following marketable products by processing 3 million metric tons/year of Kuwait crude oil:

C3 LPG
C4 LPG
93 ON (RES) premium gasoline
84 ON (RES) regular gasoline
Kerosene (for cutter stock)
Automotive grade diesel (to be maximized)
No. 6 fuel oil

The sensitivity of the premium gasoline is to be seven octane numbers. It is required to define the product scheme and to specify the size of the individual units in terms of their stream capacity.

5.1 Solution

The following is a solution and a configuration that will be considered as a candidate for further economic analysis and possible selection as the processing scheme.

Due to the different service factors (ratio of onstream time to total time) that exist for the various processes that will make up the configuration, the calculations will be made in barrels per calendar day (BPCD). This is the number of barrels processed per day over 365 days.

This refinery will be oriented principally to produce auto diesel, LPGs, and gasoline. The following units will be considered to achieve this objective:

- An atmospheric crude distillation unit
- A naphtha hydrotreater
- A naphtha splitter
- A catalytic reformer
- LPG recovery—debutanizer, depropanizer and a de-ethanizer
- A diesel hydrofiner

Parameters for these units are as follows:

(1) *The crude unit* This will be a single-stage unit (i.e. no vacuum distillation). A full-range naphtha cut will be taken off as an overhead distillate, a small kero stream as the first side-stream, and a light and heavy gas oil taken off as second and third side-streams respectively. Taking a full-range naphtha as overhead distillate enables the column to be operated at close to atmospheric pressure at a realistic overhead condensing temperature and still retain most of the LPG. The diesel is taken off as a light and heavy gas oil to minimize the size of the diesel hydrofiner. The light gas oil will be cut to meet a sulphur content acceptable to blending into diesel without further treating. Only the heavy gas oil stream will therefore require hydrofining, and the hydrofiner can remove 85% of the sulphur in the heavy gas oil. The kero stream will be blended into the atmospheric residue as far as possible. The constraint in this case will be the flash point specification of No. 6 fuel oil.

(2) *The naphtha hydrotreater* The full-range naphtha from the crude unit will be hydrotreated before being sent to the light ends recovery units. This removes the necessity of treating the light streams individually for sulphur removal downstream. This may have to be changed as a result of the economic analysis later.

(3) *The naphtha splitter* This unit's function is to prepare reformer feed from the hydrotreated naphtha. In this scheme the LPG recovery is going to be affected after the splitter. There will be some methane and some hydrogen in the feed to the splitter although the material has been stabilized. These will be removed with the overhead distillate from the splitter. To operate the column at reasonable conditions of temperature and pressure, however, requires the presence of the light gasoline components to hold these very light materials in the distillate. Should the hydrogen

and methane be vented at this point from the overhead drum they would carry some LPG with them and these would probably be lost as products. Thus the reason for debutanizing etc. after naphtha splitting in this case is to maximize LPG recovery.

(4) *The catalytic reformer* The sensitivity specification for the premium gasoline necessitates reforming to a lower octane number and meeting the gasoline octane by blending in TEL (tetra-ethyl lead). The following are the cat reformer data as supplied by the licensor:

Feed: 194 IBP to 347 °F FBP ASTM distillation
Severity: 86 ON (RES) clear
Reformate + 1.5 Mls/IG TEL: 93 ON (RES)
                                      86 ON (MOTOR)

Reformate yield:	83.2% on feed
H_2 to C_3:	5.2% on feed
Propane:	5.0% on feed
Butanes:	6.6% on feed

ASTM dist. 10% vol.: 162 °F
           30% vol.: 200 °F
           70% vol.: 295 °F
RVP (Reid vapour pressure): 4.2 psig

(5) *LPG recovery* The light naphtha distillate from the splitter will be debutanized in the light ends recovery unit debutanizer. The overhead distillate from this column will contain all the butanes and the propane for LPG recovery. The distillate stream from the cat reformer unit will also be added to this LPG-rich stream. These will then be fed to the depropanizer column for butanes recovery. The distillate from the depropanizer will then be routed to the de-ethanizer for propane LPG recovery.

(6) *The diesel hydrofiner* The heavy gas oil side-stream from the crude unit will be hydrofined for the removal of 85% of its sulphur content.

This now completes the basic parameters that will be used for this exercise. Some of the details pertaining to this process scheme must now be filled in, and this centres on a better definition of the crude unit's operation.

5.2 Crude Unit Yields (including the naphtha splitter)

NAPHTHA

The reformer feed requires the ASTM end point of the heavy naphtha to be 347 °F. From curve A in Figure A1.3 of Appendix 1 the cut point limit is 357 °F. From the crude assay this gives a total full-range naphtha yield of 24.8% vol. on crude.

On a naphtha splitter we could expect to obtain fractionation meeting 25 °F gap between the ASTM 95% vol. of the light naphtha and ASTM 5% vol. of the heavy naphtha. By drawing the heavy naphtha ASTM curve as a straight line between 0% and 100% on the probability graph at 194 °F and 347 °F, respectively, the 5% ASTM point of the heavy naphtha and 95% ASTM of light naphtha are fixed. Draw the light naphtha ASTM line on probability paper parallel to the heavy naphtha line through the 95% point at 174 °F. From the ASTM end-point of the light naphtha calculate its cut point by using Figure A1.3. This is found to be 170 °F and is equivalent to 9.0% on crude.

There is 2.8% of C_4s and lighter in Kuwait crude. Knowing this, the yields of gas, light naphtha and heavy naphtha are predicted as follows:

Gas: 2.8% vol.
Light naphtha: 9.0 − 2.8% vol. = 6.2% vol.
Heavy naphtha: 24.8 − 9.0% vol. = 15.8% vol.

Constructing the TBPs of the light and heavy naphtha from the ASTM curve and using the method described in Chapter 6 the specific gravities of the cuts and other properties can be calculated.

KERO YIELD

The front end of the kero ASTM is fixed by what can reasonably be expected by crude unit fractionation. We could expect to attain a fractionation ASTM gap of 20 °F between 5% kero and 95% naphtha in a crude unit. Thus, as we know the 95% ASTM of the naphtha, the 5% ASTM point of the kero can be fixed at 20 °F above the naphtha 95% point which is at 335 °F.

Now the back end of the kero (and therefore the yield) determines what the front end of the light gas oil will be. The flash point of the light gas oil depends on the 5% ASTM point, so let us fix the 5% point of the gas oil to meet a flash point of, say, 175 °F. (This allows some latitude in the design and operation of the unit to be able to meet specification diesel at 125 °F flash point minimum.) From the correlation of ASTM and flash point given by the equation flash point = 0.77 (5% ASTM—150 °F), the gas oil 5% point is calculated as 380 °F.

We can reasonably expect to attain a 10 °F gap between 95% ASTM of the kero and 5% ASTM of the light gas oil. Using this, the kero 95% ASTM is fixed at 370 °F. The kero ASTM is drawn on the probability curve as the straight line between 5% and 95% at 335 °F and 370 °F, respectively. From the ASTM end point of the kero the cut point is predicted using Figure A1.3. The TBP curve is that calculated from the ASTM curve. The cut point for the kero is found to be 387 °F and is equivalent to 27.8% vol. on crude. The yield is therefore 27.8 − 24.8% vol. = 3.0% vol.

From the ASTM and TBP curves the kero has the following properties:

SG at 60 °F   0.780
Sulphur (Total)   0.13% wt
Viscosity at 100 °F   1.1 cSt
Viscosity at 122 °F   0.95 cSt
Flash point   142 °F

LIGHT DIESEL YIELD

The front-end cut-point of this fraction has already been fixed and so has the 5% ASTM and flash point. Now the critical specifications to be met when the light and heavy gas oils are blended together will be pour point and sulphur. As a first guess, let the mid-boiling point of the total diesel cut be such as to give 10 °F pour point. From inspection of the crude assay this limits the cut to about 700 °F TBP cut point. Let us take the light gas oil component to about 0.8% wt sulphur, giving a light gas oil cut, say, approximately between 27.8% vol. and 45.8% vol., a yield of 18.0% vol. (387 °F – 595 °F cut points).

The 90% concept is used in this case as we are not going to be specific or design for specific fractionation between light and heavy gas oil. The end point could, in practice, be a lot higher than that derived on a straight-line probability relationship. The TBP curve is calculated from the probable ASTM curve and the following properties are defined:

SG at 60 °F   0.820
Sulphur   0.8 wt
Pour point   30 °F
Viscosity at 100 °F   2.0 cSt
Viscosity at 122 °F   1.8 cSt

HEAVY DIESEL

The 5% ASTM point of the heavy diesel will be set at 10 °F below the 95% point of the light diesel (at 570 °F). A 10 °F overlap is reasonable in this section of the crude tower. Taking a final cut point of 700 °F on crude the yield of heavy gas oil becomes 54% – 45.8% vol. = 8.2% vol. This may have to be adjusted later.

Again using the 90% ASTM/TBP cut point relationship the ASTM 90% point is fixed at 646 °F. The ASTM curve is defined as a straight line on the probability graph between the 5% point at 570 °F and 90% at 646 °F. The TBP curve is calculated from the probable ASTM curve and the following properties predicted from it:

SG at 60 °F   0.874
Pour point   40 °F
Sulphur   2.3% wt
Viscosity at 100 °F   8.0 cSt
Viscosity at 122 °F   5.3 cSt
Flash point   200 °F

5.3   Yield Summary and Crude Throughput

The total crude throughput to the refinery is calculated as follows:

Capacity required = 3 million tons/year
In BPCD (barrels per calendar day) this is

$$\frac{3\,000\,000 \times 2205}{365} = 18.1 \text{ million lb per day}$$

lb/gal of Kuwait crude = 7.22
Then $\text{BPCD} = \dfrac{(18.1)(10^6)}{(7.22)(42)} = 59\,689$ BPCD

The stream yields are now summarized from the yield calculations as follows:

	% vol. on crude	BPCD
Gas	2.8	1671
Light naphtha	6.2	3701
Heavy naphtha	15.8	9431
Kero	3.0	1791
Light gas oil	18.0	10 744
Heavy gas oil	8.2	4894
Residue	46.0	27 457
Total	100.0	59 689

From these base yields the blending recipes and unit capacities can now be determined.

PRODUCT BLENDING

Regular gasoline

The yield of reformate is given by the licensor as 83.2% on feed. Consider that the regular gasoline will be made up of all the light naphtha make and with sufficient reformate to meet 80 octane number clear and will then be blended with about 3.0 ml/IG of TEL to make the specification octane. The light naphtha octane number is 75 RES Clear.

Let x be the volume of reformate in the blend. Then:

$$x(86) + (1-x)75 = 80$$
$$x = 0.454 \text{ vol.}$$

Total light naphtha $= 3701 \text{ BPCD} = 54.6\%$ vol. of the gasoline stream

$$\text{Total regular gasoline} = \frac{3701}{0.546}$$

$$= 6778 \text{ BPCD}$$

Reformate to regular gasoline $= 3077$ BPCD

Premium gasoline

Total reformate make $= 0.832(9431) = 7847$ BPCD
Reformate to regular gasoline $\quad = 3077$ BPCD
Reformate to premium gasoline $\quad = 4770$ BPCD

This reformate when blended with 1.5 ml/IG TEL will meet 93 ON (RES)

The predicted ASTM distillation is

	Specification	Predicted
10% min. vol. received at	167 °F	167 °F
50% min. vol. received at	284 °F	250 °F
90% min. vol. received at	392 °F	370 °F

The vapour pressure can always be adjusted by allowing more C_4s in the reformer stabilizer bottom product.

Diesel blending

The blending of the straight-run light gas oil and desulphurized heavy gas oil meets specification as shown.

Pour point

The blended pour point will be as follows:

Pour point (°F)	5% ASTM (°F)	Composition % vol.	Pour index	Pour factor	ASTM 50% factor
Light diesel − 30	470	68.7	2.7	1.85	323
Heavy diesel + 40	620	31.3	26	8.14	194
Blend		100.0		9.99	517

Blended pour point is 10 °F. Therefore this sets the limit of the diesel product yield.

Sulphur

Total blended diesel sulphur is as follows:

	% vol.	SG at 60 °F	Weight factor	Sulphur (% wt)	Sulphur factor
Lighter diesel	69.7	0.820	56.30	0.8	0.45
Heavy diesel	31.3	0.874	27.4	0.35	0.095
Total	100.0		83.7		0.545

SG at 60 °F of blend = 0.837
Sulphur = 0.65% wt which meets spec.

ASTM distillation

The 90% of heavy diesel is 650 °F. This, as it is, meets spec. The 90% of the blend will be considerably lower than this, therefore the spec. is met.

Viscosity

Final viscosity of the blend will be as follows:

At 100 °F

	% vol.	Viscosity at 100 °F	Viscosity Index	Factors
Light diesel	68.7	2.0 cSt	58	40
Heavy diesel	31.3	8.0 cSt	41	12.8
Blend	100.0			52.8

Blend viscosity at 100 °F = 2.8 cSt which meets spec.

At 122 °F

	% vol.	Viscosity at 122 °F	Viscosity Index	Factors
Light diesel	68.7	1.8 cSt	60	41
Heavy diesel	31.3	5.3 cSt	45	14.1
Blend	100.0			44.1

Blend viscosity = 2.4 cSt which meets spec.

Fuel oil product

Residue from crude unit is 46%
From assay data (not shown in Appendix 2)

SG at 60   = 0.972
Viscosity at 100 °F = 1800 cSt
Viscosity at 210 °F =   58 cSt

Blending with total kerosene yield gives the following properties:

Flash point

	% vol.	Flash point (°F)	Index	Factor
Residue	93.9	350	0.65	0.61
Kero	6.1	142	510	31.11
Blend	100.0			31.72

Flash point = 188 °F which meets spec.

Viscosity at 100 °F

	% vol.	Vol. at 100 °F	Viscosity Index	Factor
Residue	93.9	1800	16	15
Kero	6.1	1.1	70	4
Blend	100.0			19

Viscosity = 700 cSt at 100 °F which meets spec.

	% vol.	Viscosity at 122 °F	Viscosity Index	Factor
Residue	93.9	700	19	17.8
Kero	6.1	0.95	74	4.5
Blend	100.0			22.3

Viscosity at $122\,°F = 260$ cSt which meets spec.

LPG Products

Assume that the LPG streams will meet specification if 70% of the C_3 and 80% vol. of C_4 are recovered as LPG.

		Actual volume	Recovered as product
From straight crude	C_4s	1134.1 BPCD	907 BPCD
	C_3s	477.2 BPCD	334 BPCD
	Gas	59.7 BPCD	430 BPCD
From reformer	C_4s	622 BPCD	498 BPCD
	C_3s	472 BPCD	330 BPCD
	Gas	490 BPCD	756 BPCD

The refinery scheme can now be shown diagramatically as a schematic flow sheet (Figure 5.1).

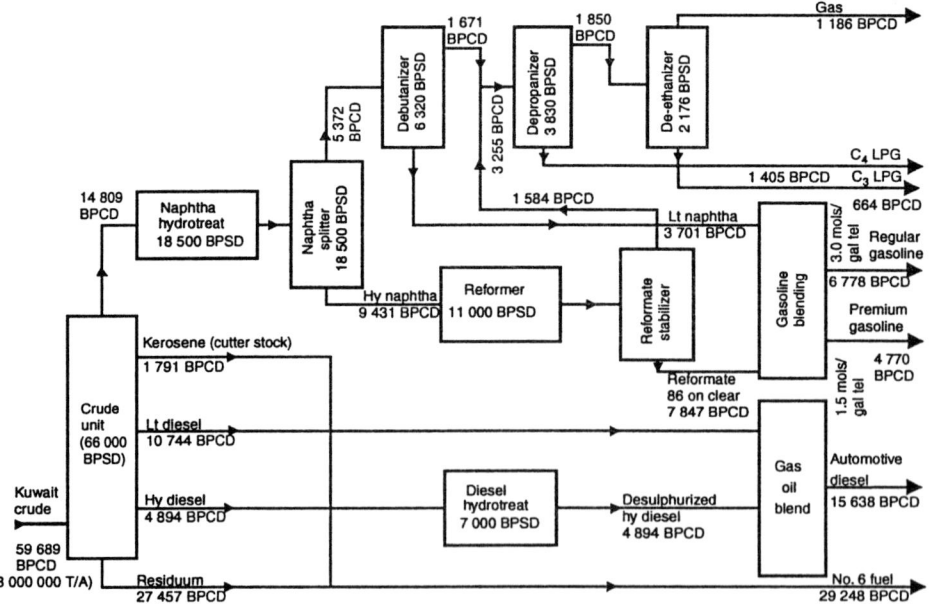

Figure 5.1. Crude oil and its products—a refinery block flow diagram

Plant capacities

The individual units will have the following service factors and their capacities are calculated by dividing BPCD by the service factor:

	Service factor	Capacities (BPSD)
Crude unit	0.9	66 000
Naphtha hydrotreater	0.80	18 500
Naphtha splitter	0.80	18 500
Platformer	0.80	11 000
Debutanizer	0.80	6320
Depropanizer	0.80	3830
De-ethanizer	0.80	2176
Diesel hydrofiner	0.70	7000

5.4 Conclusion

The basic refinery configuration and unit sizes are then established by the methods shown. The example, of course, reflects a simple exercise in utilizing the crude assay and process characteristics. Very often this type of study work encompasses the need for including conversion units to meet shortfalls in a refinery marketing programme. Such studies require complex optimization of an investment and technical nature. Some of these can be so complex that true solutions may only be provided by the use of linear programming and high-speed computers. Nevertheless, even the most complex studies commence with the definitions and methods described here.

Part 2

THE DISTILLATION PROCESSES

6 THE ATMOSPHERIC CRUDE DISTILLATION UNIT

6.1 Process Description (Typical Crude Distillation Unit)

The first process encountered in any conventional refinery is the atmospheric crude distillation unit. In this unit the crude oil is distilled to produce distillate streams which will be the basic streams for the refinery product slate. These streams will either be subject to further treating down stream or become feedstock for conversion units that may be in the refinery configuration.

THE PROCESS

Crude oil is pumped from storage to be heated by exchanges against hot overhead, product and pumparound streams in the crude unit (see Figure 6.1). At a temperature of about 200–250 °F the crude enters the desalter. Free salt water contained in the crude is washed out and separated by means of an electrostatic precipitation contained in the desalter drum. The water phase from the drum is sent to a sour water stripper to be cleaned before disposal to the oily water sewer. The oil phase enters a second surge drum where some of the light ends are flashed off and routed directly to the distillation tower (they do not pass through to the heater). The crude distillation booster pump takes suction from this drum and delivers the desalted crude under flow control to the fired heater via the remaining heat exchange train. The crude oil is heated in the heater to a temperature that will vaporize the distillate products in the crude tower. Some additional heat is added to the crude to vaporize about 5% more than is required for the distillate streams. This is called overflash and is used to ensure good reflux streams in the tower.

The unvaporized portion of the crude leaves the bottom of the tower via a steam stripper section. This is the atmospheric residue and is either further distilled, or becomes feed to a thermal cracker, or is routed to fuel oil product storage.

The distillate vapours move up the tower counter current to the cooler liquid reflux streams coming down. Heat and mass transfer take place on the trays contained in the tower. Distillate products are removed from the various sections of the tower,

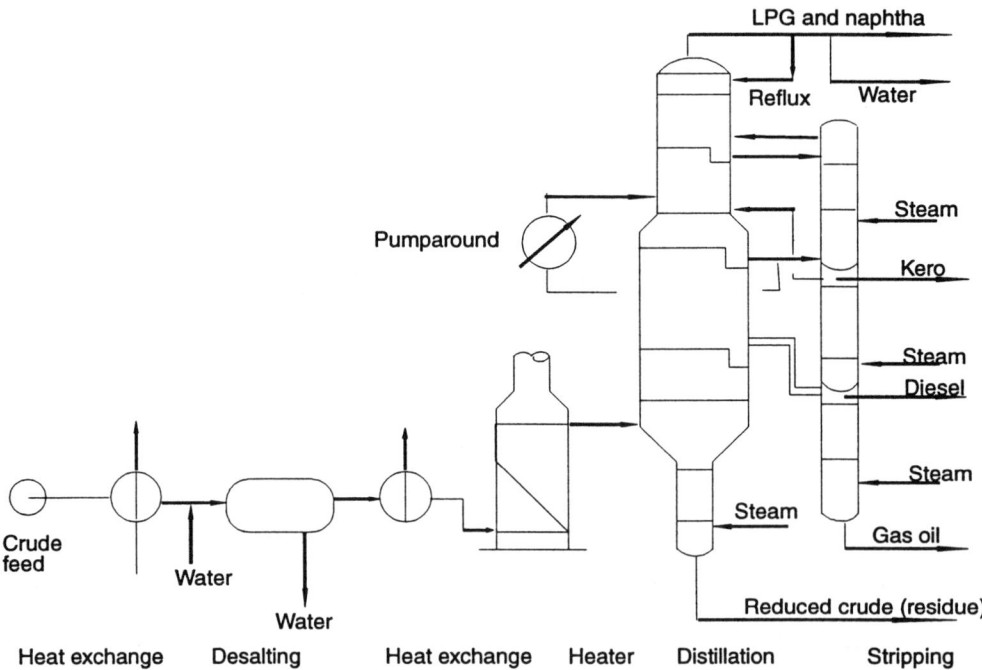

Figure 6.1. An atmospheric crude distillation unit (CDU)

steam stripped and sent to storage. The full naphtha vapour is allowed to leave the top of the tower to be condensed and collected in the overhead drum. A portion of this stream is returned as reflux while the remainder is delivered to the light end processes for stabilizing and further distillation.

The side-stream distillates shown in Figure 6.1 are:

● Heavy gas oil (has the highest boiling point)
● Light gas oil (will become diesel)
● Kerosene (will become jet fuel)

A pumparound section is included at the light gas oil draw-off. This is simply an internal condenser which takes heat out of that section of the tower. This in turn ensures a continued reflux stream flow below that section.

The overhead system described here is perhaps the most common system encountered in this process. There are other configurations that are used by many refineries. Indeed, the one used as an example of overhead calculation methods in this book uses an external pumparound stream from the top tray to provide the internal reflux required. All these methods are well proven in refinery operation and have their advantages and disadvantages.

6.2 Fractionation

The degree of fractionation in a crude distillation unit is measured by the temperature difference between the 95% vol. ASTM of the lighter product and the 5% vol. ASTM temperature of the adjacent heavier one. When the 95% point of the lighter product is less than the 5% point of the heavier the difference is called an ASTM gap. When the 95% point is higher than the adjacent product's 5% point the difference is called an ASTM overlap.

A relationship exists between the number of trays, the internal reflux ratio and the gap or overlap between the products (5). This relationship is given in Figure 6.2. This is in the form of a family of four curves, two curves at no steam present and two at maximum stripping steam. Two of the curves refer to the separation between side-streams only while the other two are for the separation of the overhead product and the first side-stream.

Obviously, the best fractionation performance is when there is an ASTM gap between the products. As fractionation deteriorates and gaps become overlaps more and more of the components of the adjacent products are not separated. It must be stressed that good fractionation can be expected between naphtha and kero and also between kero and LGO (i.e. ASTM gaps). Lower in the column, however, and in vacuum columns such separation is not possible. In this level of distillation high overlaps will be experienced.

The following example illustrates the use of this method to establish the internal reflux required for a kero and gas oil side-stream separation.

STRIPPING STEAM

Stripping steam is used in the crude distillation unit to remove the entrained light ends from the draw-off products. When the side-stream product is drawn off the

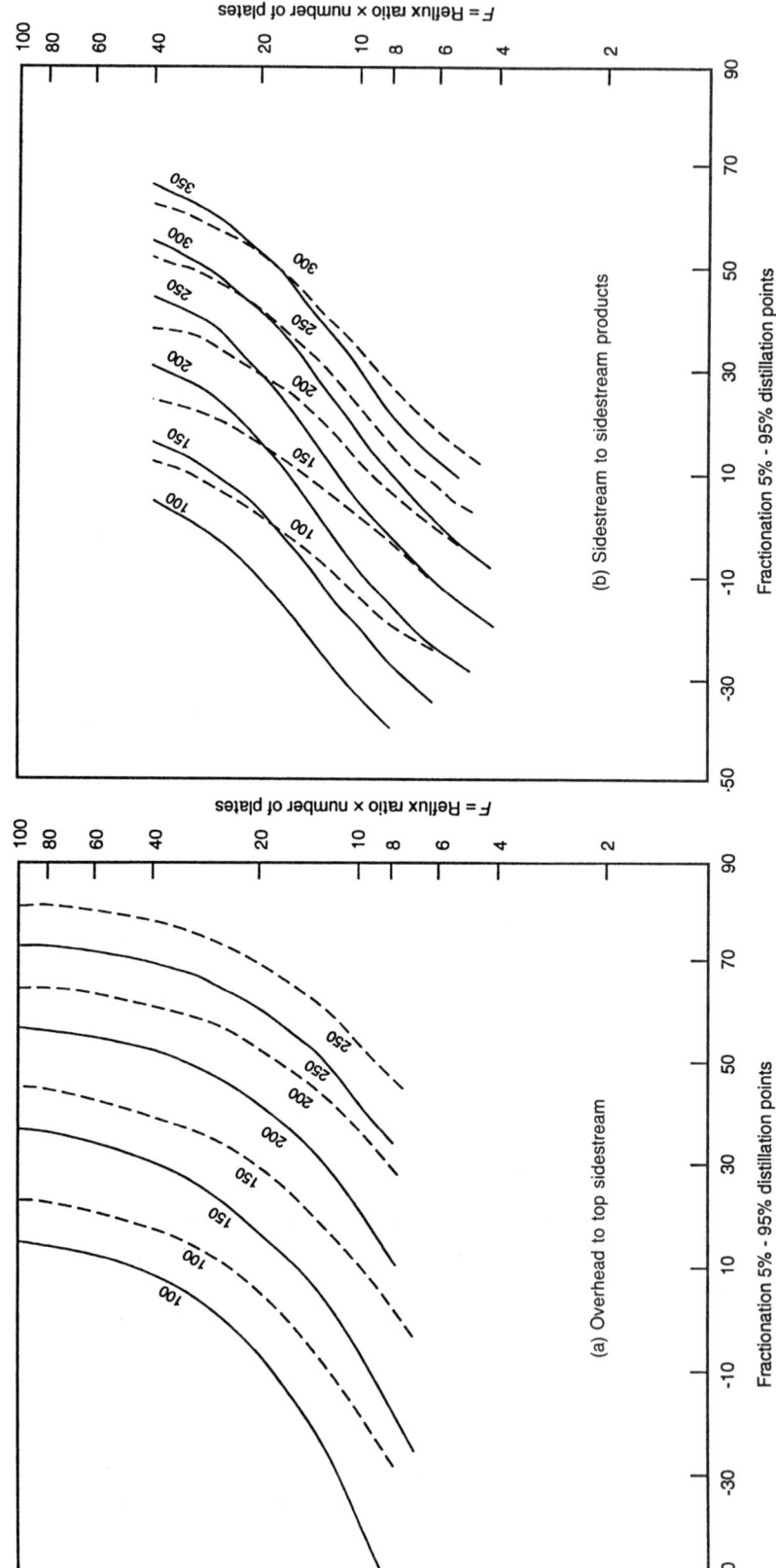

Figure 6.2. Crude unit fractionation curves for (a) overhead to top side-stream products and (b) side-stream to side-stream products. —— no steam, — — max. stripping steam generally used. In (a) the numbers on the curves represent the °F difference in 50% distillation points between overhead and top side-stream. In (b) they represent the °F difference in 50% distillation points between overhead draw-off plate and liquid drawn off at the lower side-stream; Reflux ratio = gallons hot overflow/total gallons product vapours entering the upper side-stream draw-off plate and top side-stream draw-off plate in (a) and top side-stream draw-off plate in (b)

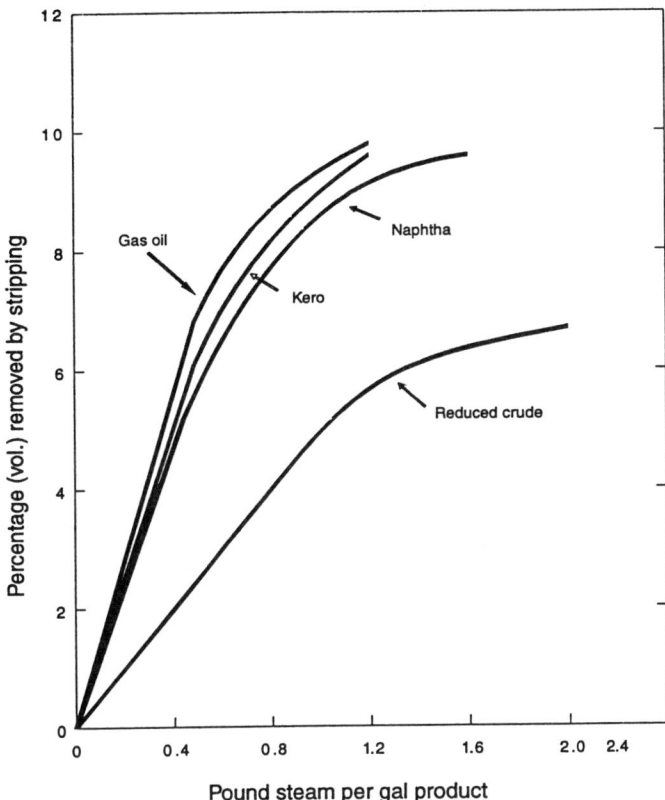

Figure 6.3. Stripping steam

main column it will contain entrained light components from the vapours passing through the draw-off tray. This draw-off product is routed to a side-stream stripper and steam is introduced to the bottom of the tower to flow counter-currently to the liquid. The stripper usually contains four trays which promote the contact between oil and the steam. The entrained light ends are removed by the steam and they and the steam re-enter the main column. The residue from the bottom of the main distillation tower is also steam stripped before leaving the tower. The quantity of steam and the percentage stripout is given in Figure 6.3.

Example

Total vapour flow through kero draw-off tray is naphtha plus kero products.
Based on Kuwait crude the 50% TBP point for this stream is 290 °F.
The 50% point of the LGO stream in this case is about 550 °F.

$$\text{Therefore } \Delta T 50\% = 260\,°\text{F}$$

In this tower there are eleven trays between kero and LGO, and we would expect a 0 °F gap between the products. Maximum stripping steam will be used.

From the Packie (5) curves the reflux $\times$ tray product is 4.4

$$\text{Therefore reflux ratio is } \frac{4.4}{11} = 0.4$$

Knowing the total gallons of hot product vapour passing through the kero draw-off tray (this will be overhead naphtha plus kero products), the hot volume of internal reflux can be calculated:

Assume total products $= 18\,000$ gal/h

The required overflow from kero tray will be

$$\frac{x}{18\,000} = 0.4$$

$$\underline{x = 7200 \text{ hot gal/h}}$$

6.3   Developing the True Boiling Point Curve (TBP) and the Equilibrium Flash Vaporization Curve (EFV) from Lab ASTM Distillation

Two items of data must be available to produce a meaningful TBP curve and therefore an EFV curve:

● The ASTM distillation curve for each product.
● The flow rate for each product and the crude feed flow rate.

The flow rates will be used to determine the yield of each product on crude.

CONVERTING THE PRODUCT ASTM DISTILLATION TO TBP

Most crude distillation units take a full-range naphtha cut as the overhead product. This cut contains all the light ends ethane through pentanes and, of course, the heavier naphtha range. All the light ends are gases in solution, therefore it is not possible to prepare a meaningful ASTM distillation on this material directly. Usually the laboratory data give the ASTM distillation on a stabilized naphtha, and a separate analysis is used to determine the composition of the light ends.

There are two well-proven methods to develop the TBP curve from ASTM data. The first is that by Wayne Edmister given in his book *Applied thermodynamics* (6) and the second by Maxwell in *Databook on Hydrocarbons* (3). The correlations from both these sources are given in Appendix as Figures A1.1 and A1.2. The Edmister method is used in this book for developing the TBP curve.

The ASTM distillation is tabulated as the temperatures for IBP, 10%, 30%, 50%, 70%, 90%, FBP. The IBP is the initial boiling point of a sample (equivalent to 0% over) and the FBP is the final boiling point (equivalent to 100% over). The temperatures between the IBP and the FBP are read off at the respective volume percentages distilled. Using Figure A1.1 the 50% vol. TBP is read off from the 50% point of the ASTM distillation.

Figure A1.1 is again used to determine the TBP temperature differences between the IBP and 10%; 10% and 30%; 30% and 50%; 50% and 70%; 70% and 90%; and 90% and FBP. From the established value of the TBP 50% point and the temperature differences determined from Figure A1.1 the TBP temperatures for IBP, 10%, 30%, 70%, 90%, FBP can now be tabulated.

IDENTIFYING AND USING PSEUDO-COMPONENTS TO MAKE AN OVERALL TEMPERATURE VERSUS RELATIONSHIP (TBP) CURVE

Each product TBP curve is split into pseudo-components and the ranges of these components established on a temperature basis. Any number of these components may be used. However, when a component is identified in terms of its temperature it must be used in all product distillation curves where this specific range exists.

The volume percentage of each pseudo-component is listed for each product. Multiplying these percentages by the yield of the respective product on crude provides the percentage on crude of each pseudo-component. The accumulated volume of each can now be related to a cut point temperature and thus a cumulative TBP curve.

The very heavy end of the TBP curve is not calculated but is usually determined by extrapolation. It plays no major role in any future calculations for the processes described in this book.

DEVELOPING THE EQUILIBRIUM FLASH VAPORIZATION CURVE

The Maxwell curves given in Figure A1.2 are used to develop the EFV curve from the TBP. The EFV curve gives the temperature at which a required volume of distillate will be vaporized. The distillate vapour is always in equilibrium with its liquid residue and the developed EFV curve is always at atmospheric pressure. Other pressure and temperature-related conditions are determined using vapour pressure curves or by constructing a phase diagram.

The TBP reference line (DRL) is first drawn as a straight line through the 10% point and the 70% point on the TBP curve. The slope of this line is determined as temperature difference per volume per cent. This slope is then used to determine the 50% volume temperature of a flash reference line (FRL). The curve in A1.2 which relates ΔT_{50} (DRL − FRL) to the DRL slope is used for this. Finally, the curve in Figure A1.2 relating the ratio of temperature differences between the FRL and the flash curve (EFV) and the ratio for the TBP to DRL is applied to each percentage volume of the FRL to arrive at the atmospheric EFV curve.

A sample calculation for the compilation of both TBP and EFV curves follows. Note that TBP curves are used to define product yields while EFV curves are used to define temperature/pressure conditions for distillation.

Example calculation

Building a crude TBP curve from ASTM lab data

This example is based on Kuwait crude:
Top product is full-range naphtha and its yield is gas to 25% vol.

Second product is kerosene and its yield on crude is 10% vol. (plant data).
Third product is light gas oil with a yield of 12% vol. on crude.
Fourth product is heavy gas oil which is 7.0% vol. on crude.

(a) Converting lab ASTM data to TBP

The overhead naphtha cut

It will not be possible to test this whole cut using ASTM distillation. The front
end is very light hydrocarbons ranging from C_2s ethane to C_7. The front end
will be analysed separately using the POD (Podblinka) apparatus. The result of
this was:

$$
\begin{array}{lll}
C_2 & 0.11 & \text{\% vol. on crude} \\
C_3 & 0.84 & \text{\% vol. on crude} \\
iC_4 & 0.40 & \text{\% vol. on crude} \\
nC_4 & 1.53 & \text{\% vol. on crude} \\
\left.\begin{array}{l}iC_5 \\ iC_5\end{array}\right\} & 3.02 & \text{\% vol. on crude} \\
C_6 & 3.6 & \text{\% vol. on crude} \\
C_7 & 4.5 & \text{\% vol. on crude}
\end{array}
$$

Total   14.0   % vol. on crude

This represents $\dfrac{14.0}{0.25}$% of the naphtha cut

$$= 56\% \text{ vol.}$$

The ASTM distillation on the heavy end of the full-range naphtha is as follows:

$$
\begin{array}{ll}
50\% \text{ vol. over} & 300\,°F \\
70\% \text{ vol. over} & 310\,°F \\
90\% \text{ vol. over} & 330\,°F \\
FBP & 358\,°F
\end{array}
$$

We will now convert this to TBP using a method by Edmister. See Figure A1.1 in
Appendix 1.

This is as follows:

Vol.	ASTM		TBP	
	°F	Δ°F	ΔT°F	°F
50%	300			305
70%	310	10	14	319
90%	330	20	27	346
FBP	358	28	32	378

The kero cut

Converting this cut from ASTM to TBP is as follows:

Vol. %	ASTM (lab test)		TBP	
	°F	ΔT°F	ΔT°F	°F
IBP	345			287
10	376	31	57	344
30	400	24	44	388
50	419	19	32	420
70	436	17	27	447
90	460	24	33	480
FBP	495	35	38	518

The light gas oil cut

	ASTM (lab test)		TBP	
	°F	ΔT°F	ΔT°F	°F
IBP	495			457
10	520	25	48	505
30	532	12	25	530
50	543	11	20	550
70	559	16	26	576
90	572	13	19	595
FBP	595	23	27	622

The heavy gas oil cut

	ASTM (Lab test)		TBP	
	°F	ΔT°F	ΔT°F	°F
IBP	595			568
10	615	20	40	608
30	626	11	25	633
50	633	7	14	647
70	640	7	12	659
90	658	18	25	684
FBP	675	17	20	704

This completes the conversion of the atmospheric distillates to TBP and these are plotted in Figure 6.4.

We now require to define the TBPs on the vacuum distillates. These are developed similarly to the atmospheric one except that the ASTM tests are done at 10 mmHg. These, however, are converted and quoted at atmospheric pressure.

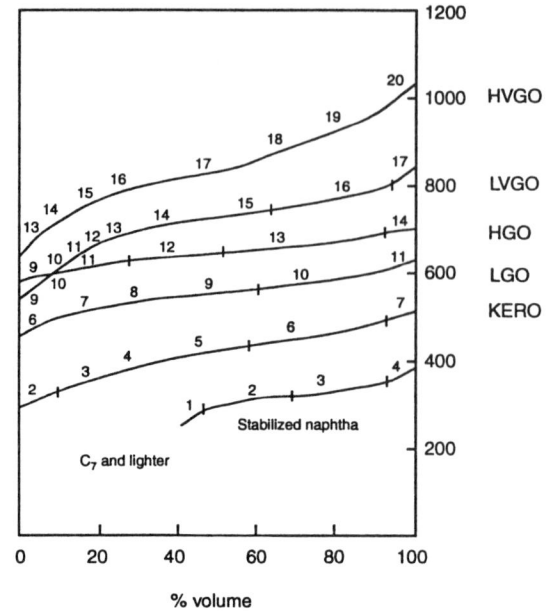

Figure 6.4. Pseudo-components

Light vacuum gas oil cut (LVGO)

	ASTM (lab test)		TBP	
	°F	$\Delta T°F$	$\Delta T°F$	°F
IBP	618			605
10	643	25	49	654
30	668	25	47	701
50	682	14	25	726
70	700	18	27	780
90	720	20	27	753
FBP	760	40	43	823

Heavy vacuum gas oil (HVGO)

	ASTM (lab test)		TBP	
	°F	$\Delta T°F$	$\Delta T°F$	°F
IBP	640			637
10	702	62	95	732
30	740	38	61	793
50	760	20	32	825
70	805	45	59	884
90	856	51	59	943
FBP	925	69	76	1019

This completes the conversion of the vacuum distillate ASTMs to TBP. The yield of LVGO was 11% vol. on crude and HVGO was 20%.

(b) Making the component balance

Yields vol. % on crude

Components	Cut points of		Naphtha	Kero	LGO	HGO	LVGO	HVGO	Total yields Vol. frac.	Cumul.
C_2	—		0.11						0.11	0.11
C_3	—		0.84						0.84	0.95
iC_4	—		0.40						0.40	1.35
nC_4	—		1.53						1.53	2.88
iC_5	—	}	3.02						3.02	5.90
nC_5	—									
C_6	—		3.60						3.60	9.50
C_7	—		4.50						4.50	14.00
Pseudo 1	to 280		5.17						5.17	19.17
Pseudo 2	320		2.53	0.5					3.03	22.20
Pseudo 3	365		2.86	1.5					4.36	26.56
Pseudo 4	405		0.44	2.0					2.44	29.00
Pseudo 5	440			2.5					2.50	31.50
Pseudo 6	490			2.8	0.6				3.40	34.90
Pseudo 7	515			0.7	1.8				2.50	37.40
Pseudo 8	550				3.6				3.60	41.00
Pseudo 9	575				2.4	0.14	0.55		3.09	44.09
Pseudo 10	600				2.4	0.35	0.44		3.19	47.28
Pseudo 11	625				1.2	1.26	0.33		2.79	50.07
Pseudo 12	650					1.75	0.44		2.19	52.26
Pseudo 13	675					2.59	0.44	0.40	3.43	55.69
Pseudo 14	725					0.91	3.08	1.20	5.19	60.88
Pseudo 15	750						2.20	1.40	3.60	64.68
Pseudo 16	800						3.08	3.80	6.88	71.36
Pseudo 17	850						0.44	5.20	5.64	77.00
Pseudo 18	900							3.00	3.00	80.00
Pseudo 19	950							3.40	3.40	83.40
Pseudo 20	1020							1.60	1.60	85.00
Resid.	+ 1020								15.00	100.00
Totals			25.0	10.0	12.0	7.0	11	20	100	

This curve is shown as Figure 6.5.

(c) Developing the equilibrium flash vaporization curve

The Maxwell curves given in Figure A1.2 in Appendix 1 will be used.

The distillation reference line (DRL) is drawn through the 10% and 70% vol. of the TBP.

Slope of this line:

$$50\% \text{ vol.} \quad 600\,°F$$
$$40\% \text{ vol.} \quad 500\,°F$$

$$\text{Slope} = \frac{100}{10} = 10\,°F/\%$$

From the top curve in Figure A1.2 slope of the flash reference line (FRL) $= 6.8\ °F/\%$

From the middle curve in Figure A1.2:

$$\Delta T_{50\%}\ (DRL - FRL) = 39.5$$

The 50% point of FRL then equals temperature at 50% DRL $- 39.5\ °F$.

$$= 600 - 39.5 = 560.5$$

$$\text{Say } 561\ °F$$

The FRL can now be drawn and is given in Figure 6.5.

Using the bottom curve of Figure A1.2 the EFV curve can be developed thus:

	(ΔT TBP/DRL)	ΔT (flash/FRL)*		EFV
% Vol.	°F	ΔT (TBP)/DRL °F	ΔT flash/FRL	°F
0	-100	0.2	-20	200
10	0	0.4	0	290
20	0	0.38	0	355
30	$+25$	0.37	9	434
40	$+40$	0.37	15	510
50	$+25$	0.37	9	569
60	$+20$	0.37	7	637
70	0	0.37	0	700
80	0	0.37	0	770
90	100	0.37	37	872
100	∞	0.37	—	∞

* From Figure A1.2

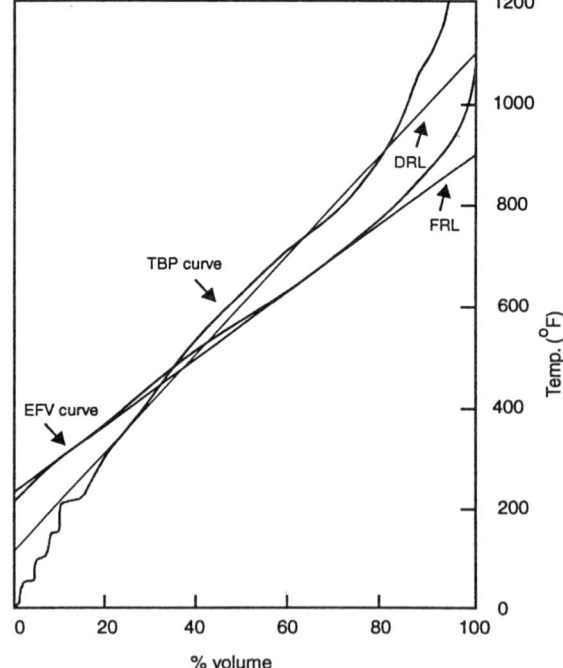

Figure 6.5. TBP and EFV curves for whole crude

6.4   Procedure for Establishing Flash Zone Conditions

Flash zone conditions, particularly flash zone temperature, is difficult to measure accurately. This is due to the profiles set up in the flash zone by turbulence of the feed entering and vapour disengaging. Knowledge of flash zone conditions is, however, essential when evaluating the unit performance and troubleshooting the unit. This section therefore deals with a calculation procedure to establish these conditions.

THE FLASH ZONE

The flash zone is the area in the crude distillation tower where the distillate vapours are allowed to separate from the unvaporized liquid. The transfer line from the heater enters the flash zone. The vapours rise up through the tower to be condensed by cold reflux streams coming down. Steam enters the flash zone from the bottom product stripper section located below the flash zone.

Steps to calculate flash zone conditions

- *Step 1*   Establish the total flash zone pressure. This is done by setting the tower-top reflux drum pressure, then by adding condenser pressure drop, tower and condenser line pressure drop to the reflux pressure. A total pressure in the flash zone is obtained.
- *Step 2*   Set the quantity of stripping steam to be used for bottoms stripping. This was discussed in Section 6.2. Normally a quantity of about 1.2 lb steam per gallon of product is used.
- *Step 3*   From the material balance calculate the moles per hour of vapour leaving the flash zone. Now this will be the total moles of products plus a quantity called overflash. The overflash is never drawn off as a product but is the mechanism whereby additional heat is transferred to the tower to provide reflux.
- *Step 4*   Set the amount of overflash. Normally this will be 3–5% volume on crude above the total product cut point. Thus if it is intended to vaporize 54% of the total crude as distillate products we would actually operate at a temperature sufficient to vaporize 57–59% of the crude. The amount of overflash depends on operating economics and the degree of fractionation required.
- *Step 5*   Construct the equilibrium flash vaporization curve. This is described and discussed in Section 6.3. Set the cut point to include overflash, remembering that this is at atmospheric conditions.
- *Step 6*   From a knowledge of the amount of hydrocarbon vapour moles, the moles of steam in the flash zone and the flash zone's total pressure, calculate the partial pressure of hydrocarbons. This is

$$\frac{\text{moles HC vapour}}{\text{moles steam} + \text{moles HC vapour}} \times \text{total pressure}$$

- *Step 7*   Adjust the EFV curve to the calculated partial pressure. This is accomplished by selecting a temperature on the atmospheric EFV curve.

Then use the vapour pressure charts given in Appendix 1 (Figure A1.6). Take the selected temperature at atmospheric pressure and move along parallel to the vapour pressure line until the required partial pressure axis is intersected. Read off the new temperature. Plot this on the percentage volume axis on which the original temperature was selected. Draw the new EFV through this point parallel to the atmospheric EFV. Locate the required cut point on this new EFV and read off the temperature. This will be the flash zone temperature.

The following example illustrates this procedure.

Example calculation

Using the following material balance and the TBP for Kuwait crude calculate the flash zone temperature and pressure. Use 3% vol. overflash and 1.2 lb/h steam per gallon for bottoms stripping.

Material balance

Stream	TBP cut range (%)	% Volume Cut	% Volume Cumu-lative	Volume BPSD	Volume GPH	Volume SG	*/Gal	Weight lb/h	Weight % wt	Moles Mw	Moles Moles/h
Whole crude	—	100	100	30 000	52 500	0.8685	7.23	379 575	100.0	225.3	1684.8
Overhead	Gas – 380	25.0	25.0	7500	13 125	0.702	5.84	76 850	20.2	95.2	805.1
Kero	– 480	10.0	35.0	3000	5250	0.798	6.64	34 880	9.2	161.6	215.7
Light gas oil	– 810	12.0	47.0	3600	6300	0.843	7.02	44 226	11.7	215.6	205.1
Heavy gas oil	– 690	7.0	54.0	2100	3675	0.875	7.28	26 754	7.0	270.2	99.0
Residuum	+ 890	46.0	100	13 800	24 150	0.957	8.16	197 085	51.9	547.8	359.9
Total products	—	100		30 000	52 500	0.8685	7.23	379 575	100.0	225.3	1684.8

Flash zone material balance

Stream	TBP cut range (%)	% Volume Cut	% Volume Cumu-lative	Volume BPSD	Volume GPH	Volume SG	*/Gal	Weight lb/h	Weight % wt	Moles Mw	Moles Moles/h
Overflash	690 – 725	3.0	57.0	900	1575	0.891	7.4	11 655	3.0	295	39.5
Product vapour	Gas – 690	54.0	54.0	16 200	26 350	0.773	6.43	182 490	48.1	137.7	1324.9
Total vapour	Gas – 725	57.0	57.0	17 100	29 925	0.780	8.49	194 145	51.1	142.3	1364.4
Liquid phase	+ 725	43.0	43.0	12 900	22 575	0.988	8.22	185 430	48.9	578.7	320.4
Total	1	100	100	30 000	52 500	0.8685	7.23	379 575	100.0	225.3	1684.8

*Does not include liquid overflow from bottom wash tray

Flash zone calculations

(a) Total pressure at flash zone:
　　　Establish overhead drum pressure　　　　at　5 psig
　　　Establish air condenser pressure drop　　at　5 psi
　　　Establish crude/overhead exchanger　　at　2 psi (shell side)
　　　　　Total overhead pressure　　　　　　12 psig

Assume 40 trays between flash zone and tower top.
Let ΔP per tray be 0.25 psi
　　　　Total tower pressure drop　　= 10 psi
Then total flash zone pressure　　= 22 psig

Call it 25 psig for design

(b) Calculate partial pressure of vapour in flash zone

From Figure 6.3 we will take bottom stripping steam requirement at 1.2 lb/gal

$$= 1.2 \times 24\,150 = 28\,980 \text{ lb/h steam}$$
$$= 1610 \text{ moles/h steam}$$

Partial pressure of vapour is

$$\frac{\text{Moles HC vapour}}{\text{Total moles vapour}} \times \text{Pressure}$$

$$= \frac{1364.4 \times 25 \text{ psig}}{1364.4 + 1610}$$

$$= 11.47 \text{ psig} = 26.17 \text{ psia}$$

(c) Calculate EFV of whole crude at atmospheric pressure

Slope of whole crude ref line = 11.8 °F/% vol. (Figure 6.5)

(DRL)

Slope of flash ref line (FRL) (from Figure A1.2)

$$= 8.5 \text{ °F/\%}$$

ΔT_{50} (DRL − FRL) = 40 °F
T_{50} DRL   = 667 °F
FRL 50%   = 667 − 40 = 627 °F

% Vol.	Actual	TBP DRL	ΔT	Ratio	ΔT	Flash FRL	Flash
0	−127	75	−202	0.24	−48	200	152
10	190	190	0	0.4	0	285	285
30	420	430	−10	0.34	−3	450	447
50	645	667	−22	0.34	−8	627	619
70	900	900	0	0.34	0	795	795
90	1235	1140	95	0.34	32	915	947
FBP	2192	1250	942	0.34	320	1040	1360

The atmospheric EFV curve is plotted in Figure 6.6. The 50% volume temperature is 619 °F. Using the vapour pressure curves in Appendix 1 locate 619 °F on the atmospheric pressure line. Then move parallel with the nearest sloped line to a pressure of 26.17 psia. Read off this temperature and plot it on Figure 6.6. Draw a line through this point and parallel to the atmospheric EFV curve. This is the EFV curve at 26.17 psia pressure. Read off the temperature for 57% volume distilled. This is the flash zone temperature and is *720 °F*.

6.5 Estimating Tower Draw-off Temperatures

TOWER BOTTOM TEMPERATURE

The tower bottom temperature (residue draw-off) is calculated by a simple heat balance when the flash zone conditions and the quantity of stripping steam are known. Using

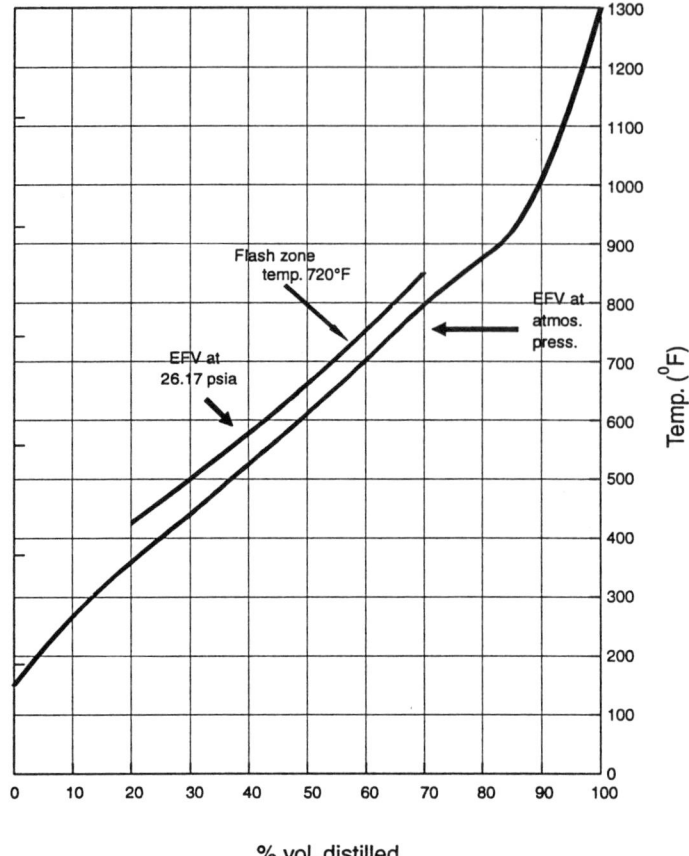

Figure 6.6. Whole crude EFV curves

the material balance given, for example, in Section 6.4 and the calculated flash zone condition, the bottoms temperature can be calculated as follows:

Consider the material on the top stripping tray. This will consist of product residue plus the stripout as liquid:

Product residue is 197 085 #/h
Stripout will be 6% by vol. = 1449 gal/h
Let stripout be 7.5 #/gal and have a mol. wt of 305
The stripout percentage is read from the curves given in Section 6.2 and using the enthalpy data in terms of BtU/lb the heat balance can be closed to give a value to the outlet temperature as shown below. Note that the enthalpy data used in this item have been taken from the author's own files. It is recommended that engineers use enthalpy data that can be obtained in standard data books such as Maxwell's *Hydrocarbon Data* or the GPSA *Engineering Data Book* or some other recognized source.

	V/L	°API	K	°F	lb/h	BTU/lb	Enthalpy BTU/h × 10⁻⁶
In							
Residue	L	13	11.5	720	197.085	396	78.046
Stripout	L	11.5	11.5	720	10.868	410	4.456
Steam	V			450	28.980	1290	37.384
Total					236.933		119.886
Out							
Residue	L	12	11.5	T	197 085	X	197 085 X
Stripout	V	25.5	11.5	715	10 868	490	5.325
Steam	V			*715	28 980	1383	40.079
Total					236 933		197 085X + 45.404

*At partial pressure of FZ = 13.53 psig

$$197\,085\,X + 45.404 = 119.886$$

$$X = \frac{74\,482\,000}{197\,085}$$
$$= 378 \text{ BTU/lb}$$
$$= 704\,°F \text{ (from enthalpy)}$$

The outlet temperature is 704 °F.

SIDE-STREAM DRAW-OFF TEMPERATURE

A method developed by J. W. Packie (5) has been used extensively for this purpose. This method uses the partial pressure of the side-stream product FRL (flash reference line) initial temperature. The calculation to obtain the FRL has already been dealt with in Section 6.3. In arriving at the partial pressure on the tray all material lighter than the draw-off product and the steam passing through the tray are considered inert. Carry out the calculation using the following steps:

● *Step 1* Fix the amount of steam passing through each draw-off. Use the curves in Section 6.2
● *Step 2* Fix the total pressure for each draw-off tray (pressure profile). Again use the method given in Section 6.4 and an estimate of the tray location.
● *Step 3* Calculate the FRL and establish initial boiling points. Use the TBP cut point for this (not the TBP end points).
● *Step 4* To obtain the total moles of vapour passing through the tray an estimate of overflow from the draw-off tray must be made. For this purpose use a rule of thumb which gives the following ratio of overflow to product moles:

HGO overflow	2.9
LGO overflow	1.2
Kero	0.9–1.0

● *Step 5* Calculate the partial pressure of side-stream draw-off at each tray. Using the vapour pressure curves in Appendix 1 relate the FRL initial boiling

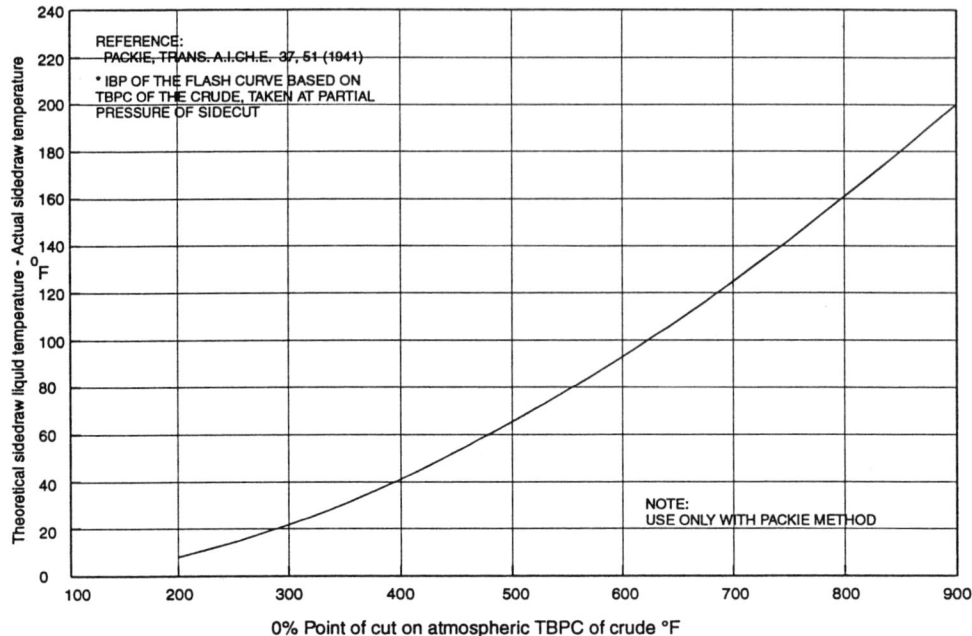

Figure 6.7. Temperature correction for calculating side-stream draw-off temperature—crude distillation

point (which is at atmospheric) to that for the partial pressure. This has been described in Section 6.4.

- *Step 6*   Now the temperature at the partial pressures evaluated in Step 5 is corrected using Figure 6.8. The resulting temperature is the draw-off temperature from the main column.

Consider the following example using the material balance given for Section 6.4 and the steam rates provided by Figure 6.3;

Example calculation

- *Steam rates to be used in side-stream stripping*

  All strippers have four actual trays.

  HY gas oil will use 0.3 # steam per gal
  LT gas oil will use 0.5 # steam per gal
  Kero will use 0.65 # steam per gal

  Steam to

  Heavy gas oil - 3675×0.3 $= 1838 \, \text{lb/h}$
  Light gas oil - 6300×0.5 $= 3150$
  Kero   - 5250×0.65 $= 3413$

● *Pressure profile in tower*

This will be 40 trays in the fractionating side.

Summary of tray locations and pressures

	Tray no.	Pressure (psig)
HGO draw-off	32	23.4
Bottom PA return	29	22.7
LGO draw-off	22	19.5
LGO PA return	19	18.8
Kero draw-off	12	15.5

● *Calculate theoretical initial boiling points*

We will use the Packie method for this. From cut TBP produce the FRL for each cut.

HGO   50% point TBP is 648 °F
Slope of TBP is 0.6 °F/%
Slope of FRL is 0.2 °F/%
Δ50 (DRL − FRL) is 7
50% FRL is 641
From FRL curve IBP is 626 °F at atmospheric

LGO   50% point TBP is 550 °F
Slope TBP is $\dfrac{575 - 505\,°F}{60} = 1.16\,°F/\%$
Slope FRL is 0.3 °F/%
50% FRL is 550 °F
IBP is 527 °F at atmospheric

Kero   TBP 50% point is 420 °F
TBP slope is $\dfrac{448 - 345}{60} = 1.72\,°F/\%$
FRL slope is 0.5 °F/%
FRL 50% is 420 °F
IBP FRL is 395 °F at atmospheric

● *Calculate approximate partial pressures and draw-off temperature*

Internal reflux is assumed as follows:

to HGO tray   290 moles
to LGO tray   250 moles
To kero tray   200 moles

HGO tray partial pressure

$$\frac{\text{Moles HC}}{\text{Total moles vapour} + \text{steam}} \times \text{Total pressure}$$

$$= \frac{1614.9}{3224.9} \times 38.1 = 19.1 \text{ psia}$$

Theoretical temperature is 660 °F (using vapour pressure curves)

From Figure 6.7:

Theoretical temperature − actual temperature = 93
actual temperature = 567 °F

LGO draw-off temperature

Steam to tray is: Bottoms stripping   1610 moles/h
                       HGO stripping   102 moles

                       Total   1712 moles/h

Partial pressure of HC

$$\frac{1225.9 + 250}{3187.9} \times 34.2 \text{ psia}$$

$$= 15.8 \text{ psia}$$

FRL at 14.7 0% = 530 °F
FRL at 15.8 psia = 545 °F

From Figure 6.7 actual temperature is 545 − 52

$$= 493 \text{ °F}$$

Kero draw-off temperature

Steam to tray is: Total from   HGO tray   1712
                                  LGO stripping   175
                                  Total   1887

Partial pressure of HC

$$\frac{1020.8 \times 200}{3107.8} \times 30.2 = 11.9 \text{ psia}$$

0% FRL at 14.7 psia is 395 °F
0% FRL at 11.9 psia is 380 °F

Figure 6.7 at actual temperature is   380 °F − 36
                                        = 344 °F

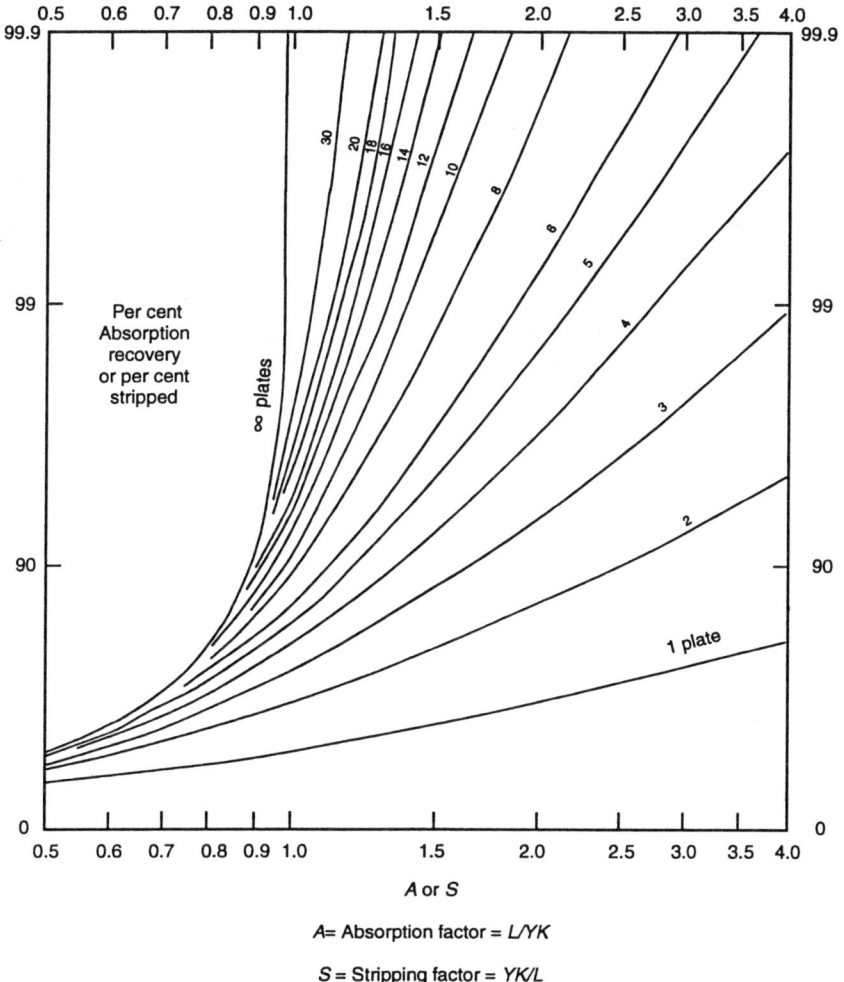

Figure 6.8. Absorption and stripping factors (Reproduced by permission of Fluor Daniel Inc)

6.6 Calculating Tower-top Conditions

Before proceeding with this section it is necessary to define and establish some fundamental thermodynamic definitions relating to fractionation in general:

- *Bubble point* This is the temperature and pressure conditions at which a hydrocarbon liquid begins to vaporize.
- *Dew point* This is the temperature and pressure conditions at which a hydrocarbon vapour begins to condense.
- *Flash vaporization* This is the quantity and composition of a liquid and vapour in equilibrium at any given condition of temperature and pressure. Note that the composition of liquid and vapour in equilibrium will be different to one another.

The tower-top conditions relate to the temperature and pressure for the reflux drum and above the top tray of the tower.

REFLUX DRUM CONDITIONS

The temperature condition for this drum is set by either ambient air conditions, if the tower overheads are air cooled, or water temperature if water cooled. Normally the reflux drum in a crude unit handles totally condensed material (i.e. no vapour phase) and is usually cooled to a point below its boiling point (or bubble point). The pressure at which the drum is operated is calculated from the bubble point of the overhead material. Because the drum is normally subcooled the pressure is maintained by a fuel gas stream. Consider the following example and calculation.

Example calculation

The overhead from the material balance given in Section 6.5 has the following component:

Mole fraction composition:
C_2 0.008
C_3 0.054
iC_4 0.021
nC_4 0.084
C_5s 0.143
C_6 0.155
C_7 0.175
Comp 1   0.124
Comp 2   0.124
Comp 3   0.075
Comp 4   0.037
Total   1.000

The reflux drum temperature will be fixed at 100 °F. The pressure will be calculated at the bubble point of this material at 100 °F.

Bubble point is defined as the sum of all Ys = sum of all KXs
K is the equilibrium constant. This can be read from the equilibrium curves or can be considered (rough and not be used for definitive design) as:

$$\frac{\text{Vapour pressure}}{\text{Total pressure}}$$

This relationship will be used for this calculation.

The calculation is iterative (trial and error) as follows:

		At 100 °F			
		1st trial at 5 psig		2nd trial at 10 psig	
	Mol fract. X	K	$Y = KX$	K	$Y = KX$
C_2	0.008	40.6	0.325	32.4	0.259
C_3	0.054	9.3	0.502	7.42	0.401
iC_4	0.021	3.55	0.075	0.38	0.05
nC_4	0.084	2.54	0.213	0.03	0.171
C_5	0.143	0.89	0.127	0.71	0.102
C_6	0.155	0.254	0.039	0.20	0.031
C_7	0.175	0.084	0.015	0.067	0.011
Comp 1	0.124	0.023	0.003	0.020	0.002
Comp 2	0.124	Neg	Neg		
Comp 3	0.075	Neg	Neg		
Comp 4	0.037	Neg	Neg		
	1.000		1.299		1.027

For 2nd trial (estimate)

Take the K value of the highest 'y' fraction (in this case C_3) where $K = 9.3$.

Take this $\dfrac{K}{1.299} = 7.16$ (new K)

Make the 2nd trial with KC_3 at 7.10

$\dfrac{pC_3}{P} = 7.10$ pC_3 at 100 °F = 190 psia then total pressure $= \dfrac{190}{7.16} = 26.5$ psia

2nd trial pressure = 26.5 psia = 11.8 psig

   Let us set it at 10 psig

The second trial gives sum of Ys = 1.027 as this is considered close enough. Then drum will be operated at 100 °F and at 10 psig.

TOWER-TOP CONDITIONS

The temperature and pressure conditions that will exist at the top of the tower will be the dew point of the overhead hydrocarbon product at its partial pressure with the steam. This can be calculated in a similar manner as that for the reflux drum. This time, however, the relationship for the condensing phase will be used. This is sum of all xs = sum of all Y/K.

   The total pressure in the top of the tower is fixed by the reflux drum pressure. The resulting temperature is calculated using the steps given in the following example. The quantities (moles/h) of hydrocarbons follow those given in the material balance Section 6.5. The steam figures used are those previously used.

Tower-top temperature conditions

The crude tower configuration in this example uses a pumparound to condense reflux overflow. Many crude units condense product and an external reflux stream using an overhead condenser (see Section 6.1).

In this example only the product vapour and total steam need to be considered for partial pressure conditions at the tower top. Assume that total pressure is 15 psig at the top of the tower:

$$\text{Total vapour moles} = 805.1 \text{ moles/h}$$
$$\text{Total moles steam} = 2076$$

$$\text{Partial pressure} = \frac{805.1 \times 29.7 \text{ psia}}{2881.1}$$
$$= 8.3 \text{ psia}$$

The following dew point calculation will be carried out at 8.3 psia.

| Comp | Mole frac. Y | 1st trial at 220 | | 2nd trial at 225 °F | | Mol wt | Weight factor | SG | Vol. factor | Liquid prop. |
		K	X = Y/K	K	X = Y/K					
C₂	0.008	—	Neg		Neg					
C₃	0.054	84.3	0.001	93.9	0.001	44	0.044	0.508	0.009	
iC₄	0.021	38.6	0.001	39.8	0.001	58	0.058	0.563	0.010	mol. wt = 130.7
nC₄	0.084	29.52	0.003	30.1	0.003	58	0.174	0.584	0.030	
C₅	0.143	12.53	0.011	12.65	0.011	72	0.792	0.629	0.126	SG = 0.766
C₆	0.155	4.70	0.033	4.94	0.031	85	2.635	0.675	0.390	
C₇	0.175	2.17	0.081	2.19	0.080	100	8.00	0.721	1.110	°API = 53
Mid-BP 260	0.124	1.00	0.124	1.16	0.107	114	12.198	0.743	1.642	
Mid-BP 300	0.124	0.506	0.245	0.518	0.239	126	30.114	0.765	3.936	K = 12
Mid-BP 340	0.075	0.229	0.328	0.253	0.293	136	39.848	0.776	5.135	
Mid-BP 382	0.037	0.108	0.343	0.126	0.294	152	44.688	0.788	5.671	
Totals	1.000		1.170		1.06	130.7	138.551	0.767	18.059	

$$K = \frac{\text{Vapour press at selected temperature}}{\text{Total systems pressure}}$$

$2nd\ trial = 0.108 \times 1.170$ (K from mid-BPT 362 component)

New $K = 0.126$ then $VP = 8.3$ psia $\times 0.126 = 1.05$ psia $= 225$ °F

2nd trial is close enough actual temperature:

$0.126 \times 1.06 = 0.134$ VP $= 1.11 = 229$ °F

Notes

(a) In estimating for the 2nd trial and final temperature the K of the highest X component is multiplied by the total value of the X function. Then vapour pressure curves are used to give the component temperature corresponding to this new vapour pressure.

(b) The molal composition of the final X is the composition of the liquid in equilibrium with the product vapour. In other words, the composition of the pumparound liquid and the internal reflux.

The tower-top conditions are 229 °F at 15 psig.

6.7 Fractionation and Internal Reflux at the Tower-top Section

This crude tower condenses internal reflux using a pumparound section at the very top of the tower. This is a four-tray section and these trays are used for heat transfer only. However, a credit of one tray can be taken for separation. Consequently, the total number of trays between tower top and kero draw-off for mass transfer is nine.

Using the procedure described in Section 6.2, internal overflow is established by the family of curves given in Figure 6.2 for max. steam and the overhead to first side-stream. This is illustrated in the following example.

Example calculation

The material balance given in Section 6.5 is used as the basis. An ASTM Gap of 30 °F is required between the 5% point for kero and the 95% point for naphtha.

The 50% TBP difference between overhead and 1st side-stream is approximately $420 - 265 = 155$ °F

Fractionation factor is 17.4 (read from the curves in Figure 6.2)

The total number of trays between kero and the tower top is nine.

Therefore reflux ratio is $\dfrac{17.4}{9} = 1.933$

Reflux ratio is $\dfrac{\text{Gals of hot overflow}}{\text{Gals of total product vapours.}}$

Then gals of hot overflow required

$$= 1.933 \times 13\,125$$
$$= 25\,271 \text{ gal/h}$$

Gal/h overflow at 60 °F $= \dfrac{25\,371}{1.137}$ (this is expansion factor)

$$= 22\,314$$

From the dew point calculation item 1.7 lb/gal for overflow is 6.36 (53° API) overflow
$$141\,917\,\text{lb/h}$$

6.8 Calculating the Top (Naphtha) Pumparound Duty to Meet an Internal Reflux Requirement

This is a heat balance over the tower-top configuration using tower-top conditions and knowledge of the required internal reflux between naphtha and kero. The following example to illustrate this uses the material balance from Section 6.4, the tower conditions developed in Section 6.6 and the internal reflux calculated.

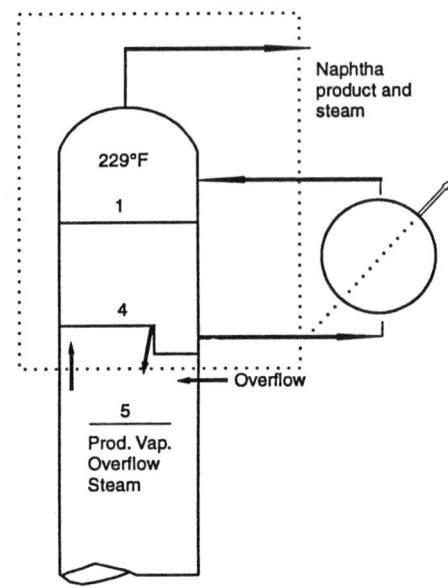

Consider the heat balance over the envelope shown in the diagram:

Heat in with
● Product vapour from tray 5
● Overflow vapour from tray 5
● Steam from tray 5

Heat out with
● Naphtha product vapour
● Steam from tower top
● Overflow liquid to tray 5
● Pumparound

The heat balance is as follows:

		V or L	API°	K	°F	lb/h	Enthalpy BTU/lb	Enthalpy BTU/h × 10⁶
In	Product vap.	V	70	12	245[1]	76 650	266	20.389
	Overflow vap.	V	53	12	245	141 917	252	35.763
	Steam	V			245	37 381	1194	44.633
	Total in					255 948		100.785
Out	Product	V	70	12	229	76 650	258	19.776
	Steam	V			229	37 381	1188	44.409
	Overflow	L	53	12	234[2]	141 917	114	16.179
	Pumparound							C
	Total out					255 958		80.364 + C

(1) Estimated from tower profile
(2) Rule of thumb—top tray temperature is tower top + 5 °F

$$\text{Heat in} = \text{Heat out}$$
$$\text{Then } 100.785 = 80.364 + C$$
$$\text{Pumparound duty} = 20.421 \times 10^6 \text{ BTU/h}$$

6.9 Calculating the Required External Reflux to the Top of the Tower

(Note that this calculation is applicable to the more conventional overhead configuration where the total product, reflux and steam are condensed in the overhead condenser.)

There is a relationship between the quantity of reflux pumped back to the tower from the reflux drum and the internal reflux. As the external reflux flow can be measured and the internal reflux cannot, it is desirable to calculate the relationship between the two reflux streams. This will be necessary because it is the internal reflux that establishes the amount of fractionation between the adjacent cuts (ASTM gap). The steps to establish this are as follows:

- *Step 1*   From the product naphtha lab analysis calculate tower-top conditions using the steam input to the tower (Section 6.6).
- *Step 2*   Establish the internal reflux using ASTM gap required (Section 6.7).
- *Step 3*   Calculate the condenser duty required to generate the internal reflux (Section 6.8). Adjust the calculation to that for total overhead condenser.
- *Step 4*   Using the condition at the tower top and reflux drum calculate a heat balance over the top of the tower. The unknown in this case will be the flow of external reflux. Equate and solve for the unknown.

Steps 1 to 3 have already been described and using the example provided in these steps the heat balance for Step 4 is given in the following example.

Example calculation

	V or L	API°	K	°F	lb/h	BTU/lb	Enthalpy BTU/h $\times 10^6$
In							
Naphtha product	V	70	12	246[1]	76 650	266	20.389
Reflux	V	70	12	246	X	266	266 X
Steam	V			246	37 381	1197	44.745
Total in					114 031 + X		65.134 + 266 X
Out							
Naphtha product	L	70	12	100	76 650	46	3.526
Reflux	L	70	12	100	X	46	46 X
Water	L			100	37 381	100	3.738
Condenser							71.360
Total out					114 031 + X		78.624 + 46 X

$$65\,134\,000 + 266X = 78\,624\,000 + 46\ X$$

$$X = \frac{13\,490\,000}{220}$$

External reflux $= 61\,318$ lb/h

Note: The tower-top temperature is now that calculated in Section 6.7 at a slightly higher partial pressure equivalent to

$$\frac{\text{Moles Prod.} + \text{moles reflux (assumed)}}{\text{Total moles HG + steam}} \times 29.7 \text{ psia}$$

6.10   Establishing the Internal
Reflux between Side-streams

This procedure utilizes the curve relationship between ASTM gaps, reflux ratios
and number of trays developed by Packie (5) and described in Section 6.2. These
steps are:

- *Step 1*   Establish the degree of fractionation required. Remember there are limits
  to this. For example, it is very unlikely that an ASTM gap will be attained in the
  heavier end of the tower (i.e. LGO to HGO).
- *Step 2*   Evaluate the 50% vol. temperature difference between the TBP of the
  two adjacent cuts.
- *Step 3*   Using the side-stream-to-side-stream family of curves for max. steam read
  off the factor corresponding to the ASTM gap (or overlap) on the Δ50% TBP
  temperature curve.
- *Step 4*   The factor is the number of trays between the cuts multiplied by the reflux
  ratio. The reflux ratio is

$$\frac{\text{Hot overflow (gal/h)}}{\text{Total product gal/h entering the highest tray}}$$

- *Step 5*   From this the hot overflow can be calculated and will be used to predict
  pumparound requirements and tower performance.

These steps are illustrated by the following example.

Example calculation

Using the quantities given in the material balance (Section 6.4) and the TBP curves
given in Section 6.3, predict the overflow required to obtain a 30 °F ASTM overlap
between LGO and HGO.

- Number of trays between LGO and HGO draw-offs are 11 (two pumparound
  trays are equal to one fractionating tray).
- The temperature difference between the 50% of *total* product passing through
  the LGO tray and the HGO product is 300 °F.
- Using 30 °F overlap and the 300 °F curve in the side-stream-to-side-stream max.
  steam curve gives a factor of 10.2.
- The reflux ratio is $\dfrac{10.2}{11} = 0.92$

- $\dfrac{\text{Hot gallons of overflow}}{\text{Gal of product vapour}} = 0.92$

Gal of product vapour = 24 675 (from material balance).
The hot gallons of overflow = 22 701 gal/h

6.11 Calculation to Establish Pumparound and Overflash that will meet Internal Reflux Requirements

Internal reflux throughout the tower is generated by heat input through overflash and its removal by pumparound or external reflux streams. All other heat to maintain heat balance is removed by the product streams. Thus, if the percentage overflash is increased there must be a comparable increase in pumparound duties. Otherwise the only effect of increase overflash will be to change the cut point.

Overflash, as described earlier, is that portion of the crude feed that is vaporized over and above that amount vaporized to provide the distillate products. Normally overflash is between 3% and 5% volume on crude. Pumparound is the method of removing this heat by an external stream drawing off a quantity of liquid from the tower. This is cooled and returned to the tower some trays (usually two to three) above the draw-off. This area in the tower acts as an internal condenser, and condenses vapours that are not removed as the product but are allowed to flow downwards from tray to tray as reflux. Normally the amount of overflash is fixed by the coil outlet temperature. The variable then is the pumparound duty at any of the relevant sections. Remember, when you change one pumparound then the other pumparounds also change to meet heat balance requirements.

Consider the following example.
It is required to calculate the bottom pumparound duty to meet the internal reflux calculated in Section 6.10. Mass balances will be as given in Section 6.4. Temperatures, flash zone conditions and steam rates will be as used previously in this chapter. In this calculation side-stream stripper bottom temperatures are going to be used instead of the calculated side-stream draw-off as a rule of thumb these are about 10 °F lower than draw off. However, they can be calculated, as will be shown later.

Overflow from tray 22 to meet fractionation of the 30 °F gap between 95% vol. ASTM of LGO and 5% vol. ASTM of HGO is 22 701 gal/h at 498 °F, or 17 487 gal/h at 60 °F. The SG at 60 is 0.843, then lb/h is 122 412. The heat balance is conducted over the envelope shown below:

Streams into the envelope
Crude feed
Overflow to tray 21
Steam to HGO stripper
Steam to bottom stripper

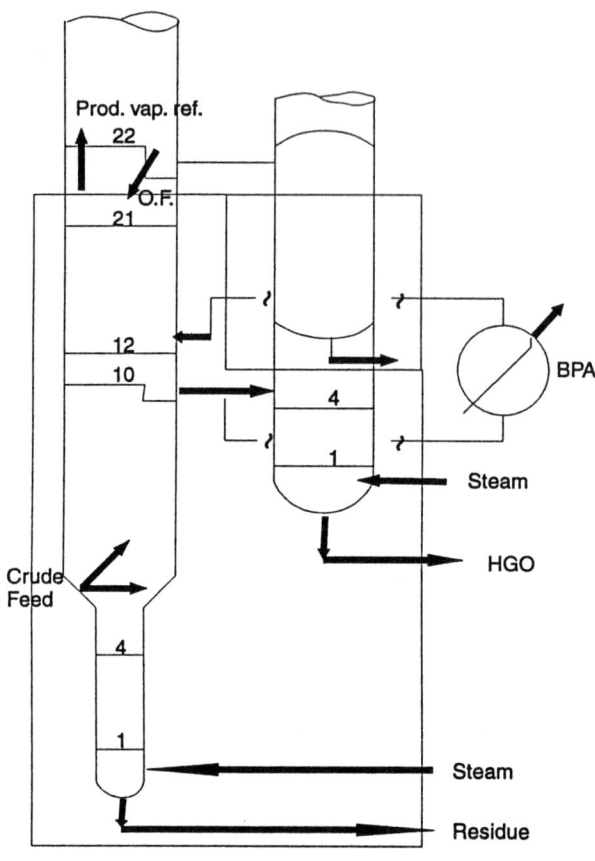

	V or L	°API	K	Temp. (°F)	lb/h	Enthalpy	
						BTU/lb	BTU/h × 10⁶
In							
Crude feed vapour	V	50	11.5	720	194 145	528	102.509
Crude feed liquid	L	11.5	11.5	720	185 430	396	73.430
Steam (total)	V			450	30 818	1290	39.755
Overflow to tray 21	L	36.6	11.5	498	122 412	263	32.194
Total in					532 805		247.888
Out							
Product distillates	V	55	11.5	500	155 736	394	61.360
Overflow	V	36.6	11.5	500	122 412	365	44.680
HGO product	L	30	11.5	557	26 754	296	7.919
Residue	L	13	11.5	704	197 085	378	74.498
Steam to LGO	V	—		500	30 818	1318	40.618
Bottom pumparound							C
Total out					532 805		229.075 + C

C = Pumparound duty
As heat in = heat out
Then 247.888×10^6 BTU/h $= 229.075 \times 10^6 + C$ BTU/h
Pumparound duty $= \underline{18.813 \times 10^6}$ BTU/h

Streams out of envelope
Product vapour plus overflow vapour (product vapour = naphtha, kero and LGO)
Steam to tray 22
HGO product (stripped)
Residue product
Pumparound duty

6.12 Overall Tower Heat Balance

When the flash zone conditions, the product draw-off temperatures, and total stream input are established it is now necessary to predict side-stream stripper product temperature to construct an overall heat balance. The purpose of the overall heat balance will be to check the total pumparound duties and establish any of those not already fixed for specific fractionation requirements.

In estimating side-stream stripper bottoms temperature the rule of thumb given in Section 6.11 can be used. However, for this exercise a heat balance over a side-stream stripper is given as an example of a temperature calculation method. The steps for preparing the overall heat balance are as follows:

- *Step 1*   Establish total heat in with crude.
- *Step 2*   Calculate stripper bottoms temperature for distillate products (an example for one product outlet is given).
- *Step 3*   Calculate bottoms temperature outlet (this has been done in Section 6.5).
- *Step 4*   Show total steam input—this will be as in the curves in Section 6.2.
- *Step 5*   Commence the heat balance as shown in the example that follows on page 81.

STRIPPER PRODUCT TEMPERATURE

The example provided in this section will be for the heavy gas oil stream using the quantities given in the material balance in Section 6.4. Steam rates and stripout will be fixed as per the curves given in Figure 6.2.

HGO stripper outlet temperature

Steam rate will be 0.5 lb/h per gal/h of product.

$$= 1838 \, \text{lb/h}$$

Stripout is 5% vol. of feed stream (Figure 6.2).

Let mol.wt of stripout be 230 (estimated average of distillates) and SG be 0.865 (again estimated). Consider the following heat balances.

Stream	V/L	°API	K	°F	lb/h	BTU/lb	Enthalpy MM BTU/h
In							
Feed Ex. HGO	L	30	11.5	567	26 754	304	8.133
Steam	V	—	—	450	1 838	1290	2.371
Stripout	L	32	11.5	567	1 390	304	0.423
Total in					29 982		10.927
Out							
Stripout	V	32	11.5	562	1 390	389	0.555
HGO	V	30	11.5	(1)	26 754	X	26 754 X
Steam	V	—	—	t	1 838	*1316	2.419
				562			
Total out					29 982		2.974 + 26 754 X

*At partial pressure of 36.3 psia
(1) Estimated at 5 °F below feed
26 754 X = 7 953 000
$\quad$ X = 297 BTU/lb
Exit temp. HGO = 558 °F

Note: The feed ex HGO tray temperature is that calculated in Section 6.6.

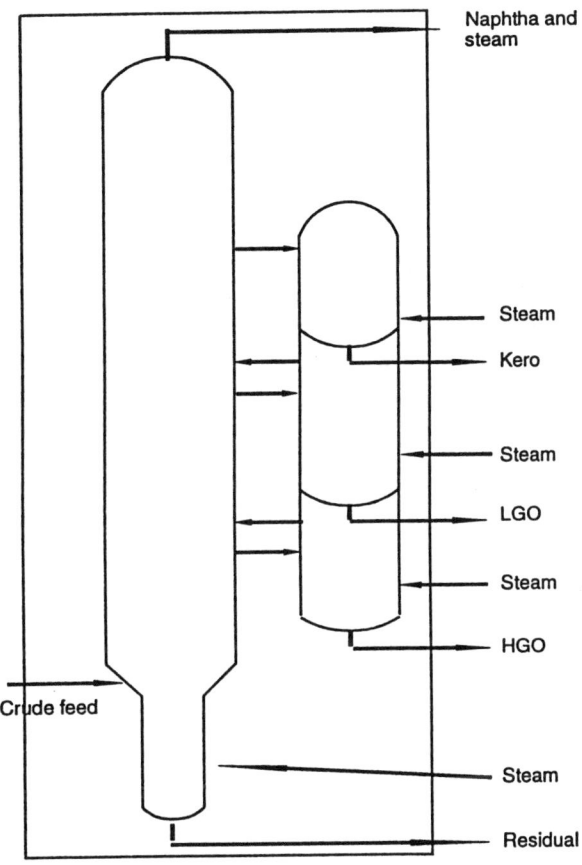

	V or L	°API	K	Temp °F	lb/h	BTU/lb	Enthalpy BTU/h × 10⁶	Section
In								
Crude feed	V+L	—	—	720	379 575	—	175.939	6.12
Steam	V			450	37 381	1290	48.221	6.9
Total in					416 936		224.160	
Out								
Residuum	L	13	11.5	704	197 085	378	74.498	6.5
HGO	L	30	11.5	558	26 754	297	7.946	6.12
LGO	L	36.3	11.5	488	44 226	255	11.278	6.5
Kero	L	45.5	12	339	34 860	184	6.414	6.5
Naphtha	V	70	12	229	76 650	258	19.776	6.9
Steam	V	—	—	229	37 381	1188	44.409	6.9
Pumparounds				(by difference)			59.839	
Total out					416 956		224.160	

Set pumparound duties as:
Top pumparound (naphtha)   20.421 BTU/h × 10⁶
Mid pumparound (LGO)   20.605 BTU/h × 10⁶
Bottom pumparound (HGO)   18.813 BTU/h × 10⁶

Total   59.839

6.13 Side-stream Stripper Performance

Side-streams from the crude distillation tower contain light material in solution. To meet most product specifications the light material will require to be stripped out. Normally, steam stripping is used for this purpose. There are some rule-of-thumb methods of predicting the amount of steam and stripout from such a process. Indeed, these rules of thumb are quite reliable and very well proven. One such method is given in Section 6.2

Steam stripping takes place in a side-stream stripping column outside the main crude distillation column. These columns contain four or five stripping trays. The unstripped oil, which is at its bubble point, enters the top of tower and onto the top tray. Steam is introduced into the tower below the bottom tray. Steam and the liquid move counter-currently in the tower. The steam moves upwards and the liquid stream downwards. Heat and mass transfer occur on the trays, taking out the light material into the steam and leaving the heavier material in the liquid phase. The amount of mass transfer that occurs depends on the ratio of vapour and liquid flow and also on the thermodynamic equilibria function of the components in both phases.

The performance of the unit may be equated to the efficiency of the trays in effecting the steamout required. Generally, stripping trays are never very efficient. In fact, usually they are below 50% even in reboiled strippers, which are more efficient than simple steam stripping. In evaluating the efficiency of the stripper or stripping trays in the steam stripping unit, the number of theoretical trays for the same amount of stripping can be calculated. The following are the steps to make such a calculation:

- *Step 1* Obtain the ASTM distillation (lab test) for the unstripped material and the stripped product. As our example is based on LGO the unstripped sample could be that for the LGO pumparound.
- *Step 2* From plant data determine the quantity of product flow and the feed to the stripper.
- *Step 3* Using the method of converting ASTM to TBP, split the TBP into pseudo-mid-boiling point components. By using assay data assign specific gravity and mol wt values to the pseudo-components. Determine total moles per hour of each component in the feed.
- *Step 4* From plant data obtain the steam rate to the stripper. From Figure 6.3 determine percentage stripout.
- *Step 5* Establish the vapour/liquid ratio for the top and bottom of the tower. Assume that all the stripout and steam enters the top tray. Assume negligible hydrocarbon in the vapour to the bottom tray. Assume all the liquid feed enters the top tray and only the product leaves the bottom tray.
- *Step 6* If available, use plant data for the temperature of feed and any streams leaving. If data are not available develop the profile by heat balance as shown in Section 6.12.
- *Step 7* Calculate the equilibrium constants K for each component in the feed at the mean temperature of the tower. The K in this case may be taken as the vapour pressure of the components divided by the total pressure. Use the vapour pressure curves in Appendix 1 for this.

- *Step 8* Calculate the stripping factor S for each component. S is equal to VK/L. Using Figure 6.8 determine the percentage stripped for each component for one theoretical plate, two theoretical plates and, if necessary, for more theoretical plates.
- *Step 9* Proceed and determine the total stripped for each case of theoretical trays under consideration. By interpolation, match the number of theoretical plates (or fractions of theoretical trays) that will give the closest to the stripout and composition based on plant data.
- *Step 10* The theoretical trays calculated divided by actual trays in the column multiplied by 100 gives the tray efficiency (overall).

The following example illustrates this calculation procedure.

Example calculation

Steam stripping

Feed rate and composition (unstripped LGO)

The basis is 6300 gal/h + 472.5 gal/h light material
  = 6772.5 gal/h

The composition of this material is as follows.

Pseudo-component mid-boiling point (°F)	% vol.	gal/h	SG at 60 °F	lb/h	Mol. wt	Moles/h	Mole fract.
180	1.2	81.3	10.705	573	93	6.16	0.022
280	3.0	203.2	0.754	1532	121	12.66	0.045
360	2.0	135.5	0.780	1057	150	7.05	0.025
410	12.8	866.9	0.793	6875	169	40.68	0.146
460	15.3	1036.2	0.801	8300	194	42.78	0.153
490	19.5	1320.6	0.836	11035	199	55.45	0.199
520	16.7	1131.0	0.844	9546	215	44.40	0.159
550	17.4	1178.4	0.846	9969	230	43.34	0.155
591	12.1	819.4	0.850	6965	254	26.89	0.096
Totals	100.0	6772.5	0.825	55852	200	279.41	1.000

The amount of steam to be used will be 0.5 lb/h per gal of product

  = 6300 × 0.5
  = 3150 lb/h steam
  = 175 moles/h steam

The Kremser equation will be used for calculations of percentage stripped. This is given in Figure 6.8.

Establish the V/L ratio

Consider the top plate
Steam $= 175$ moles/h
HC vapour at the top tray $= 7.5\%$ vol. stripout

All material lighter than 410 °F mid-BPT is stripped out and 2.36 moles/h of 410 °F BPT.

Then total moles of HC as vapour $= 28.23$
The liquid to the tray will be total moles of feed.

Top

Total vapour V $= 175 + 28.23$ moles/h
Liquid L $= 279.4 = 279.41$ moles/h

At the bottom of the stripper

The residual amount of HC vapour is 0%

The vapour is steam $= 175$ moles/h
Liquid is total moles $- 28.23$ $= 251.17$ moles/h

V/L top $= \dfrac{203.2}{279.4} = 0.73$

V/L bottom $= \dfrac{175}{251.1} = 0.70$

Ave V/L $= 0.715$

The stripping factor for each component is VK/L. Where K is the equilibrium constant.

In this case it will be $\dfrac{\text{Vapour pressure}}{\text{Total pressure}}$

Assume the total pressure for the stripper is $19.5\,\text{psig} = 34.2\,\text{psia}$ (see Section 6.6)
Then using Figure 6.8 first estimate will be based on two theoretical plates and the percentage of each component stripped out calculated thus:

Mid-BPT (°F)	Moles/h	K 493 °F 34.2 psi	VK/L	2 Theor. plates		1 Theor. plate	
				% strip	Moles strip	% strip	Moles/h
180	6.16	12.9	9.22	100	6.16	100	6.16
280	12.66	4.97	3.55	94.8	12.00	78	9.87
360	7.05	1.87	1.34	75.0	5.29	54	3.81
410	40.68	1.08	0.77	59.0	24.0	42	17.09
460	42.78	0.58	0.41	30.0	12.80	15	6.42
490	55.45	0.43	0.31	NEG	NEG	NEG	NEG
520	44.40	0.32	0.23	NEG	NEG	NEG	NEG
550	43.34	0.20	0.14	NEG	NEG	NEG	NEG
591	26.89	0.12	0.09	NEG	NEG	NEG	NEG
Totals	279.41				60.25		43.35

Temperature for K is taken as mean in the tower and is based on the heat balance and estimate method given in Section 6.2.

The amount of stripout that was achieved at the steam rates given was 28.23 moles/h.

This should have been achieved by ½ theoretical tray (interpolating). The tower had four actual trays, therefore tray efficiency was only

$$\frac{0.5}{4} \times 100 = 12.5\%$$

6.14   Prediction of Tray and Tower Performance

This section deals with analysing tray performance in terms of flood and entrainment under calculated vapour and liquid loads. This can be carried out using test run data on the plant and referring to the manfacturers' data for the tray geometry. In this case the analysis is based on percentage of critical vapour velocity as a criterion to detect possible entrainment problems. Downcomer velocities are used to detect any possible excessive downcomer back-up. A more rigorous calculation can be used to determine actual downcomer back-up.

The calculation steps used for this tray analysis, which are based on valve trays, are as follows:

- *Step 1*   From plant data, lab data and calculations described in Sections 6.8, 6.9 and 6.11, summarize tray vapour and liquid traffic and tray conditions for critical trays. Critical trays may be defined as the top tray and side-stream and pumparound draw-off trays.
- *Step 2*   Obtain the certified manufacturers' drawings relating to tray geometry and downcomer details. Make sure that this represents the actual tray installed and no modifications have been done to the tray. These drawings should also provide details of tray spacing.
- *Step 3*   Compute the liquid loading on the tray using actual GPM divided by 450 to give CFS.

Table 6.1 Valve tray design principles

Design feature	Suggested value	Alternate values	Comment
1. Valve size and layout			
(a) Valve diameter	—	—	Valve diameter is fixed by the vendor
(b) Percentage hole area A_o/A_b	12	8–15	Open area should be set by the designer. In general, the lower the open area, the higher the efficiency and flexibility and the lower the capacity (due to increased pressure drop). At values of open area toward the upper end of the range (say, 15%), the flexibility and efficiency are approaching sieve tray values. At the lower end of the range, capacity and downcomer filling becomes limited
(c) Valve/pitch/diam. ratio	—	—	Valve pitch is normally triangular. However, this variable is usually fixed by the vendor
(d) Valve distribution	—	—	On trays with flow path length $\geqslant 5$ ft, and for liquid rates > 5000 gal/h/ft (diameter) on trays with flow path length < 5 ft, provide 10% more valves on the inlet half of the tray than on the outlet half
(e) Bubble area A_b	—	—	Bubble area should be maximized
(f) Plate efficiency	—	—	Valve tray efficiency will be about equal to sieve tray efficiency provided there is not a blowing or flooding limitation
(g) Valve blanking	—	—	This should not generally be necessary unless tower is being sized for future service at much higher rates. Blanking strips can then be used. Blank within bubble area, not around periphery to maintain best efficiency
2. Tray spacing (inches)	—	12–36	Considerations are downcomer filling and flexibility. Use of variable spacings to accommodate loading changes from section to section should be considered.
3. Number of liquid passes	1	1–2	Multipassing improves liquid handling capacity at the expense of vapour capacity for a given diameter column and tray spacing. Cost is apparently no greater—at least, for lower diameters < 8 ft
4. Downcomers and weirs			
(a) Allowable downcomer inlet velocity, ft/s of clear liquid		0.3–0.4	Lower value recommended for absorbers or other systems of known high frothiness

Table 6.1 continued

Design feature	Suggested value	Alternate values	Comment
(b) Type downcomer	Chord	Chord, arc	Min. chord length should be 65% of tray diameter for good liquid distribution. Sloped downcomers can be used for high liquid rates—with maximum outlet velocity = 0.6 ft/s. Arc downcomers may be used alternatively to give more bubble area (and higher capacity) but are somewhat more expensive. Min. width should be 6 in. for latter
(c) Inboard downcomer width (inlet and outlet)		Min. 8 in.	Use of a 14–16 in. 'jump baffle' suspended lengthwise in the centre of the inboard downcomer and extending the length of the downcomer is suggested to prevent possible bridging over by froth entering the downcomer from opposite sides. Elevation of base of jump baffle should be level with outlet weirs. Internal accessway must be provided to allow passage from side to another during inspection
(d) Outlet weir height	2 in.	1–4 in.	Weir height can be varied with liquid rate to give a total liquid head on the tray (h_c) in the range of 2.5–4 in. whenever possible. Lower values suggested for vacuum towers, higher ones for long residence time applications
(e) Clearance under downcomer (in.)	1.5 in.	1 in. min.	Set clearance to give head loss of approximately 1 in. Higher values can be used if necessary to ensure sealing of downcomer
(f) Downcomer seal (inlet or outlet weir height minus downcomer clearance)	Use outlet weir to give min. ½ in. seal in plate liquid	Inlet weir or recessed inlet box	In most cases plate liquid level can be made high enough to seal the downcomer through use of outlet weir only. Inlet weirs add to downcomer build up; in some cases they may be desirable for 2-pass trays to ensure equal liquid distribution. Recessed inlets are more expensive but may be necessary in cases where an operating seal would require an excessively high outlet weir
(g) Downcomer filling (% of tray spacing)		40–50	Use the lower value for high pressure towers, absorbers, vacuum towers, known foaming systems, and also for tray spacings of 18 in. or lower

- *Step 4* Calculate the downcomer velocity check by calculating the actual downcomer area. Use Figure A1.13 for this. Calculate actual velocity in downcomer as L divided by downcomer area. Check with Table 6.1 that this is within acceptable limits:
- *Step 5* Carry out tower vapour traffic check. Start by calculating

$$A_s \qquad = \text{as total tray area}$$

$$A_{dc} \qquad = \text{inlet and outlet downcomer areas}$$

$$A_w \qquad = \text{check drawing for waste area (i.e. calming zones, area above support, etc.)}$$

$$A_b \qquad = \text{bubble area (active area taken up by the valves in this case)}$$

$$A_b/A_s \qquad = \text{ratio of bubble area to the whole}$$

- *Step 6* Calculate the flood vapour velocity. This is the critical velocity condition used for design. Good design is about 90% of flood. Serious entrainment occurs at about 120% of this number. Flood is given by the expression:

$$G = K\sqrt{\rho_v} \times (\rho_L - \rho_v)$$

where

$$G = \text{mass velocity in lb/h.sqft of bubble area AB}$$

$$\rho_v = \text{density of vapour in lb/cuft at tray conditions.}$$

$$\rho_L = \text{density of liquid in lb/cuft at tray conditions.}$$

$$K = \text{is the flood constant based on tray spacing and given in figure 6.9}$$

- *Step 7* Calculate actual vapour mass velocity (G) using vapour rate obtained from load calculations in lb\h divided by the bubble area AB.
- *Step 8* The percentage flood is determined by the expression:

$$\frac{G \text{ (actual)}}{G \text{ (flood)}} \times 100$$

If this is in excess of 110% entrainment may be occurring in this section of the tower. Also there may be some excessive back-up occurring in the downcomers due to high-pressure drop across the tray.

The following example illustrates this calculation procedure.

Example calculation

This example is to calculate the percentage flood and entrainment factor associated with the HGO and bottom pumparound tray in an atmospheric crude distillation tower.

The following is a summary of the valve tray details and the traffic flow through the tower.

Tray spacing = 24 in
Tower Diameter = 13 ft
Downcomer width = 16.5 in
Downcomer Length = 96 in
Weir height = 1 in
Number of passes = 1
Total tray area As = 132.7 sq ft
Waste area (est 15%) = 20 sq ft = Aw
Total downcomer areas ADC = 15 sq ft
Bubble area AB = As − (Aw + ADC)

Draw-off tray	Liquid from tray			Vapour to tray		
	lb/h	gal/h	hot gal/h	lb/h	Mol. wt	Moles/h
HGO draw-off (1)	Temp 627 °F (2)			Temp 632 °F (2)		
Hydrocarbons	174 648	23 990	32 760	357 138	181	1976.9
Steam				28 980	18	1610
Total	174 648	23 990	32 760	386 118	107.6	3586.9
LGO draw-off (1)	Temp 498 °F			Temp 500 °F		
Hydrocarbons	122 412	17 487	22 809	278 148	154	1802.9
Steam				30 818	18	1027.3
Total	122 412	17 487	22 809	398 996	109	2830.2
Kero draw-off	Temp 367 °F			Temp 370 °F		
Hydrocarbons	21 540	3 254	3 925	133 050	115	1157.8
Steam				33 968	18	1887.1
Total	21 540	3 254	3 925	167 018	55	3044.9

(1) These data do *not* include the pumparound liquid flow
(2) These tray temperatures are not those calculated earlier in Section 6.5 but are taken from another source.

Consider the tower at the HGO draw-off

To the liquid flow quoted above is added the pumparound liquid thus:

Pumparound duty is 17 743 000 BTU/h.
Assume pumparound return is at 300 °F
Enthalpy at 627 °F   = 347 BTU/lb
Enthalpy at 300 °F   = 140 BTU/lb
Δ Enthalpy   = 207 BTU/lb
lb/h PA stream   = 85 715 #/h
                            = 11 774 gal/h
GPH of unstripped HGO   = 3868
GPH of o/flow   = 23 990
Total liquid flow   = 39 632 gal/h at 60

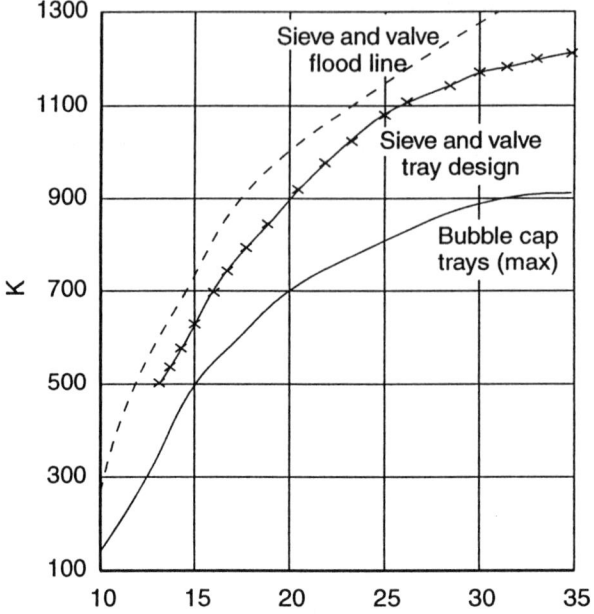

Figure 6.9. Tray spacing V_s "K" Factor

Loading data at flowing conditions

		Vapour		Liquid
Temperature	°F	632		627
Pressure	psig	23.8		—
Moles/h		3586.9	gal/hot	54 120
lb/h		386 118	lb/h	288 522
ACFs		302	gal/min	902
P_v #/CF		0.355	P_L #/CF	39.9

Tower Data

Calculating liquid loading.

$$\text{Liquid loading LL} = \frac{\text{gal/min}}{450} = \frac{902}{450}$$

$$= 2.0 \,\text{cuft/sec}$$

$$\text{Liquid velocity in d'comer} = \frac{2.0}{7.5}$$

$$= 0.27 \text{ ft/sec which is OK.}$$

$$\text{Liquid loading/ft/pass} = \frac{54120}{13}$$

$$= 4163 \text{ Gpm/ft.pass.}$$

Which is satisfactory (Should not be more than 12 000 US GPM/FT/Pass.)

Calculating vapour loading.

$$
\begin{aligned}
\text{Flood vapour loading GF} \quad &= \text{Kf} \sqrt{\rho_v \times (\rho_L - \rho_v)} \\
&= 1150 \times \sqrt{.355 \times (39.9 - .355)} \\
&= 1150 \times 3.75 \\
&= 4313 \text{ lb/h.sqft}
\end{aligned}
$$

$$
\begin{aligned}
\text{Actual vapour loading GA} \quad &= \frac{386118}{97.7} \\
&= 3952 \text{ lb/h.sqft.}
\end{aligned}
$$

$$
\begin{aligned}
\text{Percent of Flood} \quad &= \frac{3952}{4313} \times 100 \\
&= 91.6\% \text{ Which is satisfactory.}
\end{aligned}
$$

6.15 A Procedure for Evaluating Exchanger Thermal Performance

Almost all process plants in an oil refinery use shell and tube heat exchangers. Indeed, the performance of these items contribute a great deal to the standard and performance of the plant itself. Shell and tube exchangers are used to transfer unwanted heat usually in product or reflux streams with incoming feed or with another stream requiring the heat. Performance of these exchangers are therefore important in:

● Conserving energy and fuel
● Ensuring proper temperature profiles in distillation units
● Meeting storage specifications for product streams, etc.

This section deals with evaluating the performance of such equipment. This evaluation compares the overall heat transfer coefficients derived from plant test data with those accepted norms for the service. Should these derived coefficients be much lower than those listed in Table 6.2, then severe fouling may be the cause. The calculation steps for this evaluation are as follows:

- *Step 1*   Obtain the equipment data from certified manufacturer's drawing and data sheet. Check that no modifications or change of service have occurred.

- *Step 2*   From test run data, obtain the temperatures in and out of the streams. Obtain also the flow rates on the shell and tube sides.

- *Step 3*   Calculate the heat duty using the enthalpy curves given in Maxwell (3) or any other acceptable source.

- *Step 4*   Calculate the log mean temperature difference by the expression

$$LMTD = \frac{GTD - LTD}{2.3 \ \log_{10}(GTD/LTD)}$$

GTD = Greatest temperature difference
LTD = Lowest temperature difference

- *Step 5*   Correct the LMTD using the curves in Appendix 1 Figure A1.8.

- *Step 6*   From the transfer equation:

$$Heat \ transfer = \frac{Heat \ duty}{Overall \ coefficient \times LMTD}$$

Calculate a value for the overall heat transfer coefficient U_o.

Check that this is within reasonable value to those given in Table 6.2

Example calculation

Heat exchanger details (from P and I diagram)

Total area   = 2160 ft^2
2 tube passes per shell
2 shells in series
Tube side   Crude oil
Shell side   BPA

Table 6.2 Typical overall heat transfer coefficients in oil refining

Units are in BTU/h. °F. Ft^2

Fluid being cooled	Fluid being heated	U_o
Exchangers		
Naphtha pumparound	Crude	70–80
Kero	Crude	70–75
Debutanizer bottom	Debutanizer feed	75
Gas oil (inc. BPA)	Crude	40–50
Reduced crude	Crude	20–30
Light end bottoms	Light end feed	70–75
Vacuum distillates	Crude	30–40
Bitumen	Crude	20
Cat oil slurry	FCCU feed	40
CDU overheads	Crude	80–90
Coolers		
Debutanizer bottoms	CT water	75
Light naphtha	CT water	80
Gas oils	CT water	40
DEA or MEA	CT water	110
Reduced crude	CT water	30
Vapour heat exchangers		
Reformer effluent	Naphtha feed	38
Reformer effluent	Recycle gas	38
Crude tower overheads	Crude oil feed	50
Condensers		
CDU overheads	CT water	65
Splitter overheads	CT water	85
Amine stripper overheads	CT water	100
Debutanizer overheads	CT water	90
Reformer effluent	Water	65
Air coolers		
Naphtha coolers		23
Debutanizer bottoms		30
Light end overhead		40

Reboilers (use heat flux in BTU/h/ft^2)

	Hydrocarbon	Water
18 in. dia. bundles	20 000	30 000
30 in. bundles	17 500	26 500
>30 in. bundles	15 000	23 000

Plant data

Calculated duty	10.79×10^6 BTU/h
Crude temperature in	350 °F
Crude temperature out	400 °F
BPA temperature in	627 °F
BPA temperature out	412 °F

No change of phase.

(a) LMTD (Log mean temperature difference)

$$LMTD = \frac{GTD - LTD}{2.3 \log_{10} (GTD/LTD)}$$

BPA $627 \rightarrow 412$
Crude $\underline{400 \leftarrow 350}$
          $\underline{227}$ $\underline{62}$

$$LMTD = \frac{227 - 62}{2.3 \log_{10} (227/62)} = 127 \text{ °F}$$

(b) LMTD correction factor (see Figure A1.8)

$$j = \frac{400 - 350}{627 - 350} = 0.18$$

$$R = \frac{627 - 412}{400 - 350} = 4.3$$

F_n correction factor $= 0.93$
Correction LMTD $= 127 \times 0.93$
                $= 118 \text{ °F}$

Derived heat transfer coefficient U_o

$$= \frac{\text{Duty}}{\text{Area} \times \text{Corr. LMTD}}$$

$$= \frac{10.79 \times 10^6}{2160 \times 118}$$

$$= 42.3 \text{ BTU/h/ft}^2/\text{°F}$$

This is within the range of U_o for this service—performance is acceptable. See Table 6.2.

6.16   Evaluating the Heater's Thermal Performance

Fired heaters are used in a refinery to add the energy required by an oil stream to vaporize it or to raise its temperature to meet chemical reaction requirements. The mechanism of this heat input is by burning oil or gas in a firebox containing tubes through which the feed stream flows. The fired heater is a significant cost centre because it uses fuel oil or fuel gas in its operation. The proper and most effective operation of this equipment is therefore important in minimizing operating costs. This section deals with a method by which the operation of the heater can be checked. This method is described in the following calculation steps:

- *Step 1*   Obtain details of the heater from the manufacturer's data sheet or drawings. The data required are:
Tube area
Layout (is it vertical and how are the burners located?)
- *Step 2*   From plant data obtain coil outlet temperature or flow and assay data sufficient to calculate the flash zone temperature.
- *Step 3*   Again from plant data, obtain the temperature of the crude entering the heater. Also obtain the quantity of fuel fired and its properties (API gravity in particular).
- *Step 4*   From knowledge of flash zone conditions (see Section 6.4) calculate the enthalpy in the flashed crude in BTU/h.
- *Step 5*   Calculate the enthalpy in the crude entering the heater. Note the crude at this point should be all liquid. The difference between the enthalpies calculated in Steps 4 and 5 is the enthalpy absorbed by the crude feed in BTUs per hour.
- *Step 6*   Divide this absorbed enthalpy by the tube area to give the heat flux in $BTU/h/ft^2$. Heat fluxes are generally as follows:
Horizontal, fired one side 8000–12 000 $BTU/h/ft^2$
Vertical, fired from bottom on one side 9000–12 000 $BTU/h/ft^2$.
Vertical, single row, fired on both sides 13 000–18 000 $BTU/h/ft^2$ (see Figure 6.11 for the various types). If the heat flux falls outside the range given above, there could be excessive fouling. Check the pressure drop—if this is far above manufacturer's calculated value then fouling is probably the reason.
- *Step 7*   Check the thermal efficiency of the heater by giving the fuel fired a heating value. This is provided by curves in Figures A1.9 and A1.10. Use the LHV (lower heating valve) in BTU/lb and multiply it by the lb/h of the fuel.
- *Step 8*   Divide the heat absorbed by the heat released calculated in Step 7 to give the thermal efficiency. For most heaters this should be between 70% and 80%. If it falls below this range note should be taken of burner operation and the amount of excess air being used.

A calculation example now follows.

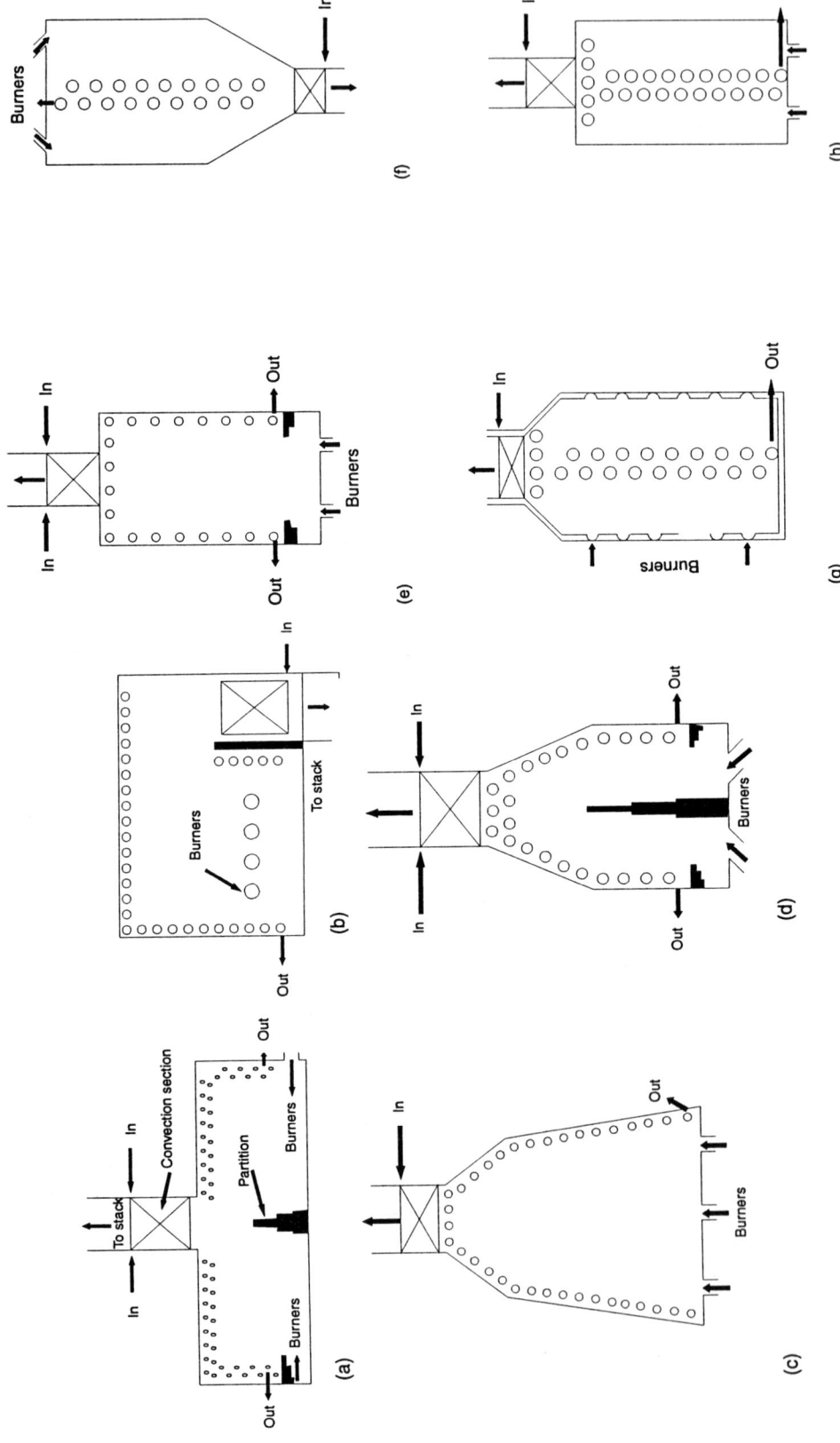

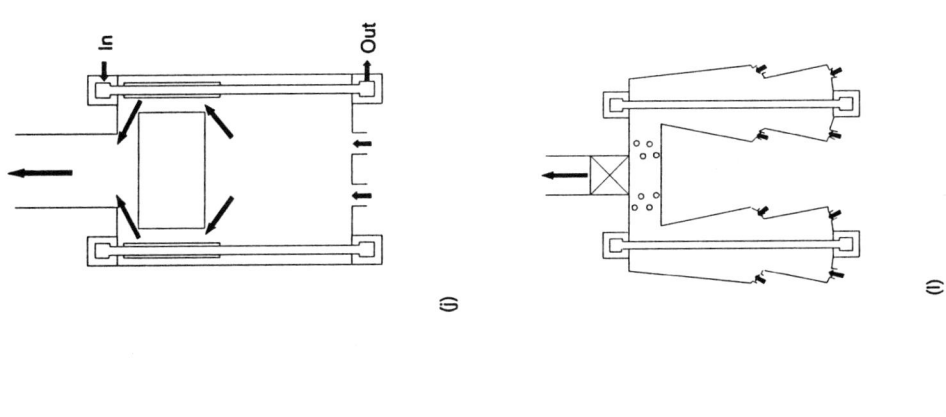

(j)

(l)

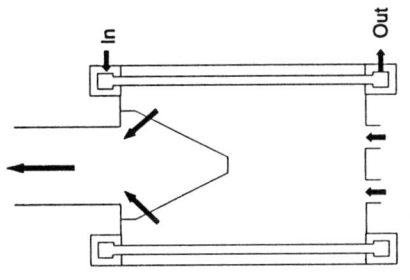

(i)

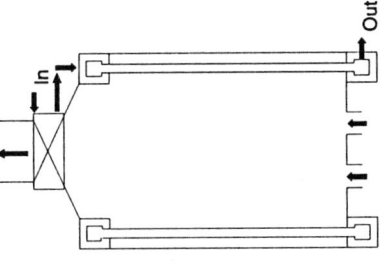

(k)

Figure 6.11. Types of fired heater configuration. (a) Box-type up connection; (b) box-type down connection; (c) A-frame type; (d) box-type double upfired; (e) box-type upfired; (f) box-type downfired; (g) box-type burners in wall; (h) box-type double fired; (i) vertical cylindrical all radiant with re-radiating cone; (j) vertical cylindrical integral convection section; (k) vertical cylindrical crossflow convection section; (l) terrace wall

Example calculation

Heater details (from manufacturer)
- Tubular area total 5200 ft^2
- Vertical fired from the bottom on all sides
- Natural draft
- No steam generation

Plant data

- Flash zone temperature calculated—720°F
- % vaporization of crude at calculated partial pressure—as per calculation in Section 6.4
- Total flow of crude
- Temperatue of crude into heater
- Fuel oil fired lb/h and °API

(A) Degree of fouling

$$BTU/h \times 10^6$$

(See section 6.11)	
Heat in crude vapour at 720 °F	= 102.509
Heat in crude liquid at 720 °F	= 73.430
Total	= 175.939
Heat in with crude (all liquid) at 515 °F	
= 379 575 × 272	= 103.24
Heater duty	= 72.699

The heat flux under these conditions is:

$$\frac{72.699 \times 10^6}{5200 \text{ ft}^2}$$

$$= 13\,980\,BTU/h/ft^2$$

This is within an acceptable limit and excessive fouling is not suspected.

Thermal efficiency

The fuel fired is 16 °API residuum

lb/h of fuel fired	= 5440 (plant data
From Figure A1.9	
LHV (heat release)	= 17 580 BTU/lb
	= 17 580 × 5440
	= 95.635 BTU/h × 10^6
Heat absorbed	= 72.699 × 10^6 BTU/h
Thermal efficiency	$= \dfrac{72.699 \times 10^6}{95.635 \times 10^6} \times 100$
	= 76.02%

Which is within acceptable limits

6.17 Calculating the Pressure Drop in the Crude Heater Transfer Line

The crude heater transfer line connects the outlet of the fired heater to the crude distillation column. This line enters the column at the flash zone and free flows from the heater to the column (that is, it contains no control valves). The size and configuration of this line is critical to the proper operation of the crude unit because the crude feed it transmits at this point is in mixed phase. That is, it contains the liquid and vapour phase of the feed. Moreover, because of the changing pressure profile along this line due to pressure drop, the ratio of liquid and vapour also changes. Under these circumstances of mixed phase feed a line not properly sized and laid out can lead to 'plug' or a similar undesirable flow pattern. Such a situation can affect the stable operation of the crude column through excessive pulsation in the flash zone.

To minimize the risk of poor flow profiles, the length of the transfer line is kept as small as possible. The line is routed to avoid, where possible, any pocketing. The size of the line must also cater for the mixed phase flow and this section deals with the sizing of the transfer line.

The procedure given in this section is based on a known line size and proceeds to prove its acceptability. The calculation starts by establishing the EFV curve for the crude. In this section the data already developed in Section 6.4 will be used together with the crude oil enthalpy calculated in Section 6.11. The calculation steps that then follow are:

- *Step 1* From the tower flash zone conditions estimate the coil outlet conditions of temperature and pressure.
- *Step 2* It is assumed that there is no heat loss over the transfer line. This is a reasonable assumption as the line will be insulated for minimum heat loss. Thus the total crude enthalpy (in BTU/h) at the flash zone will be the same as that for the coil outlet conditions. The percentage vaporization will, however, be different.
- *Step 3* Using x as the percentage vaporized at the coil outlet with the first temperature estimate equate for x constant enthalpy (BTU/h). Check the result using the EFV curve and the vapour pressure curve for the coil outlet temperature relating to the percentage vaporized.
- *Step 4* If the assumed temperature is different to that estimated in step 1, assume a second temperature and repeat. It should not be necessary to do more than three trials. Fix the percentage that best matches the calculated/assumed temperature.
- *Step 5* Using the temperature/pressure and the percentage vaporization, commence the two-phase pressure calculation for the first segment of the line. This commences with setting the K value of pounds vapour per pound liquid ratio.
- *Step 6* Estimate using the approximate mid-boiling point of the residue and the assay data for viscosity. Correct this for temperature. Vapour viscosity is estimated at 0.02 cps.

- *Step 7*   Predict the approximate gravity of the liquid from the mid-boiling point/ gravity curve of the assay. Calculate hot gal/h and f³/h. Divide total weight of liquid by ft³/h to give lb/ft³. Estimate the number of moles of vapour from the material balance (Section 6.4), calculate standard cubic feet and correct for temperature and pressure to give actual cubic feet per hour. Divide pounds of vapour by actual cubic feet to give pounds per cubic foot.
- *Step 8*   Calculate density of the mixture as follows:

$$\frac{1+K}{(1/\text{liquid density}) + (K/\text{vapour density})}$$

- *Step 9*   Calculate the *F* factor from the equation:

$$\frac{(\text{Gas viscosity}) \; R \times (\text{liquid viscosity}) \; S}{C \times \text{density of mixture}}$$

Values for *R*, *S*, and *C* are obtained from Figure 6.12.
- *Step 10*   Using Figure 6.13 read off the pressure drop per 100 ft for this section and calculate the pressure drop.
- *Step 11*   Repeat Steps 1 to 10 for the remaining section or sections of the line. The total pressure drop is the sum of those for each section calculated.

Note: The number of sections may be any number. However, the calculation is a guide and some discretion should be used concerning the time spent in carrying out this calculation.

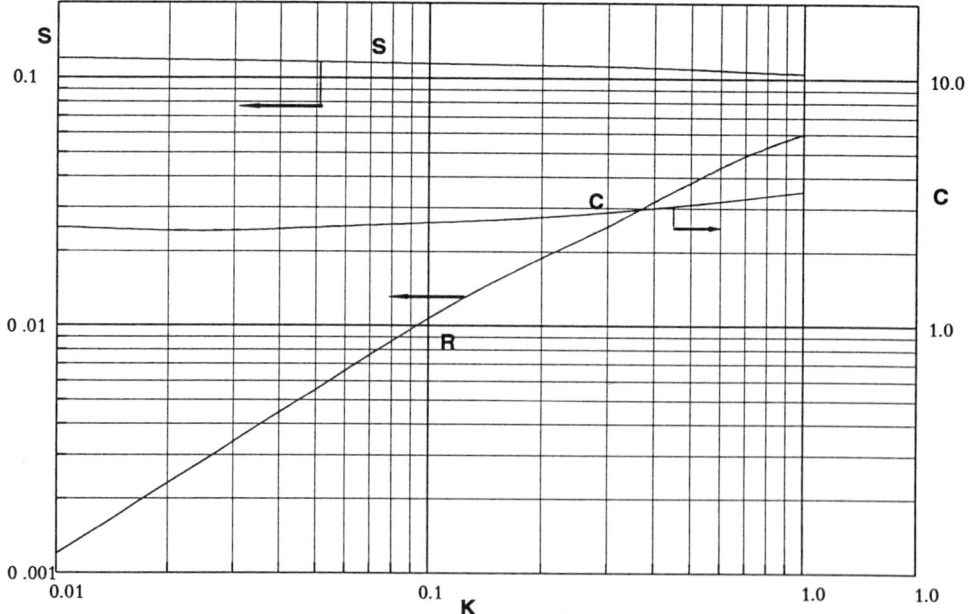

Figure 6.12. Two-phase flow functions

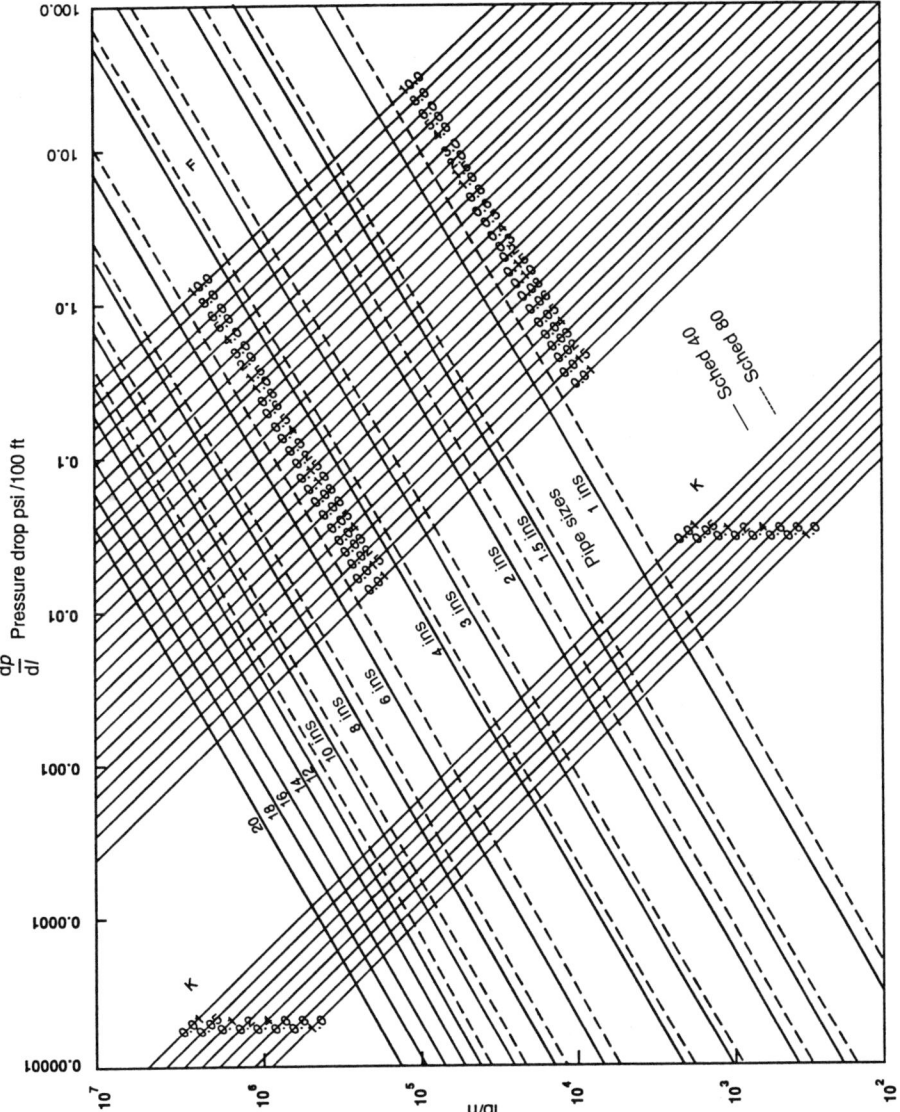

Figure 6.13. Pressure drop for two-phase flow

An example calculation now follows:

Example calculation

Calculate the pressure drop in the crude unit transfer line

This example will use data developed in Section 6.4.

Total pressure at the flash zone is 22 psig = 36.7 psia.

Allow 5 psi ΔP for inlet 'swirl', assume 5 psi ΔP for the transfer line. Coil outlet pressure = 36.7 psia + 5 + 5 = 46.7 psia, say 47 psia.

1st trial   Assume coil outlet temperature is 20 °F above flash zone = 740 °F.

There is no loss of heat in the transfer line.

Total heat in crude at the flash zone.

= 175.939 × 10⁶ BTU/h (see Section 6.11)

Then percentage vaporized at coil outlet will be as follows. Let x lb/h be crude vapour.

Heat in vapour   = (538x)   (°API = 55)
Heat in liquid   = (379 575 − x) × 414   (°API = 13)

But 538x + (379 575 − x) × 414   = 175.939 × 10⁶

$$x = \frac{18.795 \times 10^6}{124}$$

= 151 573 lb/h

55 °API = 6.3 lb/gal then gal/h = 24 059

= 45.8% vaporize

Referring to the EFV curve at atmos. (Section 6.4)

Flash temperature = 580 °F

From vapour pressure curves flash temperature at 47 psia = 690 °F

2nd trial   Assume coil outlet is 15 °F above flash zone

= 735 °F

Heat in vapour   = (532x)
Heat in liquid   = (379 575 − x) × 406

532x + 154.107 × 10⁶ − 406x = 175.939 × 10⁶

$$x = \frac{2\,183\,200}{126} = 173\,270 \text{ lb/h}$$

= 27 503 gal/h

= 52.4% vaporized

Flash temp. at atmos. $= 630\,°F$

Flash temp. at 57 psia $= 755\,°F$

From inspection estimated coil outlet temperature will be about $17\,°F$ above flash zone

$$= 737\,°F$$

From VP curve atmos. flash $= 595\,°F$

% vaporization $\qquad = 48.3\%$ vol.

Line size $\qquad = 12\,$in.

Line length including fittings $= 1.50\,$ft

Consider first 100 ft from heater outlet

% vaporization $= 48\%$ vol. or 45.6% wt

Total flow rate 379 575 lb/h

Gas/liquid ratio $= 0.838 = K$

Liquid viscosity: (mid-BPT $842\,°F$)

$$= 10\,cSt\ at\ 170\,°F$$
$$6.6\,cSt\ at\ 210\,°F$$

At, say, $735\,°F = 0.15\,$cSt $(0.650\,$Sg$)$

Cps $\qquad = 0.15 \times 0.650$

$$= 0.1\,cps$$

Vapour viscosity (estimate) $= 0.02\,$cP

Density of liquid:

lb/h $\qquad = 206\,489$
Hot gal/h $\ = 38\,310 = 5122\,$CF/h
lb/ft^3 ρ_L $\quad = 40.3$

Density of vapour ρ_v:

lb/h vapour $\qquad = 173\,086$
Approx. no. moles $= 1260.45$
ScF/h $\qquad = 1260.45 \times 378$

$$= 476\,450$$

ACFH $\qquad = \dfrac{476\,450 \times 14.7 \times 1195}{520 \times 57}$

$$= 282\,374\ ACFH$$

ρ_v $\qquad = 0.613\,$lb/ft^3

Density of mixture:

$$\rho_m = \frac{1.838}{\left(\dfrac{1}{40.3}\right) + \left(\dfrac{0.838}{0.613}\right)} = 1.321$$

$$F_{\text{factor}} = \frac{(\mu_g \times R) \times (\mu_4 \times S)}{C \times \rho_m}$$

$R = 0.054$ From Figure 6.12
$S = 0.11$
$C = 3.4$

$$F = \frac{(0.02)\, 0.054 \times (0.1)\, 0.11}{3.4 \times 1.321} = 0.14$$

From Figure 6.13

$AP/100 = 2.6\,\text{psi}/100$

Pressure at end of this section $= 44.4\,\text{psia}$

2nd section at 50 ft

Assume temperature is 735 °F

Heat in vapour	$= (532x)$
Heat in liquid	$= (379\,575 - x)\,406$
x	$= 173\,270\,\text{lb/h}$
	$= 27\,503\,\text{gal/h}$
	$= 52.4\%\,\text{vol. vaporized}$

Flash temp. at atmos. $= 630$ °F
at 44.4 psia $= 737$ °F
Then let % vaporized $= 52.4\%$

2nd 100 ft of heater coil

% vaporization	$= 52.4\%$
Liquid viscosity	$= 0.13\,\text{cP (estimated)}$
Vapour viscosity	$= 0.02\,\text{(unchanged)}$
lb/h liquid	$= 206\,000$
Hot gal/h	$= 38\,220 = 5110\,\text{Cf/h}$
ρ_L	$= 40.3\,\text{lb/cu ft.}$

Density of vapour

Moles vapour	1305
ScF/h	$= 493\,290$
ACFH	$= \dfrac{493\,290 \times 14.7 \times 1197}{520 \times 44.4}$
	$= 375\,948$

$$\rho_V \text{ lb/ft}^3 = \frac{173\,270}{375\,948} = 0.461$$

$$K \qquad = \text{lb vapour/lb liquid}$$

$$= \frac{173\,270}{206\,000} = 0.841$$

$$\rho_m = \frac{1.841}{\left(\dfrac{1}{40.3}\right) + \left(\dfrac{0.841}{0.461}\right)} = 0.996$$

$$F = \frac{(\mu_g \times R) \times (\mu_L \times S)}{(C \times \rho_M)} = 0.186$$

$$R = 0.054 \quad S = 0.11 \quad C = 3.4$$

From Figure 6.13	$= \Delta P = 7 \text{ psi}/100$
ΔP Section	$= 3.5 \text{ psi}$
Total line ΔP	$= 2.6 + 3.5$
	$= 6.1 \text{ psi}$
Pressure at coil outlet	$= 36.7 + 6.1$
	$\qquad\quad + 5.0$
	$= 47.8 \text{ psia}$

which is close enough

6.18 Evaluation of Reflux Drum Oil/Water Separation

This calculation is used to size the reflux drum in terms of oil/water separation space. In refinery operations it is required if excessive amount of oil leaves with the water phase from the reflux drum. A similar calculation is also made using the water boot dimensions and water viscosity for excessive water in the oil phase. This is normally checked if it is intended to increase throughput significantly.

The calculation is based on Stokes' law which has the expression

$$V = 8.3 \times 105 \times \frac{d^2 s}{\mu}$$

where

V = the settling rate of one of the phases from the other in inches per minute
d^2 = the droplet size, which is 0.008 in. for 35 API and lighter and 0.005 in. for heavier than 35 API.
s = the difference in specific gravity between the two phases
μ = viscosity of the phase under consideration in centipoise.

The total flow of the feed to the drum is calculated in ft³/h. The hold-up time to the NLL (normal liquid level) is obtained and the cross-sectional area to the NLL of the drum is calculated. Using the total flow in ft³/h and dividing it by the cross-sectional area to NLL gives the linear velocity of the feed. Dividing this velocity in inches per minute into the value for V gives the time required for the two phases to disengage and settle. The surge volume hold-up time is then calculated by dividing the volume to NLL of the drum by the rate in ft³/min of the feed. This hold-time must be at least twice the value calculated for disengaging time. The following calculation illustrates this method.

Example calculation

Evaluation of reflux drum oil/water separation

The overhead reflux accumulator drum is sized for the naphtha product given in Section 6.5. Dimensions of the drum are 96″ I/D × 288 in. T–T.

Separation of the oil from water

$$V = 8.3 \times 10^5 \times d^2 \times \frac{\Delta S}{\eta_{oil}}$$

Where

d = the droplet size = 0.008 in.
ΔS = the difference in SG between naphtha and water
η_{oil} = the viscosity of oil in centipoise.

$$V = 8.3 \times 10^5 \times (0.008)^2 \times \frac{(0.9931 - 0.685)}{0.247*}$$

* from average tables

$$V = 66.3 \text{ in./min}$$

Total flow of naphtha + water is: $1812 + 600$
                                                      $= 2412 \text{ ft}^3/\text{h}$

Cross-sectional area of drum to NLL $= 25.13 \text{ ft}^2$ (plant data)

Linear velocity $= \dfrac{2412}{25.13} = 96 \text{ ft/h or } 1.6 \text{ ft/min}$

                                        $= 19.2 \text{ in./min}$

The disengaging rate is 66 in./min

Time required for disengaging $= \dfrac{66}{19.2}$

                                        $= 3.4 \text{ min}$

The surge volume has a hold-up time of

$$\frac{603}{2412} = 0.25 \text{ h}$$

$$= 15 \text{ min}$$

Which is more than adequate

7 THE CRUDE VACUUM DISTILLATION UNIT

7.1   Process Description—A Crude Vacuum Distillation Unit

The atmospheric residue can be further distilled to provide the heavy distillate streams used for producing lube oil or as feed to conversion units. This distillation, however, has to be conducted under sub-atmospheric pressure conditions. The temperature required for vaporizing the residue at atmospheric pressure would be too high and the crude would crack.

The process follows very much the same pattern as atmospheric distillation. The cold feed in this case is heat exchanged against hot product and pumparound streams before being vaporized in the distillation unit heater. Thereafter the distillate vapours are condensed in the tower by heat and mass transfer with the cold reflux streams moving down the tower. The products are taken off at the appropriate sections and pumped to storage (Figure 7.1).

Neither the vacuum residue that leaves the bottom of the tower in this process nor the side-streams are steam stripped. This is called the dry vacuum distillation process. The vacuum condition is produced by steam ejectors taking suction from the top

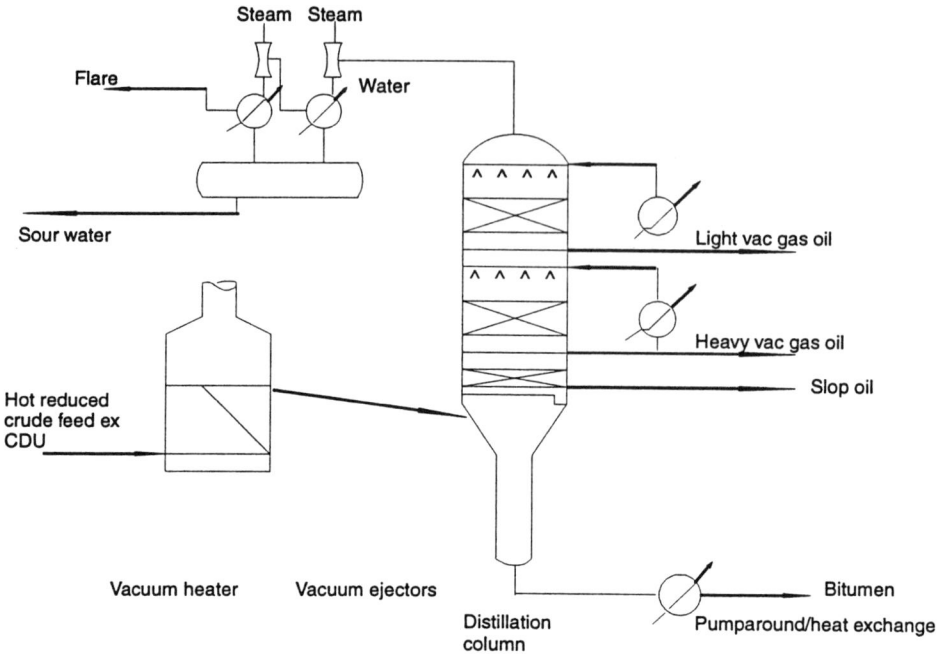

Figure 7.1. Crude vacuum distillation unit (VDU)

of the tower. These ejectors remove inert and other vapour that may exist and pull a vacuum of about 5 mmHg absolute.

7.2 Coil Outlet Temperature

As mentioned in the previous section, the distillation of the residue portion of the crude cannot be carried out at atmospheric pressure. At this condition of pressure, the flash temperature to achieve any meaningful degree of vaporization would be extremely high (say, in excess of 900 °F). At these temperatures the heavy residue will begin to break up or crack. This forms coke in the extreme and olefinic products which may not be desirable to the refiner. Effective vaporization and fractionation can be achieved, however, at reduced pressures. Under this condition reasonable flash temperature (say, 650–750 °F) can be easily obtained.

Modern vacuum distillation units handling reduced crudes operate at 3–5 mm Hg at the top of the tower and about 25–30 mm Hg in the flash zone. No steam is used for stripping. The oil can still crack, of course, if the cut point desired is so high that an excessively high flash temperature is required to meet it even at reduced pressures. Figure 7.2 graph is a guide to the critical cracking temperatures.

This graph shows a plot of a range of temperatures within which the oil will begin to crack. This is correlated to the Watson characteristic factor K. Most residuums with a 700 °F cut point for Middle East crudes have a K factor of about 11.5. From the curve, therefore, it can be seen that these residuums would begin to crack at

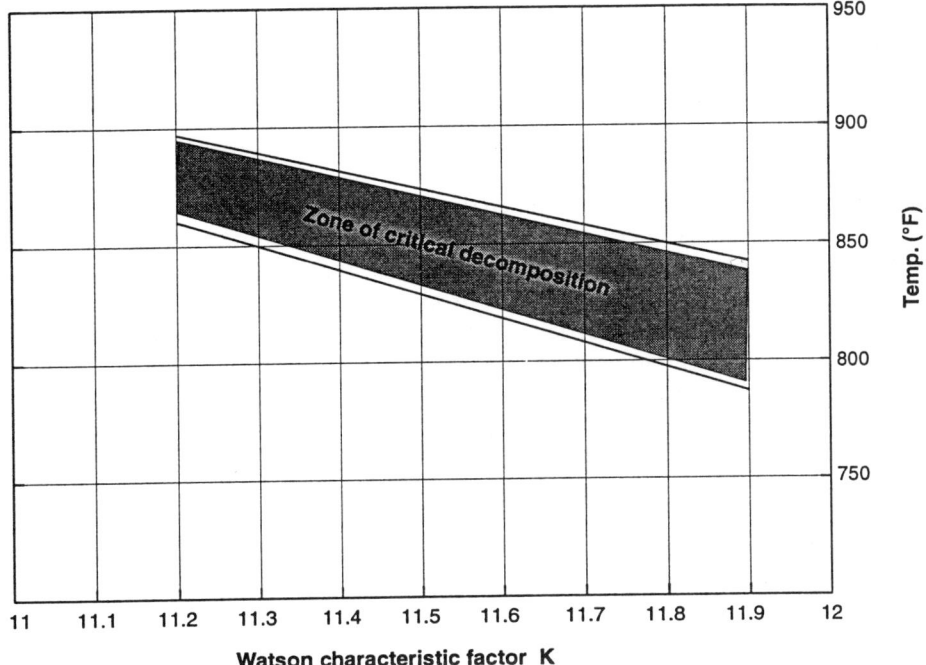

Figure 7.2. Zone of critical decomposition

temperatures between 830 °F and 855 °F. The degree of cracking at or above the 'zone of critical decomposition' will be a function of temperature and the residence time of the oil at that temperature condition.

Significant cracking of the oil in a vacuum tower causes:

- High load to the ejectors (due to the formation of light ends)
- In lube oil production, decolorization of the distillate streams
- In producing feed to hydrotreaters or hydrocrackers, high hydrogen consumption in these units due to the presence of unsaturates as the product of cracking.

It is therefore very desirable to avoid these critical temperatures in a vacuum unit.

7.3   Tower Vacuum Criteria and Pressure Drop

As discussed in Section 7.2 it is highly desirable to establish a high vacuum and a low pressure drop in a crude vacuum unit to avoid excessive oil cracking. During the 1960s a considerable amount of development work went into designing units that would achieve precisely this condition. Better vacuum-producing equipment such as the steam ejector were developed and, perhaps more importantly, the low-pressure internals were developed.

Up to the 1960s most vacuum tower internals were conventional trays designed to provide as low a pressure drop as possible. This, however, was not very low compared with present standards and trays with very low pressure drop tend to lose out to mass and heat transfer. Then in the early 1960s Glitch developed a grid packing which provided very low pressure drops and also a high equivalent tray efficiency.

Up to the development of the grid packing, flash zone temperature reduction in vacuum units was enhanced by steam stripping of the bottoms. This is similar to the operations of the atmospheric unit. The trouble here, of course, is that each pound of steam at 18 mol. wt creates a very high vapour load in the tower under vacuum conditions. This in turn increases pressure drop in the tower. With the new grid packing (low pressure drop) and high efficient steam ejectors the steam introduced into the tower to enhance flash temperature can be eliminated. Thus developed the modern 'dry vacuum tower' with vacuum condition as low as 3 mmHg in the top and about 28 mmHg in the flash zone. Most units built today are 'dry' vacuum units.

7.4   A Method to Determine Vacuum Ejector Performance

The calculation procedure that is described here relates to the 'dry' vacuum unit where no steam is used in the distillation process itself. This method can be used to determine the efficiency of the installed ejector set under test run conditions or to specify the design of an ejector set properly. The efficiency may be determined by the actual quantities of steam used to that calculated by this method. The calculation presented here reflects the procedure to evaluate the ejector performance of a system shown in Figure 7. To commence this calculation, the following plant data need to be obtained:

- Quantity of inerts—measured at the exhaust side of the last stage
- Tower-top temperature and pressure
- Intermediate stage outlet temperatures and pressures of the process streams
- Total steam flow or steam flow to each stage ejector.

The calculation proceeds using the following steps:

- *Step 1*   Determine the quantity of inerts entering the system from the tower. If this cannot be measured, a rule of thumb is that total inerts is 0.5%–1.0% by weight on feed. This is made up of air leaking into the system and some light ends. Again by rule of thumb, light ends will be about 25% of total inerts.
- *Step 2*   Calculate the 'equivalent dry air load' to the first ejector stage. Using the equation:

$$W_a = \frac{W_i}{R_{mi} \times R_{ti}} + \frac{W_s}{R_{ms} \times R_{ts}}$$

where
W_a + Total air in o/heads = Equivalent air flow in lb/h
R_{mi} and R_{ti} = Ratio factors for component i from Figures 7.3 and 7.4.
W_s = weight flow of steam in lb/h
R_{ms} and R_{ts} = ratio factors for steam from Figures 7.3 and 7.4.
In this case W_s will be zero, as no steam will be used in the distillation.

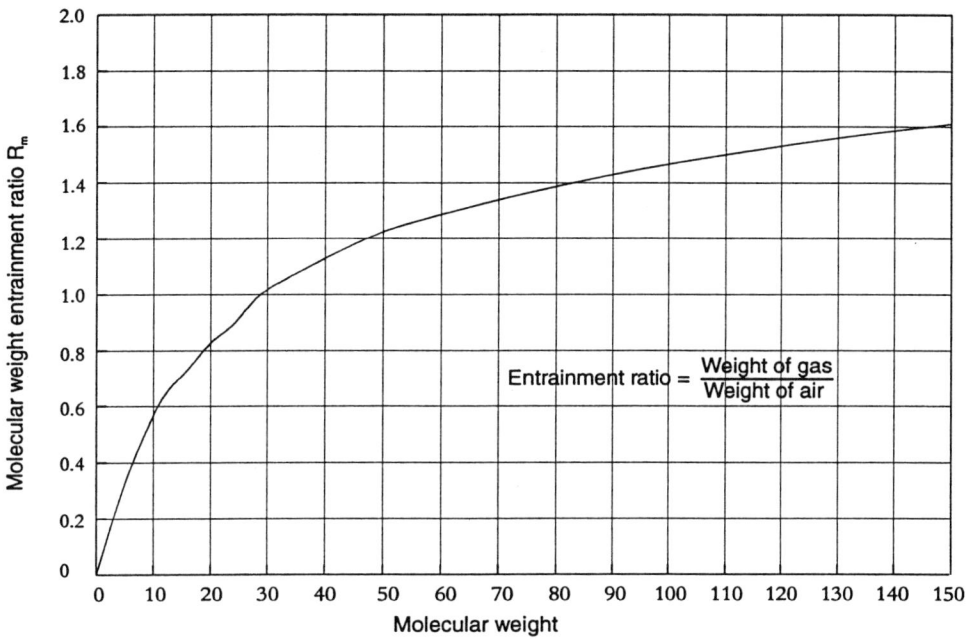

Figure 7.3. Entrainment ratios RML or RMS versus mol. weight of inerts

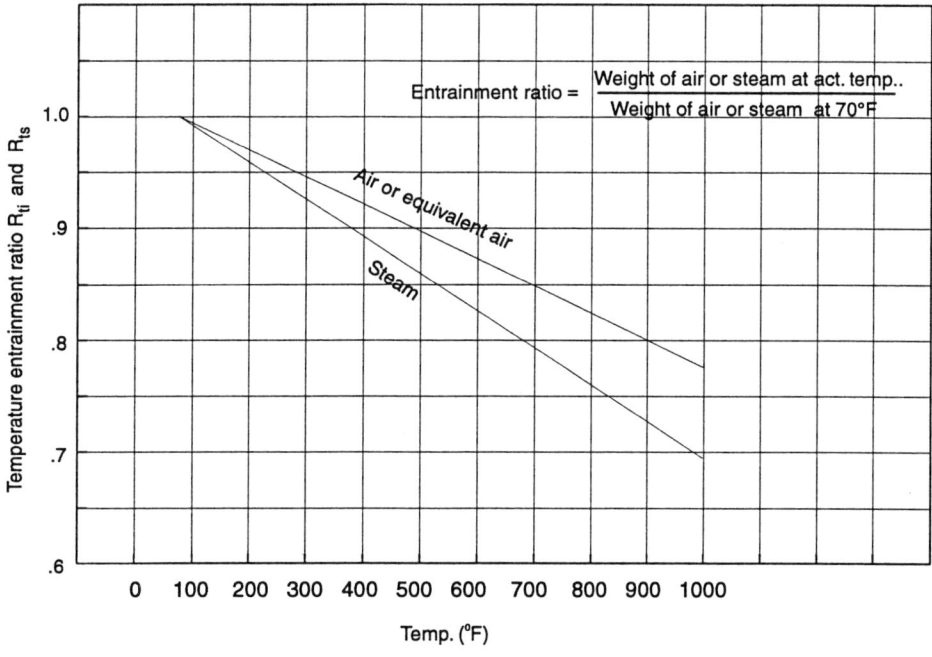

Figure 7.4. Temperature entrainment ratios R_{ti} or R_{ts} versus temperature

- *Step 3*   Calculate the steam consumption to the first ejector using tower-top pressure as the suction pressure. The consumption is calculated by:

$$W_{ms} = R_a \, M_p \, W_a$$

where

W_{ms} = Weight flow of motive steam in lb/h
R_a = ratio of lb motive steam/lb air equivalent using Figure 7.5
M_r = steam usage multiplier from Figure 7.6
W_a = Air equivalent flow in lb/h

- *Step 4*   Calculate the partial pressure of steam at the condenser pressure. This pressure is the first/second-stage intermediate pressure and is the suction pressure to the second ejector. It is read on the plant *or* it may be assumed. In the calculation it is assumed to be 50 mmHg.
- *Step 5*   From the partial pressure of steam the condensing temperature in the condenser is read from steam tables. Assume 90% steam is condensed in this first condenser.
- *Step 6*   (Optional) Calculate heat balance across the condenser and arrive at the condenser duty.
- *Step 7*   Repeat Step 2 now for 'equivalent dry air load' to stage 2. Remember that there will be steam present in this stream.
- *Step 8*   Repeat steps 3 to 5 for the second stage, again assuming a condenser pressure or reading off the actual pressure. Assume also the presence of steam condensed in the second stage condenser. This will be high around 98%.
- *Step 9*   Repeat steps 6 and 7 for the last (third stage) condenser and ejector. This condenser will be at atmospheric or some 1 to 2 psi above atmospheric.
- *Step 10*   Summarize results and compare steam consumed as calculated with those read off plant data.

An example calculation now follows.

Example calculation

Ejector sizing

Tower-top pressure to be 5 mmHg

Air leakage + light ends about 0.5 to 1% by weight on feed. For a well-designed and operated tower light ends will be about 25% of total leakage + light ends.

This calculation is based on a 'dry vacuum' tower—that is, no steam is present in the tower itself. Capacity of the unit is 13 800 BPSD from Section 6.5 of this book. lb/h = 197 085.

Total overhead draw	= say, 0.75% of 197 085
	= 1478 lb/h
Air	= 1109 lb/h
Light ends	= 369 lb/h (assume C_7s)
MW	= 100

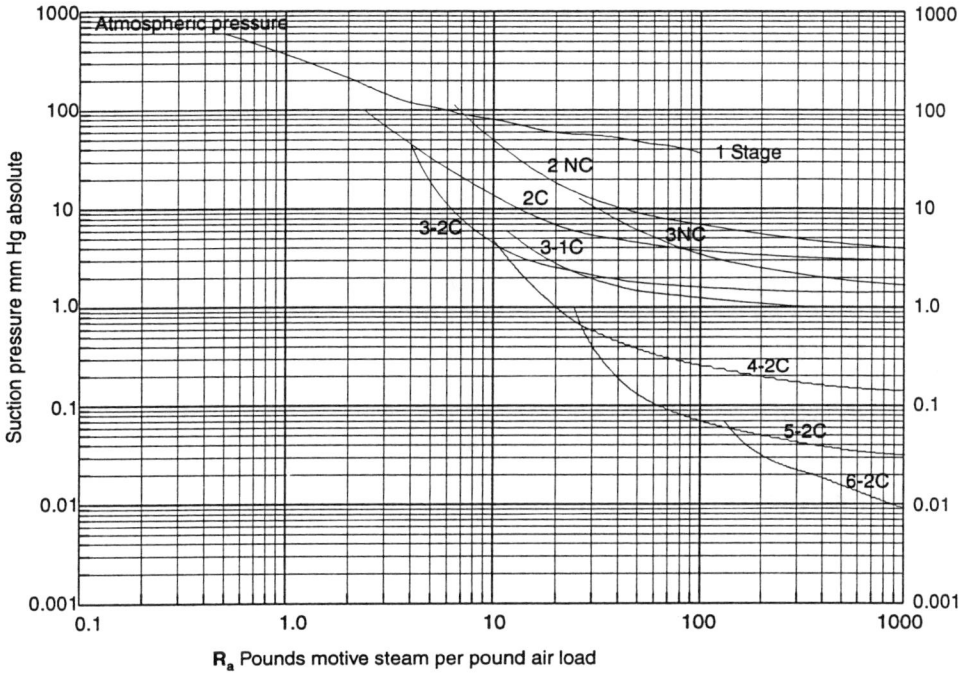

Figure 7.5. Suction pressure versus weight of motive steam (basis 100 psig motive steam)

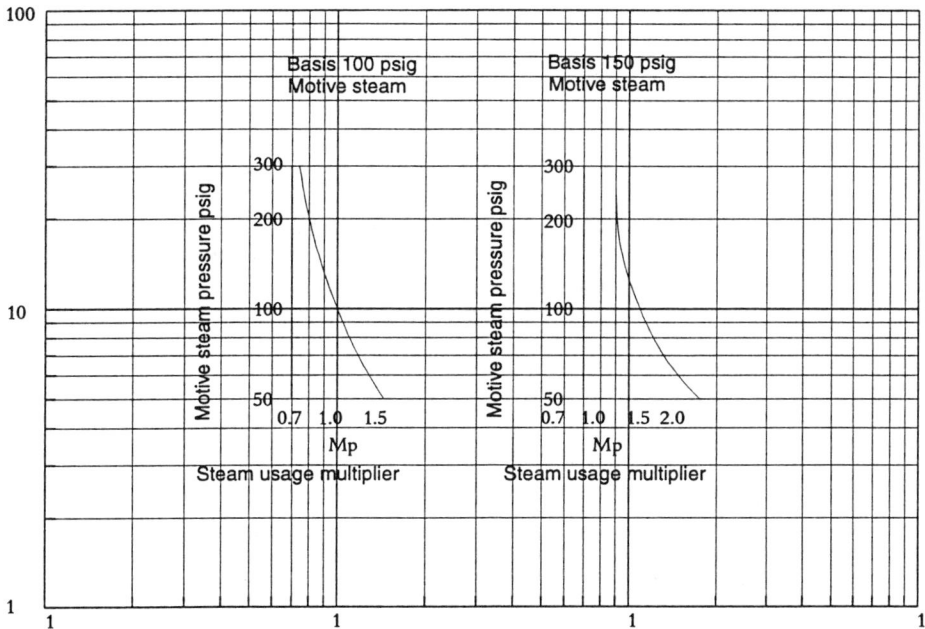

Figure 7.6. Motive steam usage correction factor

Load to first ejector

Calculate equivalent dry air load. This is given by the expression

$$W_a = \frac{W_i}{R_{mi} \times R_{ti}} + \frac{W_s}{R_{ms} \times R_{ts}}$$

where

W_a = equivalent air flow (lb/h)
W_i = weight flow of component i in lb/h
R_{mi} = molecular weight entrainment ratio of i from Figure 7.2
R_{ti} = temperature entrainment ratio of i from Figure 7.3
W_s = weight flow of steam (lb/h)
R_{ms} = molecular weight entrainment ratio of water vapour (Figure 7.2)
R_{ts} = temperature entrainment ratio water vapour (Figure 7.3)

There is no steam present. Therefore

$$W_a = 1109 + \frac{369}{1.48 \times 0.955} + 0$$

Dry air equivalent = 1370 lb/h
Overhead temp.   = 284 °F

First stage: Calculate the steam consumption

Refer to Figure 7.5

First stage   Steam consumption is given by the expression

$$W_{ms} = R_a \, M_p \, W_a$$

where

W_{ms} = weight flow of motive steam (lb/h)
R_a = lb motive steam/lb dry air equivalent (Figure 7.5)
M_p = steam usage multiple (Figure 7.6)
W_a = air equivalent flow (lb/h)

Suction pressure = 5 mmHg

$$W_{ms} = 9 \times 0.84 \cdot 1370$$
$$= 10\,357 \text{ lb/h}$$

Condenser duty

Assume shell side 50 mm/Hg
Motive steam   = 10 357 lb/h   = 575 moles/h
Air   = 1109 lb/h   = 38.2 moles/h
HC (assume C_7) = 369 lb/h   = 3.7 moles/h

Total moles = 616.9 moles/h

Partial pressure of steam $\qquad = \dfrac{575}{616.9} \times 50\,\text{mmHg}$

$\qquad\qquad\qquad\qquad\qquad\qquad = 46.6\,\text{mmHg} = 0.9\,\text{psia}$

Condensing temperature $\qquad = 99\,°\text{F}$

Assume 90% of the steam condenses.

Total heat in $\qquad\qquad\qquad\qquad\qquad\qquad\qquad\qquad$ BTU/h $\times 10^6$

With motive steam at 378 °F	$= 10\,357 \times 1229$	$= 12.733$
With air at 248 °F	$= 1109 \times 0.5 \times 248$	$= 0.138$
With HC at 248 °F	$= 369 \times 265$	$= \underline{0.098}$
Total in		$= \underline{12.969}$

Total heat out

With motive steam at 99 °F	$= 1036 \times 1136.6$	$= 1.179$
With air at 99 °F	$= 1169 \times 0.5 \times 99$	$= 0.055$
With HC at 99 °F	$= 369 \times 210$	$= 0.077$
With condensed steam at 99 °F	$= 9321 \times 99$	$= 0.923$
With condenser duty	$= \text{(by difference)}$	$= 10.735$

Second stage   Calculate new dry air equivalent

$$W_a = \frac{W_i}{R_{mi} \times R_{ri}} + \frac{W_s}{R_{ms} \times R_{rs}}$$

$$W_a = 1109 + \frac{369}{1.48 \times 0.955} + \frac{1036}{0.8 \times 0.987}$$

$$= 1109 + 261 + 1321$$

$$= 2682\,\text{lb/h}$$

Calculate steam consumption

Ejector suction	$= 45\,\text{mmHg}$
Set discharge	$= 250\,\text{mmHg}$
W_{ms}	$= R_a\,M_p\,W_a$

where

$R_a \quad = 4$

$M_p \quad = 0.84$

$W_a \quad = 2682$

$W_{ms} = 4 \times 0.84 \times 2682$

$\qquad\quad = 9012\,\text{lb/h}$

Condenser duty

Assume 98% steam condenses

Motive steam   = 9012 lb/h = 500 moles
Air   = 1109 lb/h = 38.2 moles
HC   = 369 lb/h = 3.7 moles
Steam from stage 1= 1036 lb/h = 57.5 moles
                                _____
Total   = 599.4 moles

Partial pressure of steam $= \dfrac{557.5}{599.4} \times 250$ mm

$\qquad\qquad\qquad\qquad = 233$ mmHg $= 4.5$ psia

Condensing temperature $= 158\,°F$

With 80 °F water shell side can be taken down to 100 °F comfortably

Total heat in

Motive steam at 378 °F	= 9012 × 1229	= 11.065
Steam from 1st stage at 102 °F		= 1.179
HC from 1st stage at 102 °F		= 0.077
Air from 1st stage at 102 °F		= 0.057
Total		= 12.389

Total heat out

Motive steam at 100 °F	= 180 × 1137	= 0.205
Condensed steam at 100 °F	= 9868 × 100	= 0.987
HC at 100 °F	= 369 × 208	= 0.077
Air at 100 °F	= 1109 × 0.5 × 100	= 0.055
Condenser duty (by difference)		= 11.065

Third stage   Calculate new dry air equivalent

$$W_a = \frac{W_i}{R_{mi} \times R_{ri}} + \frac{W_s}{R_{ms} \times R_{ts}}$$

$$W_a = 1109 + \frac{369}{1.48 \times 0.99} + \frac{9.0}{0.8 \times 0.99}$$

$$= 1474 \text{ lb/h}$$

Calculate steam consumption

Ejector suction = 250 mmHg
Set discharge   = atmos. = 760 mmHg

$W_{ms} = R_a\, M_p\, W_a$
$\qquad = 1.8 \times 0.84 \times 1474$
$\qquad = 2229$ lb/h

Condenser duty

    Assume 100% steam condenses

Motive steam	$= 2229\,\text{lb/h}$	$= 123.8$
Residue steam ex second stage	$= 180\,\text{lb/h}$	$= 10.0$
Air and HC		$= \underline{41.9}$

Total moles	$= 175.7$
Steam partial pressure	$= \dfrac{133.8}{175.7} \times 14.7$
	$= 11\,\text{psia}$
Condensing temperature	$= 198\,°\text{F}$
Drum will be run at $100\,°\text{F}$	

Heat in $\text{BTU/h} \times 10^6$

Motive steam at $378\,°\text{F}$	$= 2229 \times 1229$	$= 2.739$
Residual steam at $100\,°\text{F}$	$= 90 \times 1137$	$= 0.102$
Air and HC at $100\,°\text{F}$		$= \underline{0.132}$
Total		$= 2.973$

Heat out

Condensed steam at $100\,°\text{F}$	$= 0.232$
Air at $100\,°\text{F}$	$= 0.077$
HC at $100\,°\text{F}$	$= 0.055$
Condenser duty (by difference)	$= 2.609$

Summary

Steam consumption (lb/h)

1st stage	10 357
2nd stage	9 012
3rd stage	$\underline{2\,229}$
Total	21 598

Suction pressures (mmHg)

1st stage	5
2nd stage	45
3rd stage	250

Condenser duties (BTU/h $\times 10^6$)

1st stage	10.735
2nd stage	11.065
3rd stage	2.609

7.5   Calculating Flash Zone Conditions in a Vacuum Unit

Flash zone conditions are easier to calculate for a 'dry' vacuum unit than for the atmospheric crude unit. Indeed, the flash zone conditions can be measured in vacuum units with a greater degree of accuracy than in the case of the atmospheric column.

    The procedure for predicting flash zone conditions in this case follows a similar route to that for the atmospheric unit. The following steps describe this procedure:

- *Step 1* Develop the EFV from the TBP curve of the reduced crude. The same method that was used in Section 6.4 will apply in this case. Remember, the EFV calculated is for atmospheric pressure.
- *Step 2* Develop the material balance for the vacuum unit. As the flash zone conditions are to be used in most calculations involving the vacuum tower it is best to develop the materials balance at this stage. To do this, determine the distillate cuts required and, by using the respective crude assay, determine the specific gravities for each cut using mid-boiling points. The mid-boiling point for each distillate product only should be determined in this case. Use the method described in the introductory section of this book to determine mol. weight. Some of these will need to be extrapolated.
- *Step 3* Set the overflash. In this area of the TBP and EFV curves the slope of the curve is quite steep. That is, there is a high temperature difference for each percentage of volume increase. In vacuum units therefore a 1–2% overflash would be realistic.
- *Step 4* Determine the new flash temperature to include the overflash from the EFV curve. This is the temperature at atmospheric pressure.
- *Step 5* Calculate the total pressure of the flash zone with a pressure at or below 5 mmHg at the top. Pressure drop through the tower should not be more than 25 mmHg. A well-designed offtake trays and packing should be as follows:
  Grid packing 6–7 mmHg per 10 ft of packed height
  Draw-off (chimney trays) 2–3 mmHg per tray.
- *Step 6* There will be no partial pressure calculation of hydrocarbon vapour (as in the case of the atmospheric unit) as there is no steam in the flash zone of a 'dry' vacuum tower. The total pressure calculated in Step 5 is the actual hydrocarbon flash pressure. Using the vapour pressure curves, determine the flash zone temperature at the total flash zone pressure. This is the flash zone temperature. The TBP and and EFV curves are given in Figure 7.8 for the vacuum unit feed.

The following example illustrates these calculation steps.

Example calculation

For this example we will use the TBP distillate developed in Section 6.3 and the material balance given in Section 6.4.

1. Develop the EFV at atmos. from the TBP

Slope of DRL	$= 6.3\ °F/\%$
Slope of FRL	$= 3.5\ °F/\%$
ΔT_{50} (DRL $-$ FRL)	$= 39.5$
FRL 50% PT	$= 952 - 39.5$
	$= 912.5$

say, 913 °F

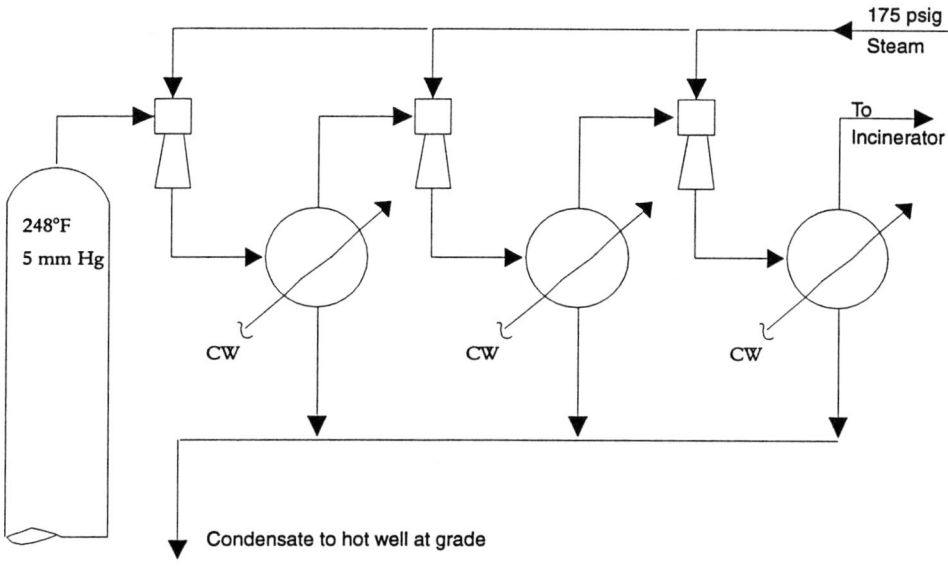

Figure 7.7. A typical three stage ejector set

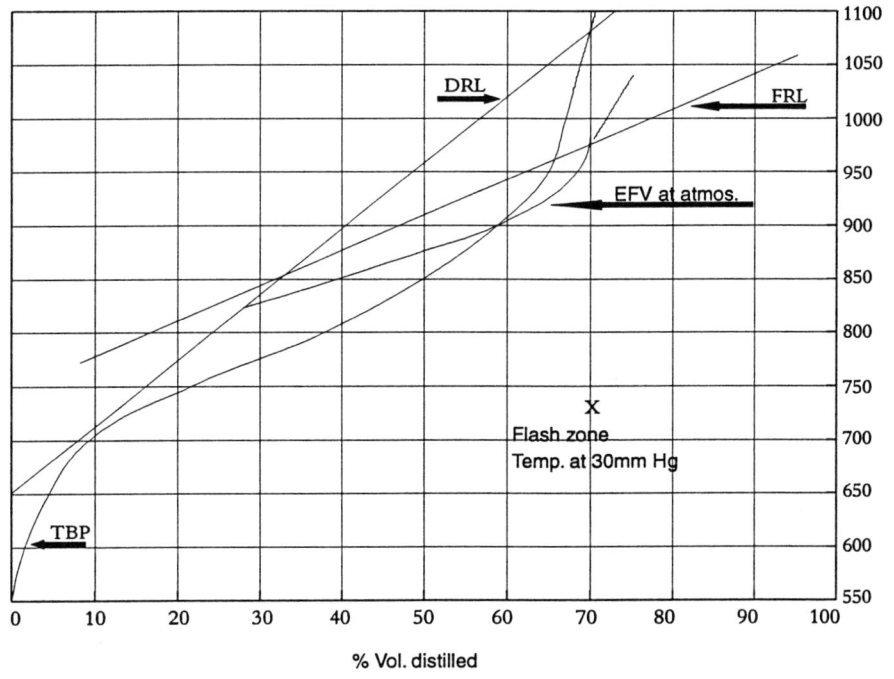

Figure 7.8. Vaccum unit TBP and EFV curves

% vol.	ΔT TBP/DRL (°F)	ΔT (Flash/FRL) ΔT (TBP/DRL) (°F)	ΔT Flash/FRL (°F)	EFV (°F)
30	− 55	0.37	− 20.4	830
40	− 88	0.37	− 32.6	847
50	− 102	0.37	− 37.7	874
60	− 104	0.37	− 38.5	908
70	− 0	0.37	0	980

2. Develop the material balance

From the TBP curve (Figure 7.8)

LVGO   = 24% vol. on reduced crude feed
HVGO   = 67% − 24% = 43% on reduced crude
Residuum = 33% on reduced crude

See table for material balance

Vacuum unit material balance: Kuwait crude

	TBP cut range	% volume		Volume		SG at 60 °F	lb/gal	Weight			Moles Moles/h	
		Cut	Cumu- lative	BPSD	GPH			lb/h	% wt	MW		
Atmos. residue	°F +690	100	100	13 800	24 150	0.957	8.16	197 085	100.0	547.6	359.9	
LVGO	IBP − 750	24	24	3 312	5 796	0.894 (26.6%°API)	7.443	43 140	21.9	325 (1)	132.7	Mid-BPT °F 715
HVGO	985	43	67	5.934	10 384	0.930 (21.°API)	7.747	80 445	40.8	542	148.4	835
Bitumen	+ 985	33	100	4 554	7 970	1.090 (0.°API)	9.20	73 500	37.3	932	78.8	
Total products	—	100	—	13 800	24 150	0.957	8.16	197 085	100.0	547.6	359.9	

Flash zone material balance

	TBP cut range	Cut	Cumu- lative	BPSD	GPH	SG at 60 °F	lb/gal	lb/h	% wt	MW	Moles/h	
Overflash	985– 1025 °F	2	69.0	276	483	0.96 (16°API)	7.99	3 859	2.0	552	7.0	Mid-BPT 1015 °F
Product vapour	985	67.0	67.0	9.246	16 180	0.915 (23°API)	7.64	123 585	62.7	439.6	281.1	
Total vapour	—	69.0	69.0	9 522	16 663	0.920 (22.5°)	7.65	127 444	64.7	442.4	288.1	
Liquid phase	+ 1025	31.0	31.0	4 278	7 487	1.10 (−4°API)	9.30	69 641	35.3	970	71.8	

(1) Extrapolated values

3. Calculate flash zone conditions

Flash zone pressure:

Tower top = 5 mmHg = 0.1 psia

There will be two beds of grid packing
(1) Section of baffle trays (top section)
(2) Chimney trays for draw-off

Pressure drops are:

Grid packing 14 mmHg
Chimney trays 6 mmHg
Baffle trays 5 mmHg

Total pressure drop   = 25 mmHg
Absolute flash zone pressure   = 30 mmHg

As there is *no* steam this will be the hydrocarbon vapour pressure.

Flash zone temperature:

From the material balance, cut point including overflash is 69% vol. distilled.

From the EFV curve at atmos. pressure this is a flash temperature of 975 °F.

From the vapour pressure charts in Appendix 1.

Actual flash zone temperature = 745 °F

7.6 Draw-off Temperature

Unlike the atmospheric crude distillation unit, the temperature of the vacuum tower bottom (bitumen) will be essentially the flash zone temperature. There will be a small difference, say 2–3 °F below actual flash zone temperature, due to overflash returning from the wash trays. Very often the overflash amount is drawn off from below the wash section and either sent to fuel or blended into the bitumen stream external to the tower. In this case the unquenched bitumen leaving the tower will be at flash zone temperature.

Side-stream draw-off temperatures are easier to calculate for a vacuum tower than was the case for the atmospheric tower. This is so because in a dry vacuum column there is no steam to influence partial pressures and, of course, there is no side-stream stripping.

A method similar to the Packie method for the atmospheric column is used for the vacuum column draw-offs. Here, however, it is only necessary to determine the initial boiling point (IBP) of the side-stream EFV curve at the tower condition to arrive at the draw-off temperature.

Note: It is the IBP of the actual EFV curve in this case *not* the IBP of the flash reference line as in the case of the atmospheric unit. Also there will be no Packie correction factors required in this case.

The calculation steps for this procedure are as follows:

- *Step 1*   Draw the EFV curve from the side-stream TBP curve using the method described in Section 6.5. Only the 0%, 10%, 30% and 50% vol. sections of the curve need to be developed. These are given in Figure 7.10 for the example.
- *Step 2*   Set the total pressure at the draw-off tray. If this is not available as plant data then use the criteria for pressure drop given in Section 7.5.
- *Step 3*   Calculate the partial pressure of the side-stream product at the draw-off tray. To do this, consider all material lighter than the draw-off side-stream to be inert. Include in the inerts estimates of air leakage and cracked hydrocarbon vapours as described in Section 7.4. The total hydrocarbon vapours will include the overflow from the draw-off tray. This is estimated and, as a rule of thumb, use overflow as:

  Top side-stream $0.8 \times$ product
  Mid-side-stream $1.0 \times$ product
  Bottom side-stream $1.5–2.0 \times$ product
- *Step 4*   Using the vapour pressure curves relate the IBP temperature of the EFV to the partial pressure determined in Step 3. This is the draw-off temperature.

The following example illustrates this stepwise procedure.

Example calculation

1.   Sidestream EFVs (using TBP data in Section 6.3)

    LVGO (see Figures 7.9 and 7.10)

    Slope of DRL $= 1.6\,°F/\%$
    Slope of FRL $= 0.5\,°F/\%$
    ΔT_{50} (DRL − FRL) $= 720\,°F$

% vol.	ΔT TBP/DRL	$\left(\dfrac{\Delta T \text{ (Flash/FRL)}}{\Delta T \text{ (TBP/DRL)}}\right)$	ΔT Flash/FRL	EFV (°F)
0	− 33	0.2	−7	685
10	0	0.4	0	698
30	+ 14	0.37	5	713
50	+ 6	0.37	2	722

IBP of EFV at atmos. $= 685\,°F$

Calculate partial pressure at draw-off

There is no steam but all material lighter than the cut is inert. In this case only air leakage and light ends are inert. From Section 7.4 let this be

    Air $= 38$ moles
    HC $= 3.7$ moles
    Total $= 42$ moles

Overflow will be about $0.8 \times$ moles cut.

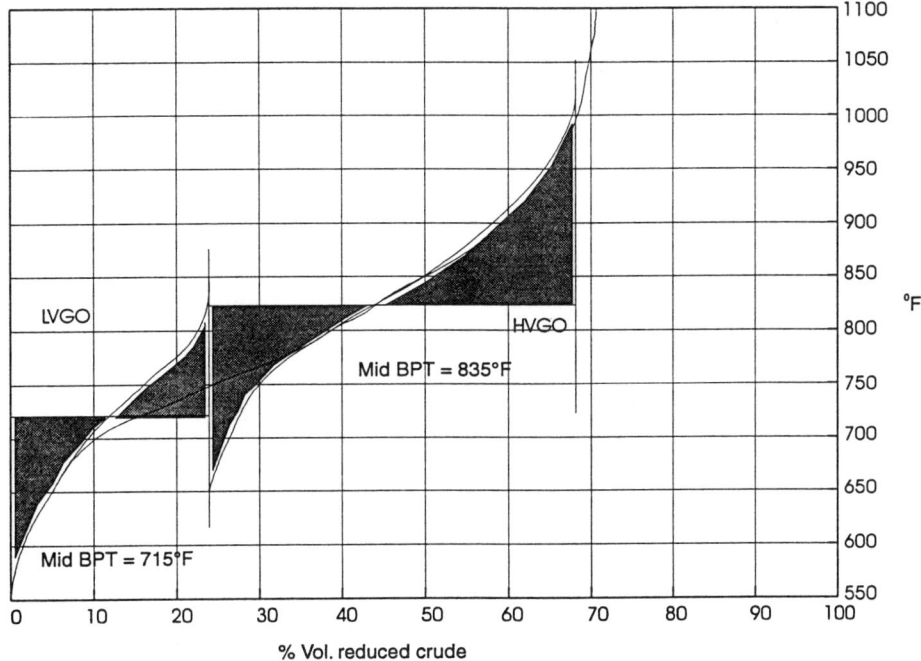

Figure 7.9. Vacuum unit distillate cut points and mid boiling points

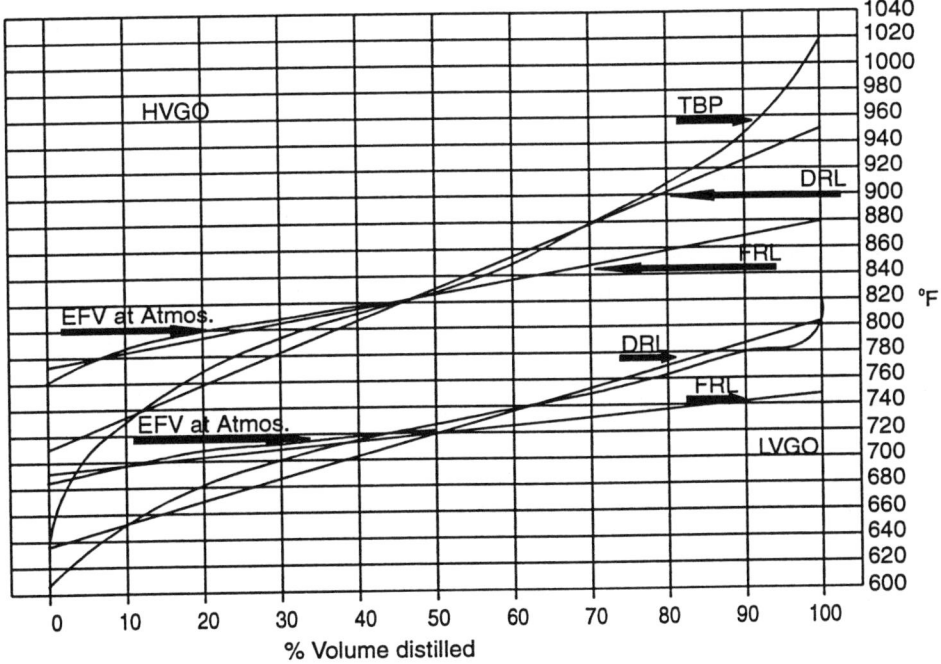

Figure 7.10. EFVs of vacuum side-streams

Then total moles LVGO + overflow vapour

$$= 239 \text{ moles/h}$$

Partial pressure $\quad= \dfrac{239}{(239 + 42)} \times 10 \text{ mmHg}$

$$= 8.5 \text{ mmHg}$$

From VP curves draw-off temperature at 8.5 mmHg

$$= 440 \, °F$$

There will be no correction in this case as there is no stripout or steam.

HVGO

Slope of DRL	$= 2.6 \, °F/\%$
Slope of FRL	$= 1.0 \, °F/\%$
ΔT_{50} (DRL − FRL)	$= 4.0 \, °F$
FRL 50%	$= 831 - 4 = 827 \, °F$

% vol.	ΔT TBP/DRL	$\left(\dfrac{\Delta T \text{ (Flash/FRL)}}{\Delta T \text{ (TBP/DRL)}} \right)$	ΔT Flash/FRL	EFV ($°F$)
0	− 70	0.2	− 14	760
10	0	0.4	0	784
30	+ 10	0.37	+ 37	808
50	+ 4	0.37	+ 1	827

EFV IBP = 760 °F at atmos.

Calculate partial pressure at draw-off

Total pressure	= 23 mmHg	
Inerts in this case are LVGO + air (etc.)		
Total moles inert	= 132.7 + 42	= 174.7 moles/h
Let overflow be 2.0 × HVGO product		
Total product + overflow vapour		= 3.0 × 148.4
		= 445.2 moles

Partial pressure $\quad= \dfrac{445.2}{445.2 \times + 174.7} \times 23$

$$= 16.5 \text{ mmHg}$$

Draw-off temperature is 520 °F at 16.5 mmHg (from VP curves)

7.7 Determining Pumparound and Internal Flows for Vacuum Tower

Now that the cut points and tower conditions of temperature and pressure are established, the internal flow and pumparound duties can be calculated. In general, fractionation requirements are not as strict in a vacuum crude unit as in the case for the atmospheric unit. Nevertheless, proper wash streams are required in vacuum towers to protect distillates which will become feed to cracking units from entrained undesirable components such as metals. Test runs on vacuum units therefore should include the determination of reflux streams and, in turn, tower loadings. Obviously, for unit design purposes, these requirements are essential in sizing the tower, heater and the ancillary equipment, to meet the unit's objectives.

The following steps outline a calculation procedure to determine pumparound requirements and overflow (reflux) in the wash section of the tower. The wash section is that section between the flash zone and the bottom distillate draw-off:

- *Step 1* Set the overflow requirement for the LVGO draw-off tray using the rule of thumb given in Section 7.6. Alternatively, if this can be measured on the plant use those data.

- *Step 2* From plant data, or such data as can be developed from Sections 7.4–7.6, calculate the heat balance below the LVGO draw-off tray.

- *Step 3* In this heat balance the bottom pumparound duty will be the unknown. Equate heat in equals heat out to determine the duty of the pumparound required to produce the set overflow.

- *Step 4* This pumparound duty can be checked on the plant by multiplying the flow in the pumparound by the enthalpy difference over the exchangers.

- *Step 5* Carry out the overall heat balance over the tower. That is, calculate the difference between the total heat in with the feed and the total heat out with all the products. This difference gives the total heat to be removed by the pumparounds. Assuming there are two pumparounds (top and bottom), then the top pumparound duty will be the total heat to be removed minus the duty of the bottom pumparound calculated in Step 3.

- *Step 6* Usually the most critical flow in a vacuum is the wash oil flowing over the bottom wash trays or packing. This is the area where most undesirable entrainment can occur and the most vulnerable area for coking. Lack of wash oil enhances contamination of the bottom product and promotes coking in this area.

- *Step 7* Carry out a heat balance over the bottom wash section of the tower. The unknown in this case is the overflow liquid from the heavy vacuum gas oil. Equate the heat in with feed and overflow with the heat out with total product vapours, overflow vapour and bitumen to solve for the unknown.

Note: The quantity of overflow in this case is independent of pumparound duties above it. It is dependent only on the amount of overflash. As an exercise, try a 1% overflash and calculate the overflow for the heavy VAC gas oil tray.

A sample calculation illustrating this procedure now follows:

Example calculation

In the side-stream temperature calculation an overflow of $0.6 \times$ distillate product in moles was allowed. This is 25 884 lb/h.

Consider the following heat balance:

Streams into the envelope are:

- Feed + overflash.
- Overflow from LVGO tray.

Stream out of the envelope are:

- Bitumen
- HVGO product
- Vapour to LVGO tray
  Product LVGO vapour
  Overflow vapour
- Bottom pump-around duty

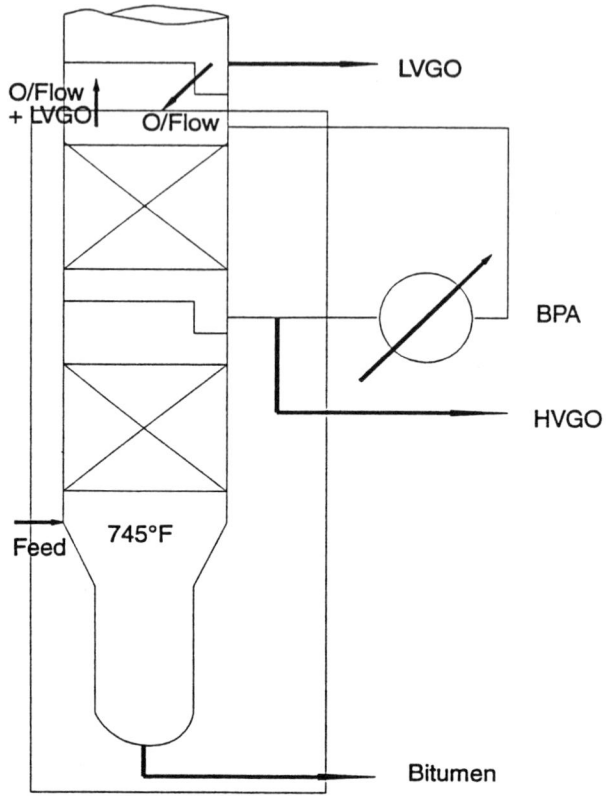

The heat balance over the envelope

	V or L	°API	K	Temp (°F)	lb/h	Enthalpy BTU/lb	BTU/h × 10⁶
In							
Distillate vapour	V	23	12	745	123 585	512	63.276
Overflash	V	12	11.5	745	3 859	494	1.906
Liquid feed	L	0	10.5	745	69 641	378	26.324
Overflow ex LVGO	L	26.6	12.0	440	25 884	212	5.487
Total in					222 969		96.993
Out							
HVGO product	L	21	11.5	520	80 445	274	22.045
Bitumen	L	0	10.5	742	73 500	374	27.489
Overflow vapour	V	26.6	12.0	450[1]	25 884	326	8.438
LVGO vapour	V	12.0	12.0	450[1]	43 140	326	14.064
Pumparound							X
Total out					222 969		x + 72.036

(1) Estimate at draw-off + 10 °F

Bottom pumparound duty to provide the required overflow from the LVGO tray is
$$(96.993 - 72.036) \times 10^6 \text{ BTU/H} = 24.957 \times 10^6 \text{ BTU/H}$$

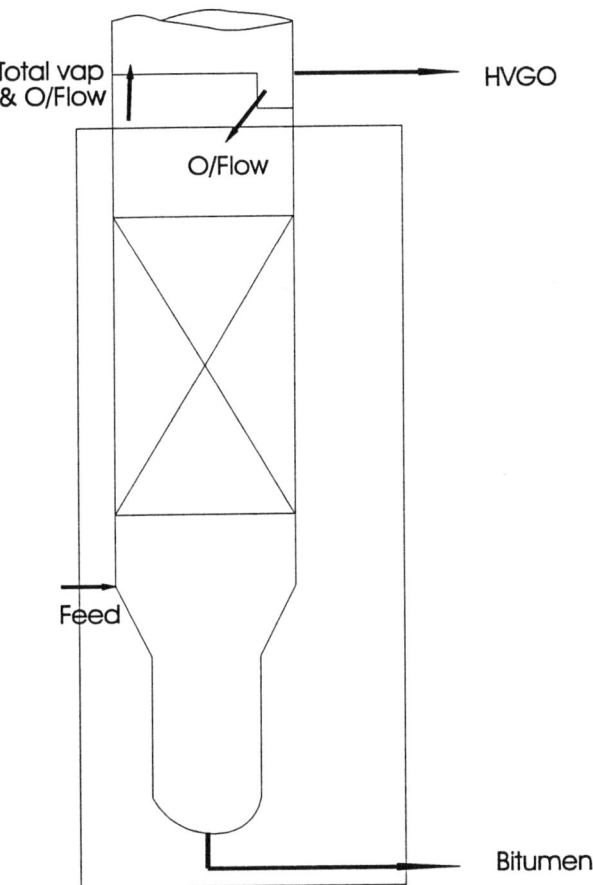

The total heat to be removed from the tower can be calculated from an overall heat balance. This was found to be 32.829 mm BTU/h.

Then the top pumparound $= (32.829 - 24.957) \times 10^6$

$$= 7.872 \times 10^6 \, \text{BTU/h}$$

Note that the overflow from the HVGO tray depends only on overflash and product cut out points. Thus consider the heat balance as follows (see diagram above):

	V or L	°API	K	(°F)	lb/h	BTU/lb	Enthalpy BTU/h × 10⁶
In							
Feed	L/V	—	—	745	197 085		91.506
Overflow	L	26.6	11.5	520	x	274	$274x$
Total in					$197\,085 + x$		$91.506 + 274x$
Out							
Product vapour	V	227	11.5	530	123 585	370	45.726
Overflow vapour	V	26.6	11.5	530	x	378	$378x$
Bitumen	L	0	10.5	742	73 500	374	27.489
Total out					$197\,085 + x$		$73.215 + 378x$

Estimated vapour temperature $=$ draw-off $+ 10\,°F$

$\qquad\qquad\qquad\qquad\qquad\quad = 530\,°F$

Solving for x above:

$91.506 \times 10^6 + 274x \qquad = 73.215 \times 10^6 + 378x$

$\qquad\quad x \qquad\qquad\qquad = \dfrac{1\,829\,100}{104}\,\text{lb/h}$

$\qquad\qquad\qquad\qquad\quad = 175\,875\,\text{lb/h}$

Ratio overflow : distillate $\quad =$

$\dfrac{175\,875}{80\,445} \qquad\qquad\quad = 2.18$

$\qquad\qquad$ Which is about right

7.8 Procedure for Evaluating Vacuum Unit Heat Exchangers

This procedure is identical to the procedure described in Section 6.16 for the atmospheric crude unit. There is, of course, the very high viscosities encountered in the vacuum unit item exchanger train and much of the flows here are laminar. However, the steps to calculate the overall heat transfer coefficient remains the same as those for the atmospheric crude unit.

Vacuum units usually generate steam from waste heat in partially cooling the bitumen stream from the tower. The thermal efficiency of this equipment is calculated in the following example. In this case the intent is to consider the exchanger as a form of reboiler with the boiler feed water in the tube side being boiled and vaporized in the tubes. Heat flux is calculated in this case and compared with those given in Table 6.2.

The same calculation can also be used to size and specify a waste heat boiler. In this case the heat flux is known (Table 6.2) and used to estimate surface area or prepare a specification sheet.

Example calculation

From the material balance item 2.5 flow of bitumen $= 73\,500\,\text{lb/h}$

From plant data the temperatures of the bitumen across the generator are:

Bitumen in$\quad = 680\,°F$

Bitumen out $= 330\,°F$

Boiler feed water enters at 220 °F

and 50 psig steam leaves at 296 °F

Calculated duty of the exchanger

$$= 14.7 \times 10^6 \text{ BTU/h}$$

From equipment data bundle size is 36 in. and total surface area is 710 ft²

Then calculated heat flux is

$$\frac{14.7 \times 10^6}{710}$$

$$= 20\,700 \text{ BTU/h/ft}^2$$

which is within the quoted typical of 23 000 BTU/h/ft² for water.

7.9 Evaluating the Heater Thermal Performance in Vacuum Unit Service

In many vacuum distillation units the feed to the vacuum heater comes directly from the atmospheric crude tower bottom. It is therefore at a high temperature, so high in fact that further heating in the convection side of the vacuum heater is not desirable. As the feed may then enter the radiant section only, a number of vacuum units supplement and generate steam in the convection side of the heater.

The same calculation steps are followed in determining thermal efficiency of this heater as those given in Section 6.16 for the atmospheric heater. In this case, however, steam will be generated in the convection side of the heater. The total duty of the heater must therefore include the heating of the oil and the heating/generation of the steam side. The surface area of the tubes must also reflect this. The following calculation illustrates these points and the step sequence for the calculation remains the same as that in Section 6.16.

Example calculation

Heater details (from manufacturer)

● Tubular area for oil	= 1097 ft
● Tubular area for steam gen	= 320 ft²
● Tubular area for steam superheat	= 189 ft²
Total tube area	= 1606 ft²

Plant data

● Calculated flash zone temp.	= 745 °F
● Observed inlet oil temp.	= 704 °F
● Flow of oil calculated from plant readings	= 197 085 lb/h
● Boiler feed water temp.	= 220 °F
● Steam pressure	= 600 psig
● Outlet steam temperature	= 700 °F
● Measured steam flow (from superheater)	= 7850 lb/h
● 16°API fuel fired	= 1995 lb/h

Calculations

Duty to oil:

Heat in with feed at 704 °F	$= 74.498 \times 10^6$ BTU/h
Heat out with feed at 745 °F	$= 91.506 \times 10^6$ BTU/h
Duty to oil	$= 17.008 \times 10^6$ BTU/h
Surface area for oil heating	$= 1097$ ft^2
Flux for oil	$= 15\,504$ BTU/h/ft^2

Which is within acceptable limits

Duty to steam generation:

Heat in with BFW at 220 °F	$= 7850 \times 220$ °F
	$= 1.727 \times 10^6$ BTU/h
Heat out with superheated steam	$= 1290.3 \times 7850$
	$= 10.129 \times 10^6$ BTU/h
Duty to steam generation:	$= 7.402 \times 10^6$ BTU/h
Total surface area to steam	$= 320 + 184$
	$= 509$ ft^2
Flux for steam generator	$= 16\,507$ BTU/h/ft^2

Which is satisfactory for steam

Thermal efficiency

Total duty of heater

To oil	$= 17.008 \times 10^6$ BTU/h
To steam	$= 81\,400 \times 10°$ BTU/h
Total	$= 251\,408 \times 10^6$ BTU/h

From Figure A1.9 in Appendix 1

LHV (heat release) for 16° API fuel

$$= 17\,580 \text{ BTU/lb}$$

Total heat release	$= 1995$ lb/h $\times 17\,580$ BTU/lb
	$= 35.072 \times 10^6$ BTU/h
Efficiency	$= \dfrac{25.408}{35.072} \times 100$
	$= 72.4\%$

which is acceptable

7.10 Calculating Tower Loading in Packed Section of Vacuum Towers

As discussed in Section 7.3 of this chapter most modern dry vacuum towers use a low pressure drop grid or stacked packing. This packing enhances heat exchange in the tower and, of course, permits the tower to operate at very low pressures. Nevertheless, this packing can become overloaded, causing a high-pressure drop in the tower and poor all-round performance.

This section describes a general method of evaluating the grid performance in terms of its pressure drop and its approach to flood conditions. Note this is a quick general method of checking performance. Proprietary grid and packing manufacturers have their own correlation which they use in their design work. The following are calculation steps used for this evaluation:

- *Step 1* From test run data determine the liquid vapour flows across the section to be evaluated. This has already been done in Section 7.7 The data calculated will be used in the example following.
- *Step 2* Manufacturers' drawings are referenced for tower details such as dimensions of the packed section.
- *Step 3* Calculate the liquid and vapour loads in terms of actual cubic feet per second for vapour and cubic feet per hour per square foot of tower for the liquid. All these will be at tray conditions of temperature and pressure.
- *Step 4* Using the liquid load calculated in step 3 read off a value for 'K' from figure 7.11. Calculate the linear velocity of vapour at flood point for the grid using the expression:

$$ `K' \quad = V_f \sqrt{(\rho_v/\rho_L - \rho_v)} $$

where

V_f = vapour velocity at flood in ft/sec.

ρ_v = density of vapour in lb/cuft at tray conditions of temperature and pressure.

ρ_L = density of liquid in lb/cuft at tray temperature.

- *Step 5* Calculate the actual vapour velocity in ft/sec from the vapour loading calculated in step 3. That is divide the load in cuft/sec by the tower cross sectional area.
- *Step 6* The percentage flood is the actual vapour velocity divided by that for flood conditions calculated in step 4 multiplied by 100.
- *Step 7* Calculate the tray (or section area) required for heat transfer by first establishing the total heat to be transferred. This is the pumparound duty as calculated in section 7.7. Then calculate the temperature of the cold external reflux stream returning to the tower. Use the draw off temperature as the inlet to the external S & T pumparound exchanger and the pumparound flow given by plant data or set in step 1.
- *Step 8* Calculate the LMTD of the packed section as if it were a heat exchanger. Then read off an overall heat transfer coefficient from figure 7.13. The height of the packed section for heat transfer can be estimated using the overall heat transfer coefficient given in Figure 7.13. This gives an answer in terms of sqft of a theoretical tray.
- *Step 9* The HETP for a well designed tray will be around 1.5 to 2.0 ft. Then calculate an estimated height based on a selected HETP between these two figures. Use the lower figure for vapour velocities below 50% of flood and the higher one for 95% and higher (to 120%).

● *Step 10* Estimate the pressure drop through the grid section using the actual vapour velocity to determine 'K', and reading off the pressure drop in inches of hot liquid from figure 7.12. Multiply this by the height of the section, to give the pressure drop over the grid. This figure is usually quoted in mm of Hg for vacuum units so convert to mm of Hg by multiplying the figure by its SG and by 1.865.

An example calculation now follows.

Example Calculation

Calculate the percent flood and the grid height for the pumparound above the HVGO draw off given in section 7.7. The flows to and from this section calculated in section 7.7 are summarized as follows:

Vapour into the section	lb/h	Moles/h
HVGO product	80 445	148.4
LVGO product	43 140	132.7
Overflow from HVGO tray	175 923	324.6
Total	299 508	605.7

Liquid into the section	lb/h	gal/h at 60
Overflow from LVGO	25 884	3478
Pumparound liquid (measured)	235 444	30 392
Total	261 328	33 870

SG at 60 $= 0.925$
SG at 350 $= 0.810$

Diameter of tower in this section is 15 feet
cross-sectional area $= 176.7$ sqft.

Mols/hr of vapour $= 605.7$
cuft per sec @ 530 °F & 20 mmHg

$$= \frac{605.7 \times 378 \times 14.7 \times 990}{520 \times 0.39 \times 3600}$$

$$= 4564 \text{ cuft/sec.}$$

$$\rho_v = \frac{83.2}{4564} = 0.018 \text{ lb/cuft.}$$

$$V_a = \text{vapour linear velocity} = \frac{4564}{176.7}$$

$$= 25.83 \text{ ft/sec.}$$

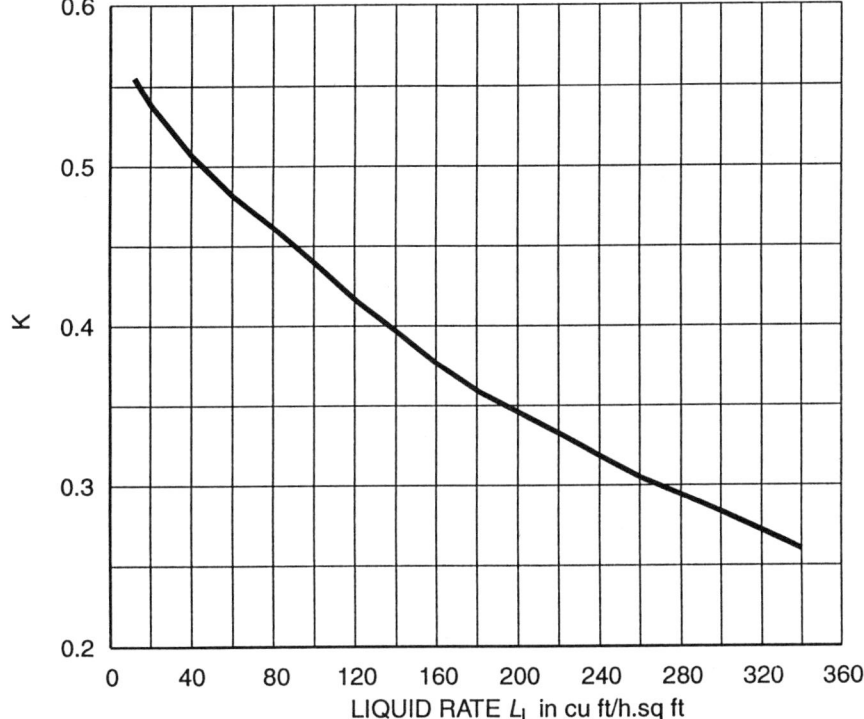

Figure 7.11. Capacity Factor 'K' V_S Liquid Rate.

Liquid loading $= 261\,328$ lb/h

$$\rho_L \quad = 62.1 \times 0.810 = 50.3 \text{ lb/cuft}$$

$$L_L \quad = \frac{261\,328}{50.3 \times 176.7} = 29.4 \text{ cuft/h.sqft}$$

From figure 7.11

$$K_f = 0.54 = V_f \times \sqrt{(\rho_v/\rho_L - \rho_v)}$$

$$V_f = \frac{0.54}{\sqrt{(0.018/50.3 - 0.018)}}$$

$$= 28.43 \text{ ft/sec}$$

$$\text{Percentage of flood} = \frac{25.83}{28.54} \times 100 = 90.5\%$$

Calculating the height of the packed section.

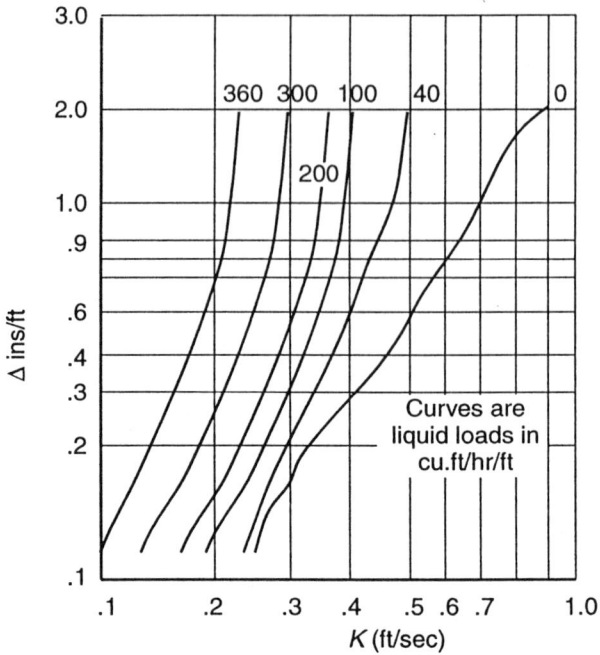

Figure 7.12. Pressure drop through grid in inches of hot liquid per foot height

Duty of the pumparound = 24 957 000 Btu/h (see item 2.7)

Temperature of cold pumparound in = 350 °F
Temperature of pumparound out   = 520 °F
Temperature of vapours in (from tray below) = 530 °F
Temperature of vapours out   = 450 °F

LMTD: 530 -------▶450
                         520◀------- 350
                         ─────────────────
                          10   100

$$\text{LMTD} = \frac{100 - 10}{2.3 \, \text{Log} \, \dfrac{100}{10}}$$

            = 40 °F

Total vapour flow in   = 299 508 lb/h
Total vapour flow out = 69 024 lb/h

$$\text{mean vapour flow} = \frac{299\,508 + 69\,024}{2}$$

            = 184 266 lb/h

$$Mv = \frac{184\,266}{176.7} = 1042.8 \text{ lb/h.sq ft}$$

From figure 7.13 Ho = 500 btu/h.sq ft °F

$$\text{total equivalent tray area required} = \frac{24\,957\,000}{500 \times 40} = 1247.9 \text{ sq ft}$$

Area of one tray = 176.7 sq ft then number of theoretical trays

$$= \frac{1247.9}{176.7} = 7 \text{ trays}$$

HETP for a well constructed grid is approx 500 mm then height of grid section

$$= \frac{7 \times 500}{25.4 \times 12} = 11.5 \text{ ft}$$

Calculating the pressure drop across the grid section

Liquid flow $L_L = 29.4$ cu ft/hr.sq ft

'K' at actual vapour flow $= Va \sqrt{(\rho_v/\rho_L - \rho_v)}$

$$= 25.83 \times \sqrt{0.018/50.3 - 0.018}$$

$$= 0.49$$

From Figure 7.12 ΔP = 0.8 in of hot liquid per ft
         height

$$= 0.8 \times 0.810$$

$$= 0.65 \text{ in of water per ft}$$

$$= 0.65 \times 1.865 \times 11.5$$

$$= 13 \text{ mmHg}$$

7.11 Tower Bottom Temperature Control

Thermal cracking is a function of temperature and the time the oil is retained at that temperature. The temperature at which cracking occurs is given in Figure 7.2 and discussed in Section 7.2. As discussed, it is important to restrain cracking as much as possible in vacuum towers. Because cracking is a function of temperature and residence time, where plants operate at or close to the critical zone of decomposition, the material should be quenched to an acceptable temperature level or disposed of as quickly as possible or both to avoid excessive cracking.

This section deals with the design for vacuum bottom product quenching and the calculation to achieve this follows:

- *Step 1*   Check the bottom temperature against Figure 7.2. If this temperature is within the decomposition zone choose a temperature below the zone to which the residue will be quenched. This should be at least 20 °F below but preferably 50 °F below the critical temperature given by Figure 7.2.

- *Step 2*   Calculate the heat to be removed to reduce the temperature to the chosen level.

- *Step 3*   Locate a cooled residue stream in the process that can be recycled for quenching. Calculate the quantity required for quenching by a simple heat balance.

- *Step 4*   Check that the bottom product pumps and the tower bottom well are large enough to handle the recycled stream as well as the product make.

An example calculation now follows.

Example calculation

Bottoms temperature in tower = 830 °F. With residuums of 11.5 K value, this temperature is within the critical zone of decomposition. Hold-up time in tower bottoms, say 5 min. Flow = 7970 gal/h at 60 = 664 gal volume. Hold-up at 830 °F = 664 × 1.33 (1.33 is expansion factor) = 883 gal = 118 ft^3. This is a 6 ft I/D section of the tower = 28.3 ft^2. Liquid height 4 ft, which is reasonable.

lb/gal at 830 °F = 6.95   lb/h = 73 500

Heat to be removed   73 500 at 830 °F

to 73 500 at 750 °F

= 73 500 (436 BTU/lb − 380 BTU/lb)

= 4.116 × 10^6 BTU/h

This is quenched by cold residue stream at, say, 360 °F

Recycle quench required   = x (380 BTU/lb − 136 BTU/lb)

= 244x BTU/lb

$$x = \frac{4.116 \times 10^6}{244} = 16\,869 \text{ lb/h}$$

= 1834 gal/h

The bottom pumps should carry this additional capacity.

7.12   Calculating Hot Well Capacity and Separation

(*Note*: This applies to ejection systems with spray condensers. Shell and tube condensers will not consider the coolant water.) When spray condensers are used as interstage heat transfer equipment in vacuum ejector systems the water picks up hydrocarbons. The effluent normally discharges into a 'hot well' which can be a weired pit (similar to an API separator) or a conventional tank with a baffle. The oil and water are allowed to separate in the 'hot well'. The water is normally stripped and returned to the cooling water system while the oil phase is sent to fuel or to crude feed.

In calculating to determine the performance or sizing the hot well Stokes' law of settling is again used. The following steps describe a calculation method to measure the performance of the unit:

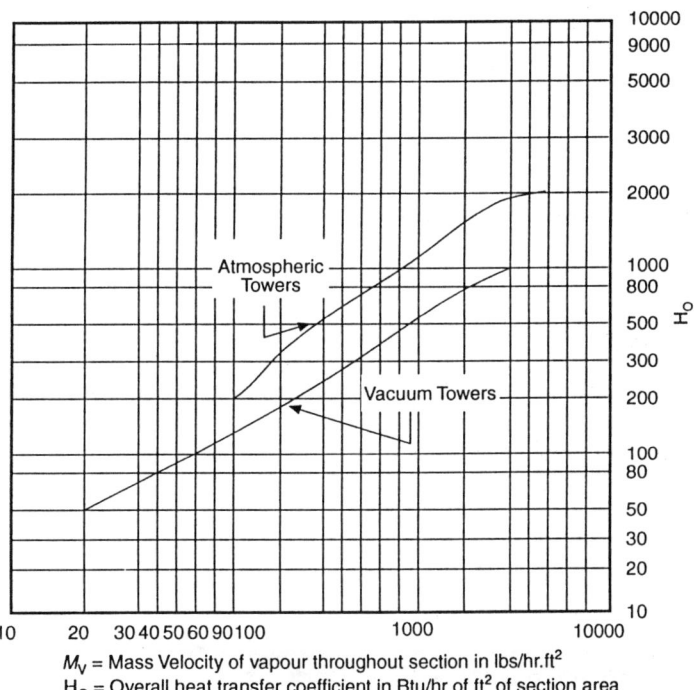

M_V = Mass Velocity of vapour throughout section in lbs/hr.ft^2
H_O = Overall heat transfer coefficient in Btu/hr of ft^2 of section area

Figure 7.13. Transfer Coefficient V_S Mass Velocity for pumparound zones

- *Step 1*   From plant data obtain the total flow rate of the effluent.
- *Step 2*   Using equipment data sheets obtain the dimensions and details of the vessel or pit.
- *Step 3*   Calculate the capacity of the vessel in cubic feet of effluent. Calculate hold-up time based on volume and flow rate.
- *Step 4*   Calculate the linear velocity of the effluent in in./min using

$$L = \frac{\text{volume flow ft}^3/\text{h}}{12 \times 60 \times \text{cross-sectional area of flow ft}^2}$$

Now the cross-sectional area of flow can be calculated using data in Table A1.1.

- *Step 5*   Using Stokes' law, which is given in Section 6.18, calculate the rate of separation of hydrocarbons from the water in in./min.
- *Step 6*   The minimum residence time required for separation will be the effluent rate in in./min divided by required rate from Step 5.
- *Step 7*   Check that the time calculated in Step 6 is less than hold-up time calculated in Step 3.

An example calculation now follows.

Example calculation

Water from the hot well is to be returned to the cooling towers after stripping.

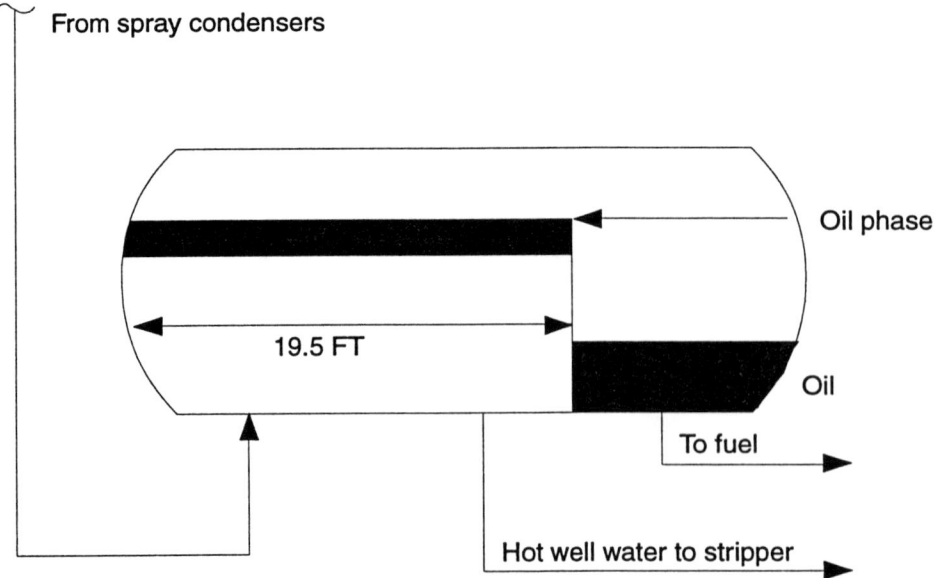

The total water flow from the condensers is

983 080 #/h at about 110 °F

Oil pick-up is about $5\% = 49\,154$ lb/h

Consider this as $C_7s = 0.688$ Sg at 60 °F and 100 MW

From data sheets the hot well drum has the following dimensions
	10 ft I/D by 25 ft tan to tan

The separator weir has a height of 8 ft which is the liquid level in the vessel. The distance from the tan on the inlet side to the weir is 19.5 ft.

$$\text{Volume of liquid} = \text{cross-sectional area of baffle} \times 19.5$$
$$= 67.3 \times 19.5 \text{ ft}^3 \text{ (see Table A1.1)}$$
$$= 1312 \text{ ft}^3$$

$$\text{Hold-up time} = 983\,080 \text{ lb/h}$$

$$= \frac{983\,080}{61.98} = 15\,861 \text{ ft}^3/\text{h}$$

$$= 264 \text{ ft}^3/\text{min}$$

$$= \frac{1312}{264} = 5 \text{ min}$$

$$\text{Linear velocity} = \frac{15861}{67.3} = 236 \text{ ft/h}$$

$$= 46.8 \text{ min}$$

From Stokes' law

$$\text{Setting rate} = 8.3 \times 10^5 \times \frac{d^2 \, \Delta S}{\text{Visc.}}$$

$$= 8.3 \times 10^5 \times \frac{(0.008)^2 \times 0.0351}{0.69}$$

$$= 23.5 \text{ in./min}$$

Minimum residence time required

$$= \frac{46.8}{23.5} = 2 \text{ min}$$

Hold-up time is 5 min, which is more than adequate.

8 STRAIGHT-RUN LIGHT ENDS UNITS

8.1 Process Description—Typical Light Ends Unit

Figure 8.1 shows the light end process configuration common to most refineries. It contains:

A debutanizer
A naphtha splitter
A depropanizer
A de-ethanizer

The normal feed to the light end unit is the overhead naphtha stream from the crude unit reflux drum. In some refineries the feed may also contain unstabilized naphtha from other units such as a hydrotreater or a cracking unit, all of which contain ethane and lighter, propane, butanes and pentanes in solution in the naphtha.

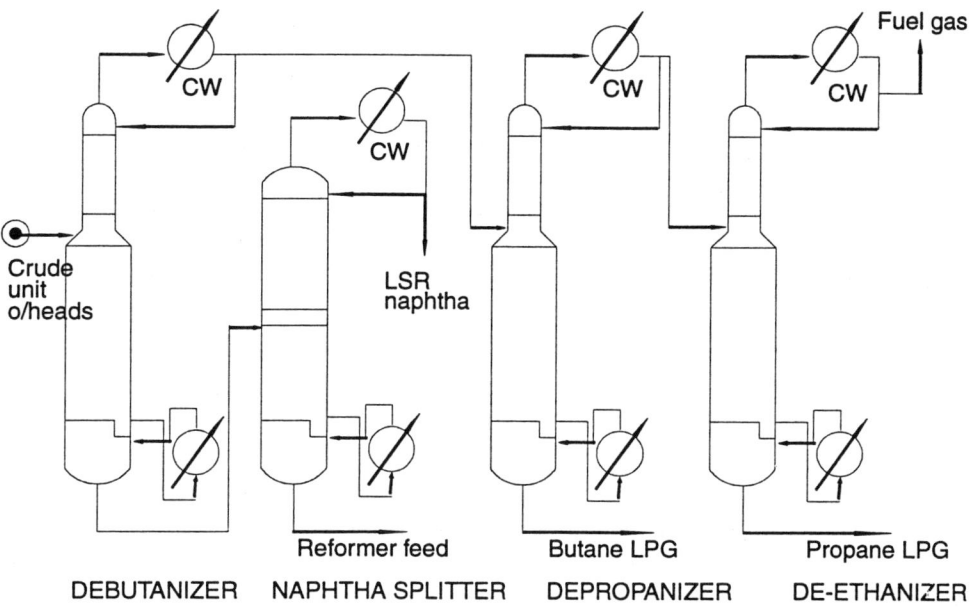

Figure 8.1. A typical straight-run light end unit

The feed stream is first routed to the debutanizer where much of the butane and lighter is taken off as an overhead stream. The debutanized naphtha leaves the bottom of the debutanizer to enter the naphtha splitter. A light naphtha is taken off as the overhead product and a heavy naphtha leaves as bottom product. This heavy naphtha is now the feed for the cat reformer after desulphurization. The light naphtha contains all the pentanes and cyclopentanes which are not required in the cat reformer.

The debutanizer overhead product enters a depropanizer. In this tower butane LPG is the bottom product and is pumped to storage. The overhead distillate is routed to a de-ethanizer. This tower removes ethane and lighter as an overhead vapour product. Propane LPG is the bottom product and is pumped to storage.

8.2 Setting Fractionation Requirements in Light End Towers

Effective separation by fractionation in light end towers obey the same laws as that in the crude distillation units, that is, the degree of separation is the product of the number of trays (or stages) and the reflux (or overflow) in the column. In the crude unit this separation was measured by the difference between the ASTM 95% point of the lighter fraction and the 5% ASTM point of the heavier fraction. This is the ASTM gap or overlap. In light end towers the degree of separation is a little more precise. This is determined by the distribution of key components in the two fractions to be separated. Key components may be real components (such as C_4s and C_5s) or pseudo-components defined by their mid-boiling points. Normally, key components are adjacent components by boiling point in the feed composition. Any two key components may be selected: a light key and a heavy key. By definition, the light key has the lower boiling point. Both key components must be present, however, in the distillate and bottoms product of the column. If a side-stream exists then these keys must also be present in the side-stream product.

There are several correlations that describe the behaviour of these key components in their distribution and relationship to one another. By far the more common of these correlations is the Fenske equation (7), which relates the distribution of key components at minimum trays with infinite reflux. The equation is relatively simple and does not require iterative calculation techniques to solve it. The Fenske equation is as follows:

$$N_{m+1} = \text{Log}\left[\left(\frac{\text{LT Key}}{\text{HY Key}}\right)_D \left(\frac{\text{HY Key}}{\text{LT Key}}\right)_W\right] \div \text{Log}\frac{K_{\text{LT}} \text{ Key}}{K_{\text{HY}} \text{ Key}}$$

where

N_m = minimum number of theoretical strays at total reflux. The +1 is the reboiler which is counted as a theoretical tray
LT Key = the mole fraction of the selected light key
HY Key = the mole fraction of the selected heavy key
D = fractions in the distillate product
W = fractions in the bottom product
K_{LT} Key = the equilibrium constant of the light key at mean system condition of temperature and pressure
K_{HY} Key = the equilibrium constant of the heavy key again at mean system conditions

The ratio of the equilibrium constants is normally called the 'relative volatilities' of the keys.

Setting fractionation requirements for light end towers is usually required to meet a product specification. More often than not, this specification is the vapour pressure of the heavier fraction or the tolerable amount of a heavy key allowed in the lighter fraction. Sometimes, however, a specification for the separation may not be given. Under this circumstance some judgement must be made in determining the most reasonable separation than can be achieved with the equipment. This section addresses calculation techniques that satisfy either premise, and the procedures for these now follow.

SETTING SEPARATION REQUIREMENTS TO MEET A SPECIFICATION

In this case it is required to determine the amount of butanes that can be retained by a light naphtha cut to meet a RVP specification. The steps are as follows:

- *Step 1* Calculate the properties of the C_5+ naphtha from a component breakdown. These properties should give weight and mole rates per hour.
- *Step 2* Carry out a bubble point calculation of the C_5+ fraction and inserting butanes as the unknown quantity x.
- *Step 3* The equilibrium constantly used (K) in the calculation will be at the temperature and pressure conditions of the RVP (i.e. normally at 100 °F, which is the test temperature).

- *Step 4* Either the equilibrium constants given in standard data books or the relationship vapour pressure divided by total systems pressure may be used.
- *Step 5* By definition, the total moles liquid given is equal to the total moles vapour (calculated) in equilibrium at the bubble point. Thus equate and solve for *x* as the quantity of butanes tolerable to meet RVP.

An example calculation is given below.

SETTING FRACTIONATION WHERE NO SPECIFICATION IS GIVEN

- *Step 1* Determine the composition of the feed in terms of real components, pseudo-components or both. Calculate this in moles/h and mole fractions.
- *Step 2* Select the key components, decide the minimum distribution of one or other key. For example, in a debutanizer C_5s allowed into LPG must not be more than 1% of the total C_4s and lighter. This is to protect the butane LPG 'weathering' test specification.
- *Step 3* Give values to the fraction of LT and HY keys in the distillate and bottoms. Use *x* as the unknown where appropriate.
- *Step 4* The Fenske equation will be used to calculate the distribution of keys. Determine the value of NM by taking the actual number of trays and using an efficiency of 70–75% to arrive at total theoretical trays. Divide this figure by 1.5 to arrive at the minimum theoretical trays NM. Don't forget to add 1 for the reboiler to use in the equation.
- *Step 5* Estimate the mean tower conditions. This can be achieved by examining past plant logs, etc. or by estimating a dew point and a bubble point from examination of the major components in the top and bottom products, respectively. Determine K values for the keys at this mean tower condition.
- *Step 6* Solve for *x* in the Fenske correlation. This will be the split of the key components and the basis for the material balance.

An example calculation is given below.

Example calculation

Normally critical specs for naphtha and LPG are vapour pressure:

To calculate C_4 content in light naphtha to meet a 5 psig RVP at 100 °F

Debutanized light naphtha composition:

iC_5 — 1.02 % vol. on crude
nC_5 — 1.85 % vol. on crude
C_6 — 3.4 % vol. on crude
C_7 — 2.8 % vol. on crude

= 9.07 % vol. on crude

Consider the component balance at 1500 BPS D

| | % vol. on crude | % | Volume on cut | | | | | | Moles /h |
			BPSD	gal/h	lb/gal	SG	lb/h	MW	
iC_5	1.02	11.25	169	296	5.19	0.624	1 536	72	21.3
nC_5	1.85	20.40	306	536	5.25	0.631	2 814	72	39.1
C_6	3.4	37.49	562	984	5.53	0.664	5 442	86	63.3
C_7	2.8	30.86	463	810	5.73	0.688	4 641	100	46.4
Total	9.07	100.00	1 500	2 626	5.50	0.661	14 433	84.9	170.1

To predict the amount of C_4s allowed in naphtha to meet 5 psig RVP.

Let C_4s be x moles/h

Carry out the bubble point calculation at 5 psig and 100 °F

	Moles/h	VP 100 °F	$K^{100\,°F}$ 5 psig	$Y = kx$
C_4s	x	61.9*	3.14	$3.14x$
iC_5	21.3	20.4	1.04	22.15
nC_5	39.1	15.6	0.79	30.89
C_6	63.3	4.96	0.25	15.83
C_7	46.4	1.62	0.08	3.71
Total	$170.1 + x$			$72.58 + 3.14x$

*Mean of iC_4 and nC_4

$$170.1 + x = 72.58 + 3.14x$$
$$x = \frac{97.52}{2.14} = 45.5 \text{ moles/h}$$

Where no specifications exist

The realistic separation will be considered where no specifications exist. Consider the debutanizing of full-range naphtha.

The composition of the feed (from Section 6.6)

	Mol. frac.	Moles/h
C_2	0.008	6.4
C_3	0.054	43.5
iC_4	0.021	16.9
nC_4	0.084	67.6
iC_5	0.100	80.5
nC_5	0.043	34.6
C_6	0.155	124.9
C_7	0.175	140.9
Mid-BT 260	0.124	99.8
Mid-BT 300	0.124	99.8
Mid-BT 340	0.075	60.4
Mid-BT 382	0.037	29.8
Total	1.000	805.1

Use the Fenske equation for minimum theoretical trays at total reflux. This equation is

$$N_{M+1} = \text{Log}\left[\left(\frac{\text{LT Key}}{\text{Hy Key}}\right)_D \cdot \left(\frac{\text{HY Key}}{\text{LT Key}}\right)_B\right] \div \text{Log}\phi$$

For a debutanizer light key will be NC_4 (normal butane) and heavy key will be iC_5 (ISO pentane). As a rule of thumb we can allow C_5s to be 1% mol. of total C_4s and lighter in the distillate product.

Note: If this puts the butane LPG off spec the split will be changed. Consider the NC_4 and iC_5 split.

	Moles/h feed	Moles/h distillate	Moles/h bottoms
nC_4s	67.6	x	$67.6 - x$
iC_5s	80.5	1.3	79.2

N_M in the equation is the minimum theoretical trays. The actual number of trays is 30, then theoretical trays = 22.5 (reasonable overall efficient 75%). Take minimum number of trays to be actual theoretical trays ÷ 1.5 = 15. Then $N_{M+1} = 16$. (The plus one is the reboiler, which is a theoretical tray.)

The expression ϕ is the relative volatility of the keys, that is, the ratio of their equilibrium constants at mean tower condition.

Say, 210 °F 110 psig.

Using K values at 800 psig conversion pressure.

$$K_{NC4} = 1.48 \text{ at } 210\,°F \text{ and } 125 \text{ psia}$$
$$K_{iC5} = 0.94$$
$$\phi = K_{NC4} \div K_{iC5} = 1.57$$

Solving for x

$$16 = \text{Log}\left[\left(\frac{x}{1.3}\right) \cdot \left(\frac{79.2}{67.6 - x}\right)\right] \div \text{Log } 1.57$$

$$16 \times 0.195 = \text{Log}\left[\frac{79.2x}{87.88 - 1.3}\,x\right]$$

$$3.12 = \text{Log}\left[\frac{79.2x}{87.88 - 1.3}\,x\right]$$

$$\frac{79.2x}{87.88 - 1.3x} = 1318$$

$$115\,826 - 1713.4x = 79.2x$$

$$x = \frac{115\,826}{1792.6} = 64.6 \text{ moles/h}$$

Thus total C_4s and lighter in distillate

$$= 131.4$$

$$\%C_5s = \frac{1.3}{131.4} \times 100 = 1.0$$

which is OK.

8.3 Predicting Operating Conditions in Light End Towers

Most light end towers for straight-run material operate with total condensing of the overheads. The exceptions are the de-ethanizer and towers whose feed contain an appreciable amount of hydrogen and methane. Setting the tower conditions for the light end towers follows those calculations used earlier in Chapter 6 for the crude tower-top conditions.

Before attempting to calculate tower conditions, a firm material balance over the tower must be prepared. If this is readily available from lab tests and plant yield data, all well and good. If not, this section provides a brief procedure of determining such a requirement.

With the material balance in place the tower conditions may be evaluated using the following procedures:

- *Step 1* Build up the material balance using plant and lab data or by the use of the Fenske equation given in Section 8.2
- *Step 2* Set the reflux drum temperature. This will depend on ambient air temperature in the case of air cooling condensers or cooling water temperature in the case of cooling water condensers. Usually 100 °F is a safe temperature for running the reflux drum.
- *Step 3* Calculate the bubble point pressure of the distillate product at the reflux drum temperature. This pressure will be the reflux drum pressure, and this sets the tower pressure profile.
- *Step 4* Determine the tower-top pressure either by plant data or by allocating pressure drops to line and equipment between tower top and reflux drum. Normally 1 psi, is sufficient for the overhead line and manufacturer's data for condenser pressure drop may be used. Previous plant logs may also be used to establish condenser pressure drop.
- *Step 5* Calculate the dew point temperature of the distillate product at the tower-top pressure. This will be the tower-top temperature.
- *Step 6* Determine the tower-bottoms pressure. This is taken as the tower-top pressure plus the pressure drop over the tower. As a rule of thumb, tower pressure drop is estimated as 0.15–0.2 psi per tray.
- *Step 7* Calculate the tower-bottoms temperature by determining the bubble point temperature of the bottom product at the tower-bottoms pressure.
- *Step 8* Join the tower-top temperature and tower-bottoms temperature on linear graph paper to give a rough estimate of tower temperature profile. Select conditions for anticipated feed inlet temperature. Do the same for pressure profile.

An example calculation is given below

Example Calculation

Consider the debutanizer operation. It is required to recover 95% of the butane in the distillate and 98% of the C_5 in the bottoms. Check to see if this is possible in a 30-tray tower again using the Fenske equation:

$$N_{M+1} = \text{Log}\left[\left(\frac{\text{LT Key}}{\text{HY Key}}\right)_D \left(\frac{\text{HY Key}}{\text{LT Key}}\right)_W\right] \div \text{Log}\phi$$

$$\phi \text{ as before} = 1.59$$

$$N_{M+1} = \text{Log}\left[\left(\frac{0.95}{0.02}\right)_D \cdot \left(\frac{0.98}{0.05}\right)_W\right]/\log 1.59$$

$$
\begin{aligned}
N_{M+1} &= 15 \\
N_M &= 14 \text{ approx no. theoretical stages} \\
&= 14 \times 1.5 \\
&= 21
\end{aligned}
$$

Tray efficiency $= 21/30 = 70\%$. The tower therefore should be able to meet this split (efficiencies $65 \rightarrow 75\%$ realistic).

Light end tower conditions are set by reflux drum temperature and pressure.

For the debutanizer the drum temperature is set at 100 °F. Calculate the component balance as follows:

Debutanizer balance

Comp.	Feed				Distillate					Bottom product				
	Mol. frac.	Moles/ h	MW	#/h	Moles/ h	Mol frac.	lb/h	lb/gal	GPH	Moles/ h	Mol frac.	lb/h	lb/gal	gal/h
C_2	0.008	6.4	30	192	6.4	0.048	192	2.97	64.6					
C_3	0.054	43.5	44	1914	43.5	0.328	1914	4.23	452.5					
iC_4	0.021	16.9	58	980	16.1	0.122	934	4.69	199.1	0.8	0.001	46	4.69	9.8
nC_4	0.084	67.6	58	3921	64.2	0.485	3724	4.87	764.7	3.4	0.005	197	4.87	40.5
iC_5	0.100	80.5	72	5796	2.3	0.017	166	5.21	31.9	78.2	0.116	5630	5.21	1080.6
nC_5	0.043	34.6	72	2491						34.6	0.051	2491	5.26	473.6
C_6	0.155	124.9	86	10 741						124.9	0.187	10 741	5.54	1938.8
C_7	0.175	140.9	100	14 090						140.9	0.210	14 090	5.74	2454.7
MP260	0.124	99.8	114	11 377						99.8	0.148	11 377	6.18	1841.0
MP300	0.124	99.8	126	12 575						99.8	0.148	12 575	6.37	1974.1
MP340	0.075	60.4	136	8214						60.4	0.090	8214	6.46	1271.5
MP382	0.037	29.8	152	4530						29.8	0.044	4530	6.56	690.5
Total	1.000	805.1	95.4	76 821	132.5	1.000	6930	4.58	1512.8	672.6	1.000	69 891	5.94	11 775.1
SG at 60 °F			=	0.6943			0.5504					0.7136			
#/gal			=	5.78			4.58					5.94			
*API			=	72.3			125.6					66.8			

The reflux drum pressure at 100 °F is calculated as the bubble point of the overhead distillate. Equilibrium constants at 800 psia convergence are used here.

	Mole frac.	1st trial at 120 psig		2nd trial at 110 psig	
		K	$Y = Kx$	K	$Y = Kx$
C_2	0.048	3.7	0.178	4.2	0.202
C_3	0.328	1.3	0.426	1.43	0.469
iC_4	0.122	0.62	0.076	0.65	0.079
nC_4	0.485	0.46	0.223	0.48	0.233
iC_5	0.017	0.20	0.003	0.22	0.004
Total	1.000		0.906		0.987

2nd trial   Use K for C_3s

$$\frac{1.3}{0.906} = 1.43$$

Corresponding pressure $= 125$ psia (110 psis)

$$\text{Actual pressure} = \frac{1.43}{0.987} = 1.45$$

Corresponding to 120 psia or 105 psig

Debutanizer reflux drum will operate at 100 °F and 105 psig. For the *total* condensation of overhead product and reflux

Debutanizer tower-top conditions

Because the total overhead is condensed the external reflux returning to the top of the tower will have the same composition as the distillate product. The tower-top temperature will therefore be the dew point of the distillate at the tower top pressure.

Predicting the tower-top pressure

Reflux drum press	=	105 psig
Overhead line pressure drop	=	1 psi
Condenser pressure drop	=	3 psi
Total tower top pressure	=	109 psig

These are experienced pressure drops for a well-designed unit operating within design parameters.

Tower-top temperature

	Mole frac. Y	1st trial at 165 °F		2nd trial at 138 °F	
		K	$x = Y \div K$	K	$x = Y \div K$
C_2	0.048	5.2	0.009	4.8	0.010
C_3	0.328	2.50	0.131	2.00	0.164
iC_4	0.122	1.38	0.088	1.00	0.122
nC_4	0.485	1.0	0.485	0.746	0.650
iC_5	0.017	0.52	0.033	0.360	0.047
Total	1.000		0.746		0.993

2nd trial Use NC_4 K value

New K $= 1.0 \times 0.746$
$ = 0.746$

Corresponding to 138 °F

This trial is close enough

Tower top conditions

Pressure 109 psis
Temperature 138 °F

Predicting tower bottoms temperature

This is the temperature of the bottoms product leaving the tower. It is the bubble point of the product at the tower bottom pressure.

Tower bottom pressure

Estimating this allow 0.17 psi pressure drop per tray (average).

Then tower bottoms will be 109 psig + (30 × 0.17 psi)
= 114 psig = 129 psia

Tower-bottom temperature

	Mole frac. X	1st trial at 320 °F		2nd trial at 380 °F	
		K	$Y = xK$	K	$Y = xK$
iC_4	0.001	3.5	0.004	4.5	0.005
nC_4	0.005	3.1	0.016	3.8	0.019
iC_5	0.116	2.0	0.232	2.8	0.325
nC_5	0.051	1.7	0.087	2.5	0.128
C_6	0.187	1.0	0.187	1.5	0.281
C_7	0.210	0.58	0.122	0.92	0.193
MP260	0.148	0.27	0.040	0.54	0.080
MP300	0.148	0.15	0.022	0.31	0.040
MP340	0.090	0.09	0.008	0.19	0.017
MP382	0.044	0.05	0.002	0.11	0.005
Total	1.000		0.720		1.099

2nd trial   Use K for iC_5

$$\frac{2.0}{0.72} = 2.8$$

Corresponding to 380 °F

Actual temperature

$$\frac{2.8}{1.099} = 2.55$$

Corresponding to 361 °F

Tower-bottom conditions

Pressure $= 114$ psig
Temperature $= 361$ °F

The temperature of the feed will be approximately 250 °F.

8.4   Calculating Tower Tray Efficiency

To evaluate the performance of a tower under test conditions it is necessary to
determine tray efficiencies under these conditions. This is accomplished by calculating
the theoretical number of trays that would give the separation provided by the test
run. The theoretical number of trays thus calculated, divided by the actual number
of trays, will then give the overall tray efficiency. Generally in light end towers this
is between 70% and 75%.

In this calculation three correlations will be used:

● The Fenske equation (7)—to determine minimum trays at total reflux
● The Underwood equation (8)—to determine minimum reflux at infinite number
  of trays
● The Gilliland correlation (9)—to determine the number of theoretical trays from
  Fenske and Underwood.

The Fenske equation has been dealt with earlier. The Underwood equation is more
complex than the Fenske and requires trial and error method for its solution. The
equation itself is in two parts. The first looks at the vapour volatility (ratio of Ks)
of each component in the feed to one of the selected keys, then, by trial and error,
arriving at an expression for a factor B that forces the equation to zero.

This first equation is written as follows:

$$\sum \frac{(\phi i)\,(XiF)}{XiF - B} = 0$$

where

ϕi = is the relative volatility of component i
XiF = is the mole fraction of component i in the feed
B = is the factor that forces the sum of the expression for each component to zero.

The second equation uses factor B calculated from the first equation to determine the mininum reflux:

$$R_{M+1} = \sum \frac{(\phi i)\,(XiD)}{XiD - B}$$

where

ϕi = is the relative volatility of component i
XiD = is the mole fraction of component i in the distillate
B = is the factor obtained from first equation
R_M = minimum reflux at infinite trays (R_{M+1} include the reboiler).

The Gilliland correlation is given in Figure 8.2, and is self-explanatory. The complete calculation follows the steps given below:

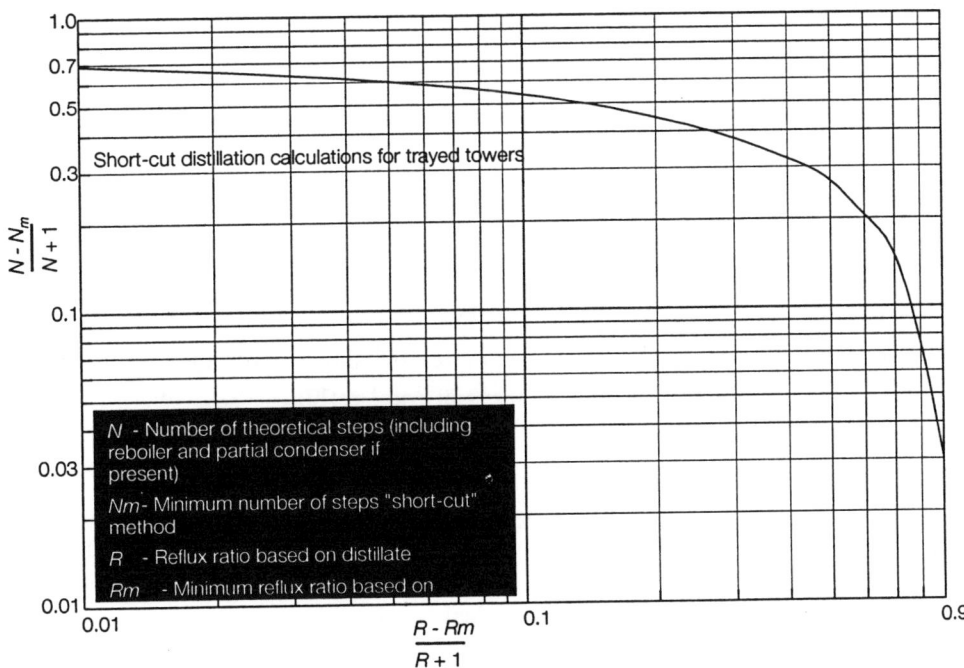

Figure 8.2. Gilliland correlation for calculating theoretical trays. From: Maxwell, Data Book on Hydrocarbons, copyright by A. K. Krieger Publishing Company. Used with permission all rights reserved

- *Step 1* Establish the material balance from lab and plant readings or by calculation as described earlier. Determine the component balance in terms of moles and mole fractions. Use real component if possible or pseudo-component.
- *Step 2* Select two key components (Section 8.2) and calculate the minimum number of trays at total reflux using the Fenske equation (see Sections 8.2 and 8.3).
- *Step 3* List the feed component and then mole fraction. Take one of the keys and calculate the average relative volatility ϕ_{ave} for each component of the feed. The equilibrium constants K are taken from the mean tower condition of temperature and pressure. List these relative volatilities also.
- *Step 4* Calculate the expression $(X_F)\,\phi_{ave}$ for each component and list them.
- *Step 5* Set a value for B in the first equation. Now this is a pure guess: try $B = 0.5$ to start and then for the next trial $B = 1.0$ to calculate for each component.

$$\frac{(X_F)\phi_{ave}}{(\phi)_{ave} - B}$$

- *Step 6* If the sum does not come to zero or close to it, try for the third time and then plot on a linear graph estimated values for B Vs the calculated $+$ or $-$. Read the value for B whose cuts result equals zero.
- *Step 7* Do final check using this value for B.
- *Step 8* Use the B factor calculated in Step 7 to arrive at R_{M+1} in the second equation.
- *Step 9* Take the R_M calculated in Step 8 and multiply by 1.5 to give the expected actual reflux ratio. Use this in the Gilliland correlation to find theoretical number of trays N.
- *Step 10* Divide N by actual number of trays in the tower to give the overall tray efficiency.

An example calculation now follows.

Example Calculation

(i) Calculating the number of theoretical trays required and hence their efficiency

Consider the debutanizer tower and the material balance given in Section 8.3.

Calculating the minimum number of trays at total reflux

Use the Fenske equation, which is

$$N_{M+1} = \text{Log}\left[\left(\frac{\text{LT Key}}{\text{HY Key}}\right)_D \cdot \left(\frac{\text{HY Key}}{\text{LT Key}}\right)_W\right] \div \text{Log}\phi$$

where

N_{M+1} = minimum number of theoretical trays + 1 for reboiler
LT Key$_D$ = light component key fraction in distillate
HY Key$_D$ = heavy component key fraction in distillate
W = bottom product
ϕ = relative volatility of key components at mean tower conditions.
                    (i.e. The ratio of their respective equilibrium constants K).

In this debutanizer the light key component will be nC_4 and the heavy key will be iC_5.

Then LT $Key_D = 0.485$ LT $Key_W = 0.005$
    HY $Key_D = 0.017$ HY $Key_W = 0.116$

ϕ at 250 °F and 126 psia.

$$= K\ nC_4 \div K\ iC_5 = 2.1 \div 1.3 = 1.615$$

$$N_{M+1} = Log \left[\frac{(0.485)}{(0.017)} \cdot \frac{(0.116)}{(0.005)} \right] \div Log\ 1.615$$

$$= Log\ (662.9) \div Log\ 1.615$$
$$= 2.85 \div 0.208 = 13.5$$
$$N_M \quad = 12.5 \text{ theoretical trays}$$

(ii) Calculating minimum reflux at infinite trays

The Underwood equation will be used for this. Thus:

Calculating for B (trial and error)

Comp.	Fract. in feed X_F	ϕ_{ave}	$(X_F)\ (\phi)_{ave}$	$\sum \dfrac{(X_F)\ (\phi)}{\phi_{ave} - B} = 0$ Trial 1 $B = 0.75$	Trial 2 $B = 0.77$
C_2	0.008	3.7	0.0296	0.010	0.010
C_3	0.054	2.3	0.1242	0.080	0.081
iC_4	0.021	1.19	0.025	0.0568	0.060
nC_4 (key)	0.084	1.00	0.084	0.3360	0.365
iC_5	0.100	0.60	0.060	−0.4000	−0.353
nC_5	0.043	0.52	0.022	−0.0956	−0.088
C_6	0.155	0.26	0.040	−0.082	−0.078
C_7	0.175	0.13	0.023	−0.037	−0.036
BP260	0.124	0.05	0.006	−0.009	−0.008
BP300	0.124	0.026	0.003	−0.004	−0.004
BP340	0.075	0.012	0.001	−0.001	−0.001
BP382	0.037	0.007	0.0003	Neg.	Neg.
Total	1.000			−0.1456	−0.052

Take K key as nC_4 at 250 °F and 126 psia.

Trial 2 is close enough to zero: let B be 0.77.

(iii) Calculate for R_{M+1}

Comp.	Fract. in dist. X_D	ϕ_{ave}	(X_D) (ϕ_{ave})	$\dfrac{(X_D)(\phi_{ave})}{(\phi_{ave}) - B}$
C_2	0.048	3.7	0.1776	0.061
C_3	0.328	2.3	0.7544	0.493
iC_4	0.122	1.19	0.1452	0.346
nC_4	0.485	1.00	0.485	2.109
iC_5	0.017	0.60	0.010	0.059
Total				2.950

$R_{min} + 1 = 2.950$
$R_{min} \quad = 1.95 \quad R = 1.95 \times 1.5 = 2.93$

Refer to the Gilliland correlation (Figure 8.2)

$$\frac{R - R_M}{R + 1} = \frac{2.93 - 1.95}{3.93} = 0.249$$

$$\frac{N - N_M}{N + 1} = 0.42 \text{ (from curve)}$$

$$\frac{N - 12.5}{N + 1} = 0.42$$

$$
\begin{aligned}
N \quad &= 0.42(N + 1) + 12.5 \\
&= 0.42N + 0.42 + 12.5
\end{aligned}
$$

$0.58N \quad = 12.92$
$N \qquad = 22.3$

Less 1 for reboiler $= 21.3$ theoretical trays
Efficiency $\qquad = 71\%$, which is reasonable

8.5 Calculating Tower-top Conditions for a Partial Condenser (De-ethanizer)

De-ethanizer towers operate close to the critical conditions at the bottom of the tower, and at critical conditions separation is not feasible by distillation processes. To minimize the pressure and temperature conditions in the tower the very light top product is not condensed completely as in most other light end towers. Indeed, only sufficient condensation occurs to supply reflux to the tower.

Predicting tower-top conditions for this kind of tower is somewhat different from the methods we have dealt with so far in this book. The steps described below gives the method used to establish these tower conditions:

● *Step 1* Develop the component balance in moles and mole fractions. Generally, the top vapour product will contain methane, ethane and some propane.

- *Step 2* Set the reflux drum temperature as described in Section 8.3.
- *Step 3* Carry out a dew point calculation for the overhead vapour product at the temperature of the drum. This dew point pressure will be the pressure of the drum.
- *Step 4* Using the drum pressure predict the tower top pressure as in Section 8.3.
- *Step 5* Set the external reflux flow in moles/h. This may be obtained from plant data or from experience. Usually, a reflux of 2:1 (reflux to overhead vapour product) will be realistic. Now the composition of this reflux will be the composition of the liquid in equilibrium with the vapour leaving the reflux drum (that is, the *x* mole fractions calculated in the dew point calculation).
- *Step 6* Unlike the conventional light end units (such as the debutanizer and depropanizer), the composition of the external reflux stream will be different from the product. Consequently, the total moles of reflux must be added on a component basis to the product to obtain the composition of the total overhead vapour leaving the top of the column.
- *Step 7* Carry out a dew point calculation on this total overhead stream to find the tower top temperature. *Note*: A partial condenser is equivalent to one theoretical tray. Thus in calculating tray efficiency (Section 8.4) it must be considered as such.

An example calculation now follows.

Example calculation

Calculating reflux drum conditions for a de-ethanizer

Consider a de-ethanizer with an overhead vapour product of:

	Moles/h	Moles frac.
C_1	1.4	0.118
C_2	6.3	0.529
C_3	4.2	0.353
Total	11.9	1.000

The partial condenser will operate at the dew point of overhead vapour and at a temperature of 100 °F

	Mol. frac.	1st trial 460 psia		Mol. wt	Weight factor	#/gal	Vol. factor	
		$K^{(1)}$	$X = Y/K$					
C_1	0.118	5.0	0.024	16	0.384	2.5	0.15	SG at 60
C_2	0.529	1.6	0.331	30	9.930	2.97	3.34	− 0.456
C_3	0.353	0.52	0.679	44	29.876	4.23	7.06	°API − 160 °
Total	1.000		1.034	38.9	40.19	3.8	10.55	

(1) As C_1s are present in greater quantity than 0.001 use 1500 psia convergence pressure.

This is close enough. Then reflux drum conditions are:
Temperature = 100 °F
Pressure = 460 psia

To calculate tower-top conditions for the de-ethanizer

Assume de-ethanizer operates with reflux twice the vapour product. Then because the reflux will have a different composition from the vapour it must be taken into consideration.

Tower top pressure

Reflux drum pressure	— 460 psia
Pressure drop for line	— 1 psi
Condenser pressure drop	— 4 psi

Tower top	— 465 psia

Tower top temperature

	Y prod.	Y reflux	Total Y	Mol Frac. Y	K	$X = Y/K$	MW	Weight factor	lb/gal	Volume factor	
					1st trial at 140 °F						
C_1	0.118	0.046	0.164	0.055	5.2	0.011	16	0.176	2.5	0.070	
C_2	0.529	0.641	1.170	0.390	2.0	0.195	30	5.950	2.97	1.97	°API
C_3	0.353	1.313	1.866	0.555	0.77	0.721	44	31.724	4.23	7.50	1.66
Total	1.000	2.000	3.000	1.000		0.927	40.7	37.75	3.96	9.54	

Actual temperature $= 0.77 \times 0.927 = 0.71$

Corresponding to 137 °F

8.6   Calculating and Evaluating Condenser Performance

When the required degree of fractionation has been established the required duty of the condenser can be calculated. This section deals with two types of condensers:

● Those that are designed to condense the overhead product and reflux completely
● Those that only partially condense the overhead stream.

Most light end units in straight-run service use total condensers in all but the de-ethanizer unit. This latter unit condenses sufficient liquid for tower reflux only. The overhead product (C_2 and lighter) is disposed of as a vapour and is not condensed.

The calculation to assess the condenser duties are different for the two types. The following steps and discussion relate to each of the two types and are as follows.

Total Condensing

The example given here relates to a debutanizer and data developed earlier in this chapter are used throughout. The steps followed are:

● *Step 1*   Establish the internal reflux or overflow from tray 1. This has been done using the Underwood equation in Section 8.4.

- *Step 2* Using the tower conditions in Section 8.3, prepare the heat balance over the top of the column. The unknown in this case is the condenser duty.
- *Step 3* To establish liquid and vapour properties sufficient to carry out the heat balance use the *x* factor (that is, the liquid composition in equilibrium with the vapour) calculated in the dew point calculation in Section 8.3. From this composition calculate the specific gravity and mole weight of the liquid.
- *Step 4* Assign a reasonable temperature for trays 1 and 2. Rule of thumb is tray 1 will be 3–5 °F higher than dew point calculated. Tray 2 vapour will be about 5 °F above tray 1 liquid.
- *Step 5* Solve for condenser duty by difference in the heat balance.
- *Step 6* Knowing the condenser duty, the overhead product rate, tower top and reflux drum conditions the external reflux may be calculated by a simple heat balance over the condenser. *Note*: Because the external reflux is sub-cooled relative to the internal reflux it will always be less in quantity than the internal stream.

Partial Condensing

The de-ethanizer operation is to be referenced for this case. Note that other partial condensers such as the platformer stabilizer usually have vapour and distillate product streams. This will be described in Chapter 9.

Unlike the total condensers the tower-top temperature/pressure cannot be established by a dew point calculation of the overhead product in a de-ethanizer. The reflux drum is itself a theoretical tray in this case and therefore the liquid from it returning to the tower as reflux, will have a different composition from the overhead product. The example given here to calculate such a condenser duty has the following steps:

- *Step 1* The tower top conditions have been calculated assuming a 2:1 reflux ratio. See section 8.5.
- *Step 2* Using these conditions carry out a heat balance over the condenser. The unknown in this case will be the condenser duty. This is found by difference.
- *Step 3* Set the tray 1 temperature at about 3 °F higher than the tower top (on dew point). Set the vapour from tray 2 at about 5 °F above tray 1.
- *Step 4* The unknown is the internal reflux. Carry out a heat balance over tray 1, condenser and the product leaving the reflux drum.
- *Step 5* Equate and solve for internal reflux. Note that this is higher than the external reflux.
- *Step 6* If the internal reflux ratio is too low for the separation required (this can be calculated using the Underwood equation, see Section 8.4), then a new external reflux rate must be selected. The new tower-top conditions must be calculated and Steps 2 to 5 recalculated to obtain a new internal reflux. This repeat calculation is not shown here.

The example calculation now follows.

Example calculation

(i) For a total condenser

The data developed earlier for a debutanizer will be used in this sample calculation:

- Overhead product will be as per the material balance given in Section 8.3
- Internal reflux (from tray 1, top tray) will be as calculated in Section 8.4
- Temperature will be as calculated in Section 8.3

Consider the following heat balance:

Heat is brought in by:

- Vapour from tray 2.

Heat is taken out by:

- Reflux overflow from tray 1
- Product distillate
- The condenser

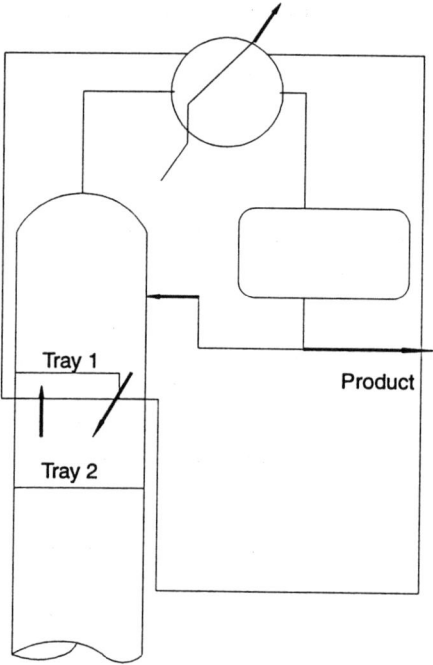

Reflux stream properties

From Section 8.3 composition of the reflux stream is the x column of the dew point calculation (liquid in equilibrium with distillate vapour).

Comp	Mole frac. x	Mol wt.	Weight factor	lb/gal	Vol. factor
C_2	0.010	30	0.3	2.97	0.10
C_3	0.164	44	7.22	4.23	1.71
iC_4	0.122	58	7.08	4.69	1.51
nC_4	0.650	58	37.70	4.87	7.74
iC_5	0.046	72	3.31	5.21	0.64
Total	0.993	56	55.61	4.75	11.70

SG   0.570
°API 116

Total internal reflux (Section 8.4)

Calculated reflux ratio	$= 2.93$
Then total reflux	$= 2.93 \times$ dist. product
	$= 2.93 \times 132.5$ moles/h
	$= 388.2$ moles/h
	$= 21\,741$ lb/h

Tray 1 temperature	$= 138 + 3\ °F = 141\ °F$ (estimated)
Tray 2 temperature	$= 141 + 5\ °F = 146\ °F$ (estimated)

The heat balance

Stream	V or L	Mol. wt.	K	°F	lb/h	Enthalpy BTU/lb*	Enthalpy BTU/h × 10⁶
In							
Vapour ref. ex 2	V	58	—	146	21 741	216	4.696
Vapour prod. ex 2	V	58	—	146	6 930	216	1.497
Total in					28 671		6.193
Out							
Liquid ref. ex 1	L	58	—	141	21 741	82	1.783
Liquid prod.	L	58	—	100	6 930	56	0.389
Condenser duty						By difference	4.021
Total out					28 671		6.193

*From Maxwell's *Data Book on Hydrocarbons*

External reflux (total condenser)

Let x lb/h be the external reflux. Then the heat balance over the condenser will be:

Heat in with external reflux + product vapour at 138 °F.
   $= (6930 + x) \times 212$ BTU/lb $= 1.469 + 212x$ MM BTU/h

Heat out with reflux liquid + product at 100 °F.
   $= (6930 + x) \times 56$
   $= 0.388 + 56x$ MM BTU/H

Heat out with condenser	$= 4.021$ MM BTU/h
Total out	$= 4.409 + 56x$
Then $1\,469\,000 + 212x$	$= 4\,409\,000 + 56x$

$$x = \frac{2\,940\,000}{156} = 18\,846 \text{ lb/h}$$

$$= 2.7{:}1 \text{ reflux ratio}$$

(ii) For a partial condenser

Consider the de-ethanizer tower used in Section 8.5.

To calculate the internal reflux generated by the external reflux of 2:1, first calculate the condenser duty thus: heat in is external reflux vapour + product vapour.

Product vapour   = 11.9 moles/h = 396.2 lb/h
Reflux vapour   = 23.8 moles/h = 925.8 lb/h

Enthalpy in vapour at 137 °F   = 230 BTU/lb (use 40 MW)

BTU/h   = 230 × (396.2 + 925.8)

                                = 304 000

Enthalpy out: Heat in product vapour at 100 °F   = 194 BTU/lb
                                                  = 194 × 396.2
                                                  = 77 000 BTU/h

Heat in reflux liquid at 100 °F   = 62 BTU/lb
                                                  = 925.8 × 62
                                                  = 57 000 BTU/h

Condenser duty by difference   = 170 000 BTU/h

Calculate the heat balance over the envelope shown in the diagram.

Heat in with vapours from tray 2

Heat out with liquid from tray 1

Heat out with product vapour from reflux drum

Heat out with condenser

Liquid reflux x lb/h will have the properties of the liquid in equilibrium at the dew point at the tower top (see Section 8.5).

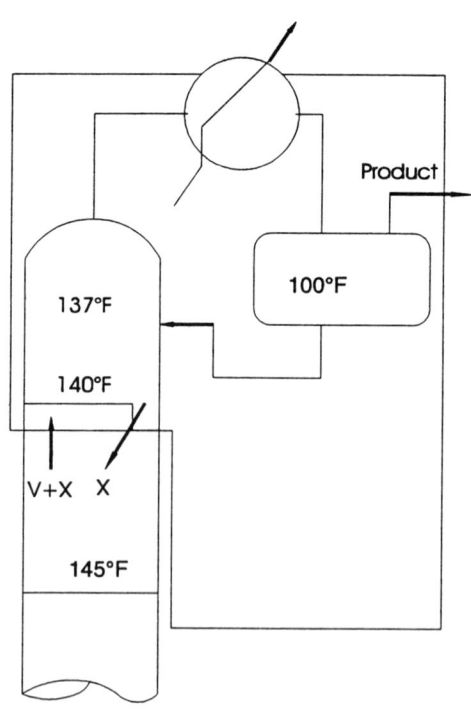

The heat balance

					Enthalpy	
	V/L	Mol. wt	°F	lb/h	BTU/lb	BTU/h
In						
Product vapour ex tray 2	V	33	145	396	210	83 000
Reflux vapour ex tray 2	V	40	145	X	200	$200X$
Total in				$396 + X$		
Out						
Product vapour ex ref. drum	V	33	100	396	194	77 000
Overflow ex tray 1	L	40	140	X	98	$98X$
Condenser duty						170 000
Total out				$396 + X$		$247\,000 + 98X$

$$83\,000 + 200X = 247\,000 + 98X$$

$$X = \frac{164\,000}{(208 - 98)} = 1608\ \text{lb/h}$$

$$= 40.2\ \text{moles/h}$$

Giving an internal reflux ratio of 3.38:1.0

8.7 Calculating and Evaluating Reboiler Performance

Calculating reboiler duty is important in evaluating the tower operation or in designing the light ends unit. This topic is dealt with more fully in Section 8.10. It is, however, highly desirable in the operation of light end towers to maintain as balanced an inlet heat load as possible. There are two sources of inlet heat to the tower:

- Heat in with the feed
- Reboiler heat input

Normally the feed to a light end tower is preheated by exchange with the bottom product and in some cases by another heating medium such as steam or hot oil. Again in most cases, the feed is introduced into the tower partially vaporized. Knowing the condition and the enthalpy of the feed, the condenser duty and the product streams enthalpy the reboiler duty can be calculated. The example given here and the following steps provide this calculation technique:

- *Step 1* The example given here utilizes the data calculated earlier in this chapter for the debutanizer. The first requirement is to establish the tower material balance.
- *Step 2* Calculate the tower conditions top and bottom. Plot the preliminary temperature profile given in Figure 8.3.

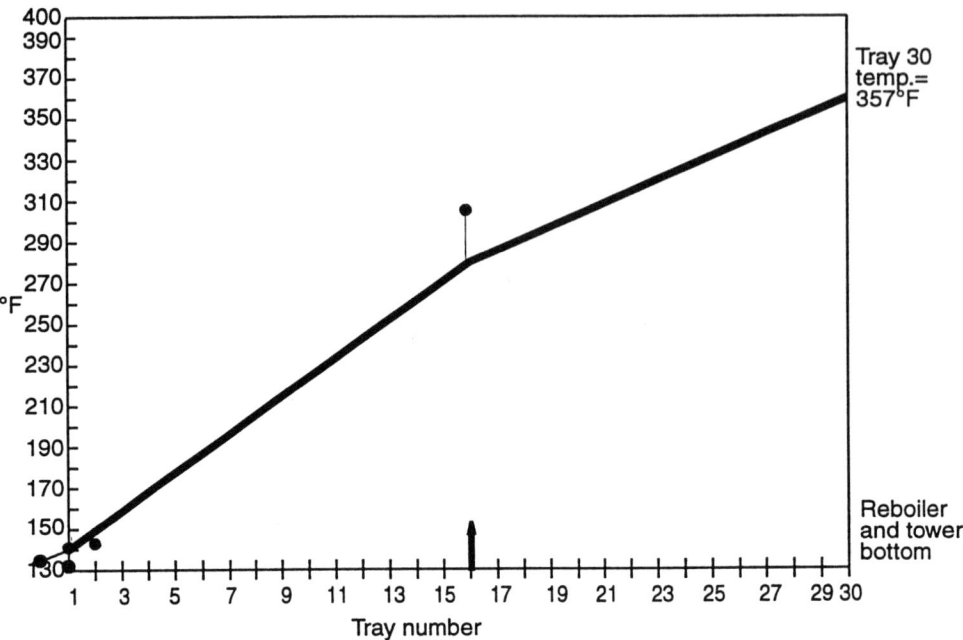

Figure 8.3. A debutanizer estimated temperature profile

- *Step 3* Unless it is known from plant data, assume a feed inlet temperature. In the calculation given here a 60 °F approach was assumed between the feed inlet temperature to the tower and the bottoms product outlet from the tower.
- *Step 4* Carry out a flash vaporization calculation on the feed at the feed inlet temperature and pressure determined in Step 3. (The pressure in this case is taken as the mean tower pressure.) This establishes the amount of vapour and liquid present under these conditions.

The flash vaporization calculation is one of 'trial and error'. A ratio V/L is first assumed, and based on this assumption a ratio V/L is calculated. The equation for such a calculation is

$$L = \frac{F}{1 + (V/L)K}$$

where

L = number of moles of each component in the liquid phase
F = number of moles of each component in the feed
V/L = the assumed vapour/liquid ratio
K = equilibrium constant for each component at the system conditions.

The sum of the L for each component is the L in the calculated ratio. The V, by definition, is the feed minus the L function. The problem converges when V/L

assumed $= V/L$ calculated. One tip: if two trials are made and plotted on LogLog paper the straight line through them will indicate the third trial, which is usually the solution.

- *Step 5* Using the calculated vapour/liquid component break down calculated density and mole weight of the vapour and liquid phases, respectively.
- *Step 6* Calculate the enthalpy for the mixed phase feed in BTU/h.
- *Step 7* Carry out the overall heat balance for the tower where the reboiler is the unknown. Equate and solve to determine reboiler duty.

An example calculation now follows.

Example calculation

For this example the data for the debutanizer given earlier will be used thus:

- Material balance as in Section 8.3
- Tower conditions as in Section 8.3
- Condenser duty as Section 8.6

(i) Calculate the feed condition and enthalpy

The feed be introduced as a mixed phase (vapour and liquid) at 300 °F and 127 psia

Calculate feed flash

	Moles /h (°F)	127 psia K 300 (°F)	1st trial $V/L = 0.5$		2nd trial $V/L = 0.2$		3rd trial $V/L = 0.1$		MW	Liquid lb/h	lb/gal	gal/h
			V/LK	$L = F/(1+V/LK)$	V/LK	$L = F/(1+V/LK)$	V/LK	$L = F/(1+V/LK)$				
C_2	6.4	9.1	4.55	1.15	1.82	2.27	0.9	3.37	30	101	2.97	34
C_3	43.5	5.0	2.50	12.43	1.00	21.75	0.5	29.0	44	1276	4.89	301.7
iC_4	16.9	3.3	1.65	44.79	0.66	10.18	0.33	12.71	58	737	4.69	157.1
nC_4	67.6	2.9	1.45	30.04	0.58	42.78	0.29	52.40	58	3039	4.87	624.0
iC_5	80.5	1.8	0.90	42.37	0.26	59.19	0.18	68.22	72	4912	5.21	942.8
nC_5	34.6	1.6	0.80	19.22	0.32	26.21	0.16	29.83	72	2148	5.26	408.4
C_6	124.9	0.85	0.425	87.65	0.17	106.75	0.085	115.12	86	9900	5.54	1787.0
C_7	140.9	0.48	0.240	113.63	0.096	128.56	0.048	134.45	100	13445	5.74	2342.3
NP260	99.8	0.212	0.106	90.24	0.042	95.74	0.021	97.75	114	11144	6.18	1803.2
NP300	99.8	0.116	0.058	94.33	0.023	97.54	0.012	98.62	126	12426	6.37	1950.7
NP340	60.4	0.083	0.032	58.53	0.126	59.55	0.006	60.04	136	8165	6.46	1263.9
NP382	29.8	0.035	0.0175	29.29	0.007	29.60	0.004	29.80	152	4530	6.56	690.5
Total	805.1			623.67		680.07		731.31	98.2	71823	5.4	12305.6
Calculated V/L				0.29		0.184		0.1			°API = 70	

lb/h vapour	= 4998
lb/gal	= °API = 100

Feed Enthalpy

Vapour	= 312 BTU/lb	= 1.559×10^6 BTU/h
Liquid	= 158 BTU lb	= 11.348×10^6 BTU/h
	Total feed	= 12.907×10^6 BTU/h

(ii) Calculate overall heat balance and reboiler duty

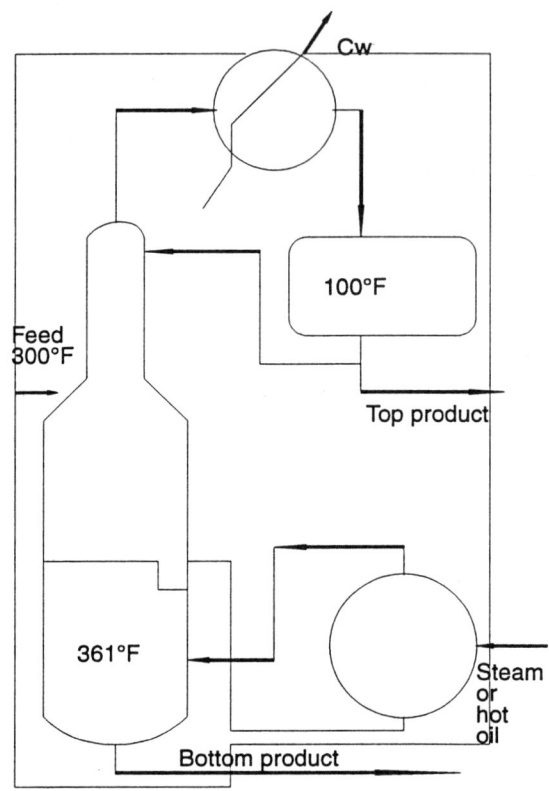

| | V/L | K | °API | °F | lb/h | Enthalpy | |
						BTU/lb	BTU/h × 10⁶
In							
Feed	$V+L$	—	—	300	76 821	—	12.907
Reboiler	—						Z
Total in					76 821		$12.907 + Z$
Out							
Top product				100	6 930	56	0.389
Bottom product				361	69 891	194	13.559
Condenser							4.021
Total out					76 821		17.969

Reboiler duty $= (17.969 - 112.90) \times 10^6$

$= 5\,062\,000$ BTU/h

8.8 Establishing Vapour and Liquid Traffic in Towers and Evaluating Tray Performance

Among the evaluation of tower performance must be the analysis of the tray performance in the various sections of the tower. The critical trays in the tower that lend themselves to this analysis are the top and bottom trays of the tower. The vapour and liquid traffic across these trays are easily calculated by heat balance and using normal plant data.

This Section describes a calculation procedure to determine tower traffic and uses these data to evaluate the tray performance. The same data is of course also used in the plant design to obtain the tower diameter. In the evaluation of the tray performance and loading the calculated tray loading is compared with the "Flood" condition allowable for this type and size of tray. It should be noted that trays can operate above the allowable flood point quite effectively. However, above the allowable flood some liquid entrainment could occur and at higher vapour and liquid rates there will be flooding by high downcomer back-up. When this occurs no separation is possible, high-pressure drops are experienced and the unit must be shut down (or at least heat input reduced) to relieve the situation.

The allowable flood point, then, is an operating and design parameter to measure tray performance. Ideally, towers should operate at between 80% and 90% of this figure. The dangers of high traffic in the tower has been discussed. Very low vapour and liquid traffic can also impair the operation and performance of distillation trays. Sieve trays in particular are affected by low tray loadings. Their performances are impaired by liquid leaking from tray to tray because of insufficient vapour up-thrust or pressure to maintain a level on the tray. Poor mixing on the trays occurs at low loadings, as very often vapour traffic becomes channelled to one section of the tray. Tower loadings should not be allowed to fall below the 40–50% of the allowable 'flood' figure.

The calculation example provided here is based on the data for the debutanizer calculated or used in this chapter. The calculation steps are as follows:

- *Step 1*   Establish the vapour/liquid loading over the top tray. This has been done in Section 8.6.
- *Step 2*   From the bubble point calculation for the bottom product (Section 8.3) develop the properties (density and mol. weight) of the vapour in equilibrium with the liquid at the bubble point.
- *Step 3*   Estimate the temperature of liquid leaving the bottom tray. The tower temperature profile given in Figure 8.3 is used for this. The liquid from the bottom trays is the feed to the reboiler (a thermosyphon shell and tube in this case). This liquid contains the bottom product and a portion that will be vaporized in the reboiler.
- *Step 4*   With the portion to be vaporized as the unknown calculate the heat balance over the bottom tray and reboiler. The reboiler duty has already been calculated in Section 8.7. Equate and solve for the unknown vapour rate in pounds per hour.
- *Step 5*   In this problem valve trays are used for the top section. Calculate the mass vapour velocity for this section of the tower using the data calculated for the top tray in step 1. The mass vapour velocity Ga is calculated by:

$$\frac{\text{lb/h of vapour}}{\text{Bubble area of tray}}$$

- *Step 6*   Calculate the liquid loading L_L in CFS at tray conditions. This is given by the expression

$$L_L = \frac{\text{Hot GPM}}{450}$$

- *Step 7* From the tray geometry calculate (Equipment Data Sheet)

  A_S = total tray area in square feet
  A_{DC} = downcomer area (both inlet and outlet)
  A_W = the waste area on the tray (estimated at 15% of total tray area).

- *Step 8* Using the data developed in Step 7, calculate the Bubble area, which is

  $$A_B = A_S - (A_W + A_{DC})$$

- *Step 9* Calculate the allowable vapour mass velocity at flood using the expression

  $$Gf = K(\rho_v \times \rho_L - \rho_v)^{0.5}$$

  where

  K = The flood constant vs tray spacing from figure 6.9
  Gf = Mass vapour velocity in lb/h.sqft at flood.

- *Step 10* Calculate the tower percentage of flood by the expression

  $$\frac{G_a}{G_f} \times 100$$

- *Step 11* Calculate the actual vapour flow and liquid flow to/from the bottom trays using data calculated in Steps 3 and 4.

- *Step 12* Although the bottom section of this tower contains sieve trays the same data contained in Figure 6.9 can also be used for this section. Repeat Steps 5 to 10 for the bottom section.

An example calculation now follows.

Example calculation

This calculation will consider the top and bottom trays of the debutanizer.
- Vapour and liquid traffic for the top trays has already been calculated (see Section 8.6)

To calculate vapour/liquid traffic to the bottom tray

Consider the heat balance over the bottom of the tower.

Estimated temperature for tray 30 from profile is 357 °F.

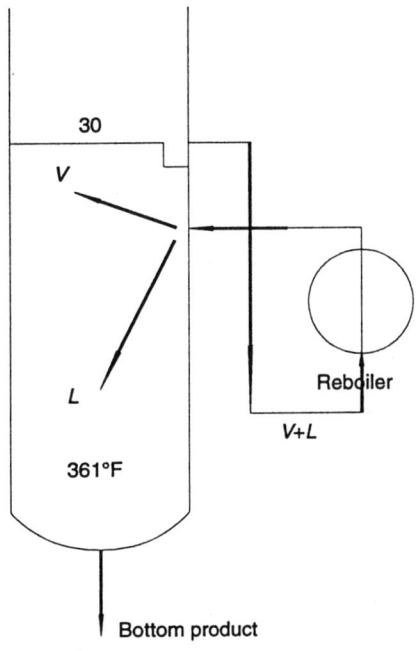

	V/L	K	°API	°F	lb/h	Enthalpy	
						BTU/lb	BTU/h × 10⁶
In							
Liquid ex tray 30	L	115	70*	357	69 891 + V	195	13.629 + 195V
Reboiler							5.062
Total in					69 891 + V		18.691 + 195V
Out							
Bottom prod.	L	11.5	66.8	361	69 891	193	13.489
Vapour to tray 30	V	11.5	74*	361	V	331	331V
Total out					69 891 + V		13.489 + 331V

*Estimated

$$V = \frac{(18.691 - 13.489)}{(331 - 195)} = 38\,250 \text{ lb/h}$$

To calculate properties of the vapour to the bottom tray

From the bubble point calculation of the bottom product (see Section 8.3) the vapour composition at equilibrium is as follows:

	$Y = XK$	MW	Weight factor	lb/gal	Vol. factor
iC_4	0.005	58	0.29	4.69	0.06
nC_4	0.019	58	1.102	4.87	0.23
iC_5	0.325	72	12.4	5.21	4.49
nC_5	0.128	72	9.22	5.26	1.75
C_6	0.281	86	24.17	5.54	4.36
C_7	0.193	100	19.30	5.74	3.36
NP260	0.080	114	9.12	6.18	1.48
NP300	0.040	126	5.04	6.37	0.79
NP340	0.017	136	2.31	6.46	0.36
NP382	0.005	152	0.76	6.56	0.12
Total	1.099	86.2	94.712	5.57	17.00

Moles of vapour to tray 30	$= 443.7$
Moles of liquid from tray 30	$= 443.7 + 672.6$
	$= 1116.3$
Volume of liquid from tray 30 at 60 °F (gal/h)	$= 6867 + 11\,775$
	$= 18\,642$
Lb per gallon of liquid ex tray 30 at 60 °F	$= 5.6 = 71$ °API and 0.698 SG
Hot SG	$= 0.48$ (at 357 °F)
Hot gallons per hour	$= 27\,109$

To calculate the tray loading at top and bottom of the tower

Assume tower dimensions are as follows:

Bottom section I/D	$= 54$ in. 4 ft. 6 in.
Top section	$= 42$ in. 3 ft. 6 in.

The top section will contain valve trays (i.e. trays 1 to 15)

The bottom section will contain sieve trays (i.e. trays 16 to 30)

Calculating the loading over the top tray section (use data for top tray)

From Section 8.6 liquid from tray	$= 21\,741$ lb/h
Cold gals per hour	$= 21\,741 \div 4.75$
	$= 4577$
At 141 °F gal/h	$= 4577 \times 1.156$
	$= 5291$ gal/h
Vapour to tray $1 = 21\,741$ lb/h	$= 388$ mole/h
$\qquad\qquad\qquad$ 6930 lb/h	$= \underline{132.5}$
Total $= 28\,671$ lb/h	$= 520.5$ mole/h
Vapour temperature	$= 146$ °F pressure 109 psig
Actual flow in CFS	$= \dfrac{520.5 \times 378 \times 14.7 \times 606}{520 \times 123.7 \times 3600}$
	$= 7.57$ CFS
Density (ρ_V)	$= 7.96 = 1.05$ lb/ft³
Liquid density (ρ_L)	$= 30.74$ lb/ft³

Section top tray geometry

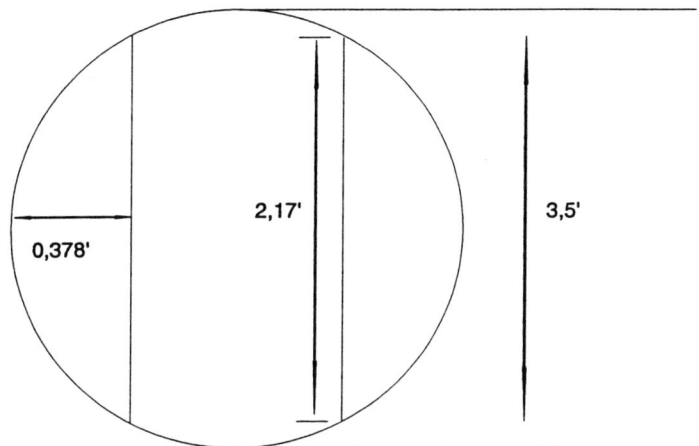

Inlet downcomer area	$= 0.56 \text{ ft}^2$
Outlet downcomer area	$= 0.56 \text{ ft}^2$
Inlet weir height	$= 1 \text{ in.}$
Outlet weir height	$= 1 \text{ in.}$
Outlet and inlet weir length	$= 2.17 \text{ ft}$
Total tray area A_S	$= 9.62 \text{ ft}^2$
Total downcomer areas A_{DC}	$= 2 \times 0.56 \text{ ft}^2$
	$= 1.12 \text{ ft}^2$
Waste area A_W	$= 1.4 \text{ ft}^2$
Bubble area A_B	$= 9.62 - (1.12 + 1.4)$
	$= 7.1 \text{ sqft}$

Top section loading

Actual vapour mass velocity $\text{Ga} = \dfrac{28671}{7.1}$

$$= 4037 \text{ lb/h.sqft}$$

Allowable mass velocity at 24 in tray spacing:

$$\text{Gf} = \text{K}\sqrt{(\rho_v \times \rho_L - \rho_v)}$$
$$= 1110 \sqrt{1.05 \times (30.74 - 1.05)}$$
$$= 6197 \text{ lb/h.sqft}$$

Then % of flood:

$$\frac{G_a}{G_f} \times 100 = 65\%$$

This section may have a problem of tray "weeping". Ideal loading would be 70–90% of flood.

Calculating Loading Over Bottom Tray Section (use data for bottom tray)

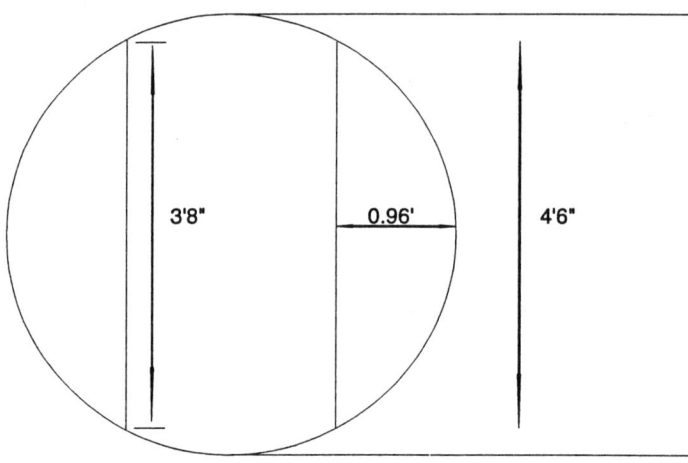

Bottom section tray geometry

Inlet and outlet downcomer area	$= 2.5 \text{ ft}^2$
Total	$= 5 \text{ ft}^2$
Inlet and outlet weir height	$= 1 \text{ in.}$
Inlet and outlet weir length	$= 3 \text{ ft } 8 \text{ in.}$
Total tray area	$= A_S = 15.9 \text{ ft}^2$
Total downcomer area	$= A_D = 5 \text{ ft}^2$
Waste area (15% total)	$= 2.4 \text{ ft}^2$

Bubble area
$$= 15.9 - (5 + 2.4)$$
$$= 8.5 \text{ sq ft}$$

Bottom Section Loading

Actual vapour mass velocity
$$= \frac{38\,250 \text{ lb/h}}{8.5}$$
$$= 4500 \text{ lb/h.sqft}$$

Allowable velocity at flood Gf
at 24 in tray spacing
$$= K \sqrt{(\rho_v \times \rho_L - \rho_v)}$$
$$= 1110 \times \sqrt{1.27 \times (29.8 - 1.27)}$$
$$= 6682 \text{ lb/h.sqft}$$

$$\text{percent of flood} = \frac{G_a}{G_f} \times 100$$

$$= \frac{4500}{6682} \times 100 = 67.3\%$$

which is acceptable.

8.9   Process Efficiency in Terms of Actual Energy Used versus Calculated

This section deals with the handling of plant utility data and comparing it with calculated data to evaluate plant performance. The example provided here is based on the debutanizer operation, the data for which are contained in the various sections of this chapter.

The calculation covers estimating the fuel consumed in the case of a fired heater reboiler. This uses an assumed heater efficiency and heats of combustion given in Appendix 1. The results of the calculation in terms of lb/h of liquid fuel used (or CF/h of gas) can be compared with actual values from a test run to compare heater efficiency.

Similarly, if a shell and tube reboiler is used the duty of the reboiler can be checked. This is accomplished by comparing calculated duty with actual duty of steam used for the heating.

Power requirements are calculated for pumps and all coolers. The air cooler consumption utilizes a known air temperature rise figure through the air cooler. This is used to determine the quantity of air flow. The brake horsepower of the fan (or fans) is determined by the expression

$$\text{BHP} = \frac{(\text{ACFM per fan} \times \text{pressure drop})}{6370 \times \text{fan efficiency}}$$

Pressure drop is taken as between 0.5 and 0.7 in. of water. Fan efficiency is taken as between 70% and 75%. Motor efficiency is about 90%.

Pump discharge pressure and capacities can be obtained with plant data or can be estimated as shown in the calculation. Pump efficiency is taken as 70% and motor horsepower calculated with an efficiency of 90%. Based on these data, the power consumption for each pump is calculated, and may be compared with actual plant data. An example calculation follows.

Example calculation

Fuel consumption

This assumes that the debutanizer is reboiled using a fired heater. Then from the example in Section 8.7 the reboiler duty is calculated as 5.062×10^6 BTU/h. This is absorbed heat. The heater should operate between 70% and 80% efficiency. As a light end unit this should be around 75% efficient. Then the BTU/h of fuel value that should be used is

$$\frac{5.062 \times 10^6}{0.75} = 36.749 \times 10^6 \text{ BTU/h}$$

From the heat of combustion curve in Appendix 1 the actual heat of combustion of the fuel can be read off. This is in BTU/lb for liquid fuels and BTU/cf at 60 °F for gases. Using these data the quantity of fuels predicted can be obtained by dividing the total fuel value by the heat of combustion. Use the net heating value.

Steam requirements

When the debutanizer is reboiled by steam the calculation for steam quantity is resolved simply by dividing the reboiler duty by the total heat of the condensing steam. Thus:

Duty of the reboiler $= 5.062 \times 10^6 \text{ BTU/h}$

600 psig steam superheated to 700 °F is the heating medium—the condensate will leave at 490 °F

Heat in steam $= 1352 \text{ BTU/h (from steam tables)}$

Heat in condensate $= (490 - 32) = 458 \text{ BTU/lb (at 32 °F datum)}$

lb/h steam $= \dfrac{5.062 \times 10^6}{(1.352 - 458)}$

                         $= 5662$

Power requirements

- Air coolers
- Pumps

(i) Air coolers

There will probably be two

- The overhead condenser
- The bottom product cooler.

The condenser

Temperature raise of air (plant readings), say 50 °F

Duty of exchanger $= 4.021 \times 10^6 \text{ BTU/h (from Section 8.6)}$

$$\text{Quantity of air flow} = \frac{\text{duty}}{\text{SP HT} \times \text{Temp rise}}$$

$$= \frac{4.021 \times 10^6}{0.24 \times 50}$$

$$= 335\,083 \text{ lb/h}$$

$$\text{Moles/h} = \frac{335\,083}{29} = 11\,554.6$$

$$\text{ACFM} = \frac{11554.6 \times 378 \times 14.7 \times 545}{520 \times 15}$$

$$= 74\,768 \text{ ACFM}$$

Static pressure drop will be between 0.5 and 0.7 in. of water. Say 0.65 in. of water

$$\text{BHP} = \frac{(\text{ACFM/FAN}) \times \Delta P}{6370 \times 0.70}$$

Fan is 70% efficient

$$= \frac{74\,768 \times 0.65}{6370 \times 0.7}$$

$$= 11 \text{ HP}$$

Say, motors are 90% efficient

$$\text{Then approx. power usage} = \frac{11}{0.9} \times 0.746$$

$$= 9.1 \text{ kW}$$

The cooler will follow a similar procedure

Pumps

There will be two pumps

- Overhead product and reflux
- Bottoms product pump

Establish the battery limit condition. For the overhead pump this will be around 320 psig and 100 °F. For the bottom pump this will be about 120 psig. In both cases it is assumed that the overheads go to the depropanizer and the bottoms to the splitter.

The overhead pump

Say reflux drum is 30 ft above grade, the suction pressure is 105 psig + static HD − friction loss. (2 ft to centreline of pump).

$$\text{Static head} = (30\,\text{ft} - 2\,\text{ft}) \times \frac{0.5504 \times 62.2}{144}$$

$$= 6.66 \text{ psi}$$

Friction loss (say, 1 psi)

Suction pressure   = 110.7 psig (say, 111 psig)

Discharge pressure will be 320 psig + control valve + friction loss.

Discharge pressure   = 320 psig + say 10 psi × 2 psi

(*Note*: this can be read at pump discharge)

	= 332 psig
Pump differential head	= 332 − 111 psi
	= 221 psi

$$= \frac{221 \times 144}{62.2 \times 0.5504} = 930 \text{ ft of liquid}$$

Capacity of pump	= product + reflux
	= 6930 + 18 846 (Section 8.6) lb/h
	= 25 776 lb/h

$$\text{BHP (assume 70\% efficiency)} = \frac{25\,776 \times 930}{60 \times 33\,000 \times 0.7} = 17.3$$

Assume motor efficiency 90%

Then power usage $$= \frac{17.3}{0.9} \times 0.746$$

$$= 14.3 \text{ kW}$$

The bottom pump will follow a similar calculation route except pressure drops over the two exchangers, i.e. feed preheat and air cooler (if any) must be taken into account.

8.10 Overall Performance Review and Checks for Light End Towers

Most light end towers are very stable in their operation. That is, once they are lined out for an operating requirement with normal unit control they maintain their stability. When performance falls off it can be attributed to one of a few reasons. This section looks at some of these reasons and how they can be evaluated and checked. By 'performance' in this case is meant the ability of the unit to make product quality at the prescribed throughput.

COLD FEED

The condition of the feed entering the tower is very important to tower operation. Ideally, the feed should enter the tower as close to a calculated feed tray temperature as possible. If the feed is well below its bubble point on entering the tower several trays below the feed tray are taken up for heat transfer before effective mass transfer can begin. This could prevent the specified product separation occurring and tray efficiency in this section of the tower falls off dramatically. Feed condition can be checked by bubble point calculation and a flash calculation (Section 8.7).

HOT FEED

This situation is probably the more serious regarding feed condition. If the feed enters at a temperature far above its bubble point its resulting enthalpy will be such as to reduce the reboiler duty. This will occur automatically, as the tower must always be in heat balance. The tower controls will maintain the product quantity and split. However, if the reboiler duty is drastically reduced, insufficient stripping vapours will be available for the stripping function. Poor separation will result.

As a rule of thumb the stripping vapour to the bottom tray must be at least 50% mole of the bottom product make. In super-fractionation such as a de-isopentanizer this figure would be at least 80–100% of bottoms make.

Heat balances as shown in Sections 8.7 and 8.8 will quickly determine the stripping vapour status.

IDEAL FEED CONDITION

Ideally, the feed should enter the tower close to feed tray temperature. Usually at the inlet pressure the feed will then be in a mixed phase with the vapour portion very close in quantity to the distillate product. As the feeds to these units are generally heated by the bottoms product heat exchange, the approach temperatures are always a consideration. To maintain good feed conditions, however, it is often beneficial to include a separate steam (or hot oil) feed preheater.

ENTRAINMENT

A common cause of poor plant performance at high throughout or high reflux rates is liquid entrainment from tray to tray. Very often in a high load and entrainment situation the problem is further agitated by increasing reflux to attempt separation improvement.

A well-designed light end tower can operate up to about 120% of allowable flood. Loading above this figure would result in some degree of entrainment.

DOWNCOMER BACK-UP AND FLOODING

If tower loadings are increased well above the allowable flood point there is a real danger that downcomers become unable to cope with the liquid load. They would fill and the tower would be in a state of flood. This will be very apparent, with very high abnormal pressure drop occurring across the tower. Separation by fractionation is not possible under these conditions. Heat input to the tower must be reduced to bring the unit back to a normal pressure drop.

LOW TOWER LOADING

Most towers have been designed with at least a 50% turndown ratio for the trays. This means that the trays should operate satisfactorily at 50% of their loadings. Nevertheless, tray performance does fall off at these low loadings. At below this turndown ratio, performance, particularly in sieve trays, is drastically reduced. This is almost certain to be due to 'weeping', where liquid falls from tray to tray. If the loads are to be for only a short time due to temporary reduced throughput, tray loading can be increased by increasing reflux. If the low throughput is to continue for an extended period a tray blanking schedule should be considered to reduce the active tray area.

OPERATING CLOSE TO CRITICAL CONDITIONS

De-ethanizers in particular operate close to critical pressure in the bottom of the tower. Careful attention should be paid to avoid any pressure surges in this unit. Feed to the unit and reflux streams should be on flow control.

No separation by fractionation can occur at pressures in excess of critical. Very often chilled water is used for overhead condensing to reduce reflux drum pressure and maintaining minimum C_3 loss in the case of de-ethanizers.

8.11 Handling Pseudo-components in Straight-run Light End Distillation Units

The pseudo-component developed in Chapter 6 for the crude unit naphtha was sufficient for the development of the feed to the light end units. Now, however, there is a need for a better definition of the pseudo-components that will be used for naphtha splitters, hydrotreating and conversion units. This is undertaken in this section and the calculation is attached.

The material heavier than C_7 has been plotted in Figure 8.4 attached. This was taken from Figure 6.4 in Chapter 6. New pseudo-components (eight of them) with

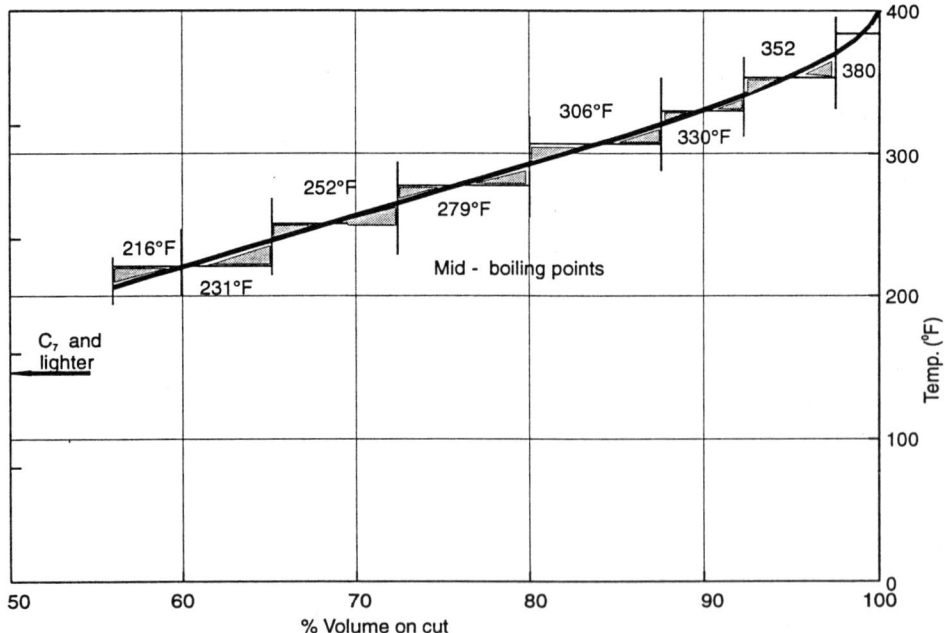

Figure 8.4. Splitter feed pseudo-components

mid-boiling points ranging from 216 °F to 380 °F are stepped out on Figure 6.4. The percentages shown here for each component are based on full-range naphtha. Using the debutanizer material balance (Section 8.3) these were calculated relating to debutanized (bottom product) naphtha stream. This will be the feed to the naphtha splitter. The bottoms from the splitter will be the hydrotreater feed.

An example calculation now follows.

Example calculation

From Figure 8.4, debutanized feed to splitter will be:

	Volume		Density (lb/gal)	lb/h	MW	Moles/h
	% vol.	gal/h				
iC_4	0.08	9.8	4.69	46	58	0.8
nC_4	0.34	40.5	4.87	197	58	3.4
iC_5	9.18	1080.6	5.21	5630	72	78.2
nC_5	4.02	473.6	5.36	2491	72	34.6
C_6	16.47	1938.8	5.54	10741	86	124.9
C_7	20.85	2454.7	5.74	14090	100	140.9
Mid-BPT 216	4.51	531.1	6.03	3203	102	31.4
Mid-BPT 231	5.64	664.1	6.06	4024	108	37.3
Mid-BPT 252	8.27	973.8	6.12	5960	115	51.8
Mid-BPT 279	8.27	973.8	6.26	6096	120	50.8
Mid-BPT 306	8.27	973.8	6.36	6193	130	47.6
Mid-BPT 330	5.64	664.1	6.43	4270	138	30.9
Mid-BPT 352	5.64	664.1	6.48	4303	147	29.3
Mid-BPT 380	2.82	332.3	6.55	2177	156	14.0
Total	100.00	117751	5.90	69421	102.7	675.9

This breakdown contains the keys and is more accurate for splitter evaluation work.

Part 3

THE CONVERSION AND TREATING PROCESSES

9 CATALYTIC REFORMING

9.1   Process Description of a Catalytic Reformer—Typical Flows

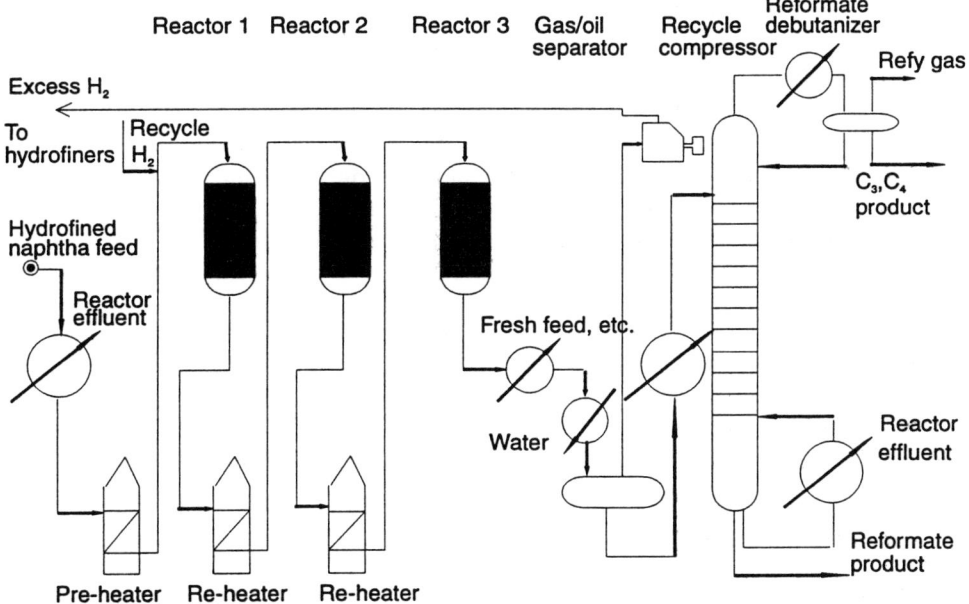

Figure 9.1. Catalytic Reforming

Catalytic reforming is a process for improving the octane quality of straight-run naphthas and of mixed naphthas containing cracked material. The principal reaction is dehydrogenation of naphthenes to aromatics, which are high in octane value. Contributing to the high octane of the product also are side-reactions such as hydrocracking of high-boiling hydrocarbons to low molecular weight paraffins, isomerization of paraffins to branched-chain structures and dehydrocyclization of paraffins and olefins to aromatics. Hydrogen recycle reduces the formation of carbon.

Feedstocks preferably are within the 220–430 °F (TBP) boiling range. Feeds with lower initial boiling point contain C_5 naphthenes which do not aromatize. Feeds with higher final boiling point:

● Cause excessive laydown of carbon on catalyst
● Necessitate product rerunning because reforming may increase the final boiling point beyond gasoline specification.

The water content of the feedstock must be under 10 ppm to prevent rapid stripping of chloride from the platinum catalyst in those processes which contain a halide-promoted catalyst. High water content is not so critical in processes which use catalysts that are not halide promoted. However, it does cause other problems in the process for all catalysts. Sulphur and sulphur compounds cause catalyst poisoning and deactivation. Poisoning of the platinum catalyst is prevented by hydrofining the feed to a sulphur content below 10 ppm. Recycled hydrogen is added to the reformer feed and the mixture is preheated and charged to the first of a series of three reactors (usually). The reactions in the first reactor are highly endothermic, and the reactor effluent is reheated before being charged to the second reactor. The reactions in this reactor are also endothermic but to a lesser degree than in the first reactor, so the effluent is again reheated before entering the last reactor. The effluent from the final reactor is separated into hydrogen-rich gas and reformate in a high-pressure separator at a temperature of around 100 °F.

A typical reformer catalyst contains 0.6 wt% platinum and 0.6 wt% chloride on alumina base. All catalysts are regenerable. The type of regeneration system and the cycle length between regeneration periods depends on feed quality and reforming severity. Good feedstock quality and moderate reforming severity favour the use of the semi-regeneration system, in which the plant is shut down after 3 months of operation to regenerate all reactors simultaneously. When reforming poorer feed and/or going into higher octane, the cyclic regenerative system is often used in which there is a spare or 'swing' reactor. A deactivated reactor can be regenerated in one day without process interruption.

The life of the catalyst is at least one year, although present catalysts have been known to exceed this time by a considerable amount. The spent catalyst may be 'reworked' to recover the platinum for use on new alumina base.

9.2 Reactor Operating Correlations

The catalytic reforming processes are proprietary to the licensors and developers of the processes. In general, the licensed processes follow a similar configuration, that is, they have usually three reactors with a preheater and two intermediate heaters. The greatest difference between the processes are the catalyst and the design and shape of the reactors themselves. It is these differences that set the operating criteria of the processes.

A typical process operating condition will be within the following ranges:

- Reactor inlet pressure (psig)   320–290
- Average reactor inlet temp (°F)   930 SOR (start of run)
                                         960 EOR (end of run)
- Space velocity (LHSV)/HR   1.6
- Hydrogen-to-hydrocarbon ratio   5.0

Licensors of the various reformer processes supply the clients with complete operating manuals on the process. These contain correlations that may be used to affect operating changes to meet the refinery's product needs. Most of these correlations show the effect of changes to operating variables such as space velocity, inlet reactor temperatures and, in some cases, reactor pressures. The correlations also relate feed quality and composition to the reactor variables. For example, a feed high in naphthenes requires far less severe reactor conditions than a feed low in naphthenes. Generally, the changes would result in operating at lower inlet temperatures and probably a higher space velocity. The term 'space velocity' is really a measure of the residence time of the reactants in the reactor.

9.3   Some Typical Crude Distillate Naphthene Contents

This section shows the relationship of the naphtha cut point to the PONA content of the fraction. PONA stands for the percentage volume of paraffins, olefins, naphthenes and aromatics making up a particular fraction. It usually refers only to the naphtha cuts of crude.

The PONA for each crude will be unique to that crude. In other words, the distribution of the paraffins, naphthenes, etc. will be different from crude to crude and vary in their relative quantities. Figure 9.2 shows the percentage volume on naphthenes for various cut points and for various crudes. For simplicity, only naphthenes are shown in this figure but the whole PONA analysis adds up to 100% of any fraction. Naphthenes content varying with cut point is also shown in Figure 9.2. This is the homologue (hydrocarbon family) that has the major effect on the catalytic reforming process.

The curves have been developed from assay data. Selective sequential cuts each with a PONA analysis have been multiplied by the relative yields of the fractions making up a C_5 to 400 °F TBP cut point naphtha. This was done for a selection of typical Middle East crudes and then again for typical North African and West African crudes. The latter were chosen to demonstrate the high quality of North and West African crude with respect to gasoline production. They are high in naphtha yields and also in naphthene content compared with most Middle East crudes (Aghajari being an exception).

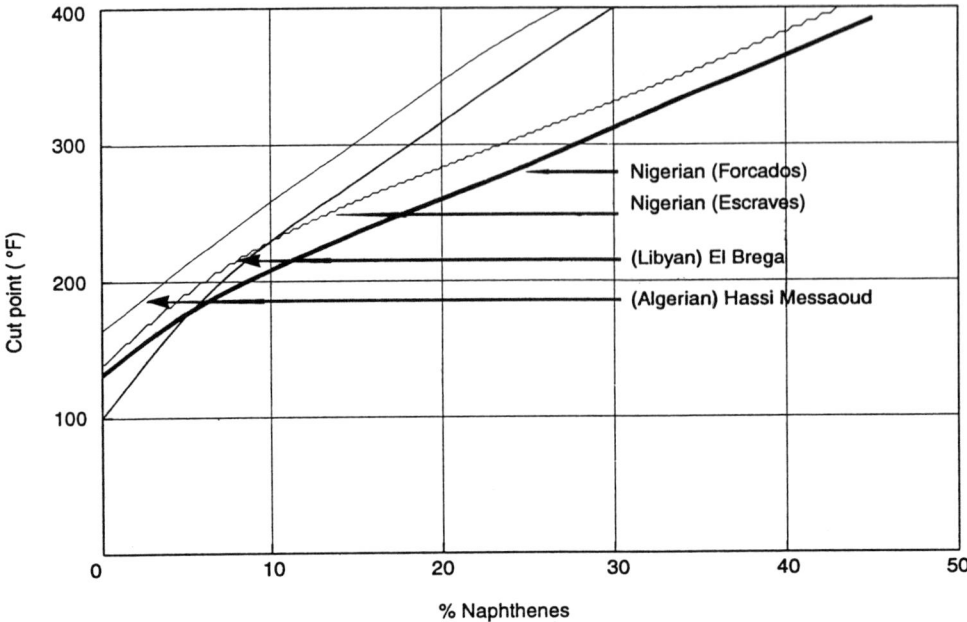

Figure 9.2. Typical naphthene contents versus cut points for African crudes

Using these curves is quite simple. Take the naphthene content of the final cut point and subtract the naphthene content of the initial cut point (this is usually very low). The result is the naphthene content of the naphtha fraction in question. Thus using Kuwait crude initial cut point is NC6 which is about 155 °F. The final cut point using the cut given in Section 6.4 is 380 °F. From Figure 9.3 naphthene content of initial cut point (155 °F) for Kuwait crude is 0.2% vol. and that corresponding to 380 °F is 20.2% vol. The naphthene content of the fraction or cut is (20.2 − 0.2) = 20% vol. This can now be used with Figure 9.3 to determine the aromatic content of the reformate.

9.4 Predicting Pseudo-components in Reformates

Products of catalytic reforming are very high in aromatic content. To use normal paraffinic properties for pseudo-components (as done in Chapters 1–3) in this case would be wrong. This section deals with a method of predicting pseudo-components and their properties for reformates.

The following calculation steps and the example calculation provides a method for predicting reformate properties such as molecular weight, specific gravity and vapour pressure:

- *Step 1* Convert the test run ASTM curve (lab result) to the corresponding TBP curve using the method given in Section 6.3.
- *Step 2* A relationship exists between the naphthene content of the naphtha feed and

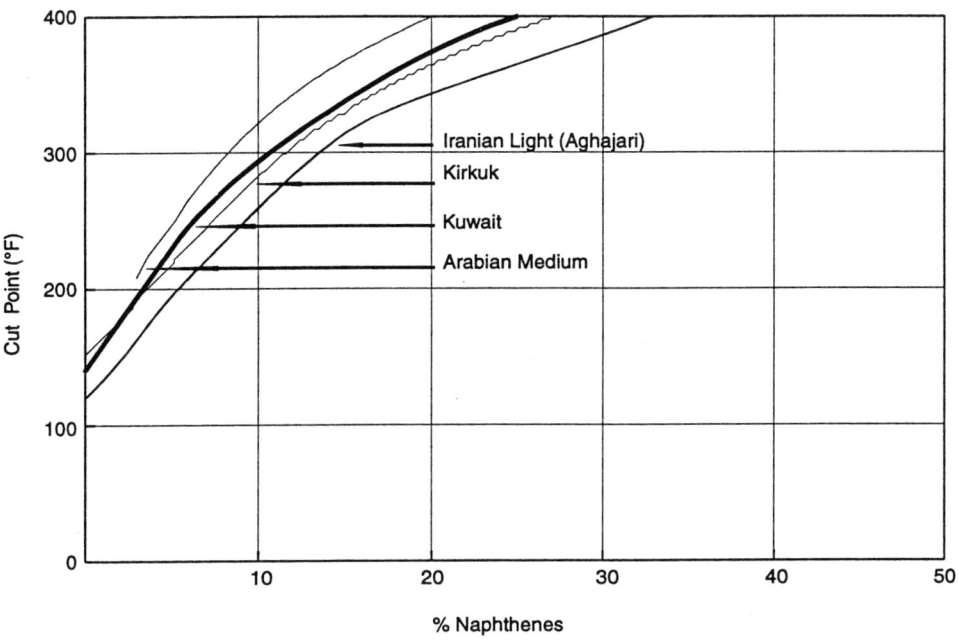

Figure 9.3. Typical naphthene contents versus cut points—Middle East crudes

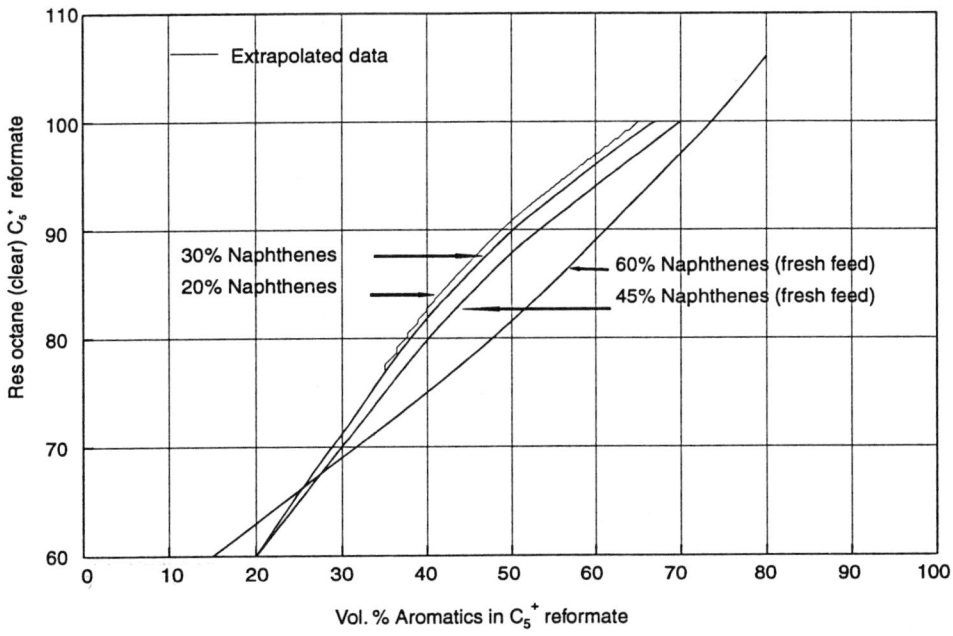

Figure 9.4. Volume of aromatics in reformate

the aromatic contents of the reformate at various octane levels. This relationship (10) is given in Figure 9.4. From lab data establish the aromatic content of the reformate.

● *Step 3*   The aromatic in this product will be benzene, toluene and some xylenes. Using the percentage aromatics step off the TBP curve (Figure 9.5) in mid-boiling points as close as possible to the aromatic, paraffin boiling points for C_6, C_7 and C_8.

● *Step 4*   Break down each mid-boiling point volume component to aromatic and other split. Table them on the volume percentage basis.

● *Step 5*   Apply real component gravities (i.e. hexane and benzene, heptane and toluene, etc.) to the paraffins and aromatics in each category. Calculate the weight factor and thus the gravity of the reformate.

● *Step 6*   Apply real component molecular weights to the paraffins and aromatics in each category. Calculate the mole factor and thus the mole weight of the reformate.

● *Step 7*   The mid-boiling points is the boiling point at atmospheric pressure. Using the vapour pressure curves in Appendix 1 the vapour pressure at any temperature for each component can now be read off.

The properties developed in this manner demonstrates good comparison with lab data. An example calculation now follows.

Example Calculation

C_5^+ reformate (lab data)

	ASTM		TBP	
% vol.	°F	ΔT °F	ΔT °F	°F
IBP	106	54	86	35
10	160	42	67	121
30	202	28	44	188
50	230	30	43	232
70	260	40	49	275
90	300	95	126	324
FBP	395			450

Calculate the composition and properties of C_5^+ reformate

From TBP curve (Figure 9.5) based on 98 RON clear:

Comp.	% vol.	% vol.	SG	Weight factor	MW*	Mole factor	Mole Frac	Mid-BPT
iC_5	15	7.5	0.6244	46.83	72	0.65	0.078	90
nC_5		7.5	0.6311	47.33	72	0.65	0.079	
C_6 P	29	8.5	0.6640	56.44	81	2.94	0.351	180
C_6 A		20.5	0.8845	181.32				
C_7 P	31	9.1	0.6883	62.64	102	2.49	0.298	250
C_7 A		21.9	0.8719	190.95				
C_8 P	25	7.5	0.707	53.03	128	1.62	0.194	330
C_8 A		17.4	0.885	154.0				
		100.0	0.793	7154.4	94.8	8.36		

Plant data SG = 0.790. Plant Data MW = 94
Total AR in reformate = 60% from Figure 9.3
*Estimated from °API, mid-BPT.

9.5   Predicting Stabilizer Feed Composition and Recycle Gas Composition

This section deals with the flash separation of the effluent from the last reformer reactor. Flash vaporization is also explained in Section 8.7. The separation occurs in the reformer flash drum which operates at a single temperature and pressure condition. Normally the temperature of this drum is maintained from 95 °F to, say, 100 °F (depending on coolant temperature) and partially condenses the cooled effluent stream. The pressure is set sufficient to provide an economic suction pressure to the compressor or some other economic factor. Normally this pressure will be around 200–230 psia.

The following steps are used in calculating the flash separation of the effluent:

- *Step 1*   Determine from lab data the composition of the effluent stream. Determine actual flash drum conditions from plant data.
- *Step 2*   Calculate the component properties of the effluent.
- *Step 3*   Using moles/h for the feed composition, construct the table having the feed column, equilibrium constant column K, and several trial columns for the term $(V/L)K$ and

$$L = \frac{E}{1 + (V/L)K}$$

(see Section 8.7).

- *Step 4*   The K value will be that for each component at the drum condition of temperature and pressure. The K values for the real components (H_2 to nC5) are obtained from standard data books. The values used here are obtained from GPSA Engineering Data at a convergence pressure of 1500 psia.
- *Step 5*   Guess at a trial V/L. Judging from the proportion of light ends and C_5^+ it will be between 5 and 10 (V/L is moles vapour divided by moles liquid). Calculate first trial to arrive at total liquid phase L. By difference, establish V and thus the ratio V/L.
- *Step 6* V/L assumed must equal V/L calculated to obtain a solution. If trial 1 does not succeed then apply a second guessed V/L and repeat Step 5.
- *Step 7*   Plot assumed V/L against calculated V/L on LogLog paper. Establish where assumed is equal to calculated and use this value in the final trial.
- *Step 8*   When an acceptable V/L ratio is obtained calculate the vapour molar composition by subtracting the calculated L for each component from the feed component.
- *Step 9*   The liquid component L is the feed to the stabilizer. The vapour component V is the reformer offgas and recycle gas.

An example calculation now follows.

Example Calculation

An analysis of the effluent is as follows; and the moles C_5^+ liquid yield from stabilizer is 451 moles per hour (liquid feed $C_6^+ = 556.6$ moles/h).

	Mole %	Moles/h	MW	lb/h	lb/gal	gal/h
H_2	62.82	3219	2	6 438		
C_1	8.43	432	16	6 912		
C_2	8.31	426	30	12 780		
C_3	7.38	378	44	16 632		
iC_3	1.81	93	58	5 394		
nC_4	2.45	126	58	7 308		
C_5^+	8.80	451	94.8	42 755	6.604	6474
Total	100.00	5125	19.2	98 219		

To carry out flash vaporization calculation

The reactor flash drum operates at:

Temperature	95 °F
Pressure	190 psia
Convergence pressure	1500 psia

Comp.	F moles/h	K 95 °F 190 psia	1st trial $V/L=10$		2nd trial $V/L=5.0$		Final $V/L=9.0$		Vapour	
			V/LK	$L=\dfrac{F}{1+(V/L)K}$	V/LK	$L=\dfrac{F}{1+(V/L)K}$	V/LK	$L=\dfrac{F}{1+(V/L)K}$	Moles/h	Mole %
H_2	3219	260	2600	1.2	1300	2.47	2340	1.37	3218	69.7
C_1	432	12	120	3.6	60	7.08	108	3.96	428	9.3
C_2	426	3.2	32	12.9	16	25.06	28.8	14.79	411	8.9
C_3	378	0.98	9.8	35.0	4.9	64.07	8.82	38.49	340	7.4
iC_4	93	0.42	4.2	17.9	2.1	30.00	3.78	19.45	73	1.6
nC_4	126	0.33	3.3	29.3	1.65	47.55	2.47	31.74	94	2.0
iC_5	35	0.089	0.89	18.5	0.445	24.22	0.80	19.44	16	0.3
nC_5	36	0.068	0.68	21.4	0.340	26.87	0.61	22.36	14	0.3
C_6	159	0.015	0.15	138.3	0.075	147.91	0.135	140.09	19	0.4
C_7	134	0.003	0.03	130.1	0.015	132.02	0.027	130.48	3	0.1
C_8	87	NEG	—	87	—	87	—	87	NIL	—
Total	5125			465.9		594.25		477.44	4616	100.0
			$VL=9.35$		$V/L=7.6$		$VL=9.1$			

Close enough.

9.6 Carrying Out a Component Balance Over the Reformer Stabilizer

This section demonstrates the development of a component balance over the stabilizer. It would be essential to perform this calculation before any meaningful design could commence. In the evaluation of an operating unit the material balance would be developed from plant and laboratory data. However, such a calculation technique may be used prior to a test run to establish conditions required to meet a set specificiation. The development of the balance is set out by the following steps:

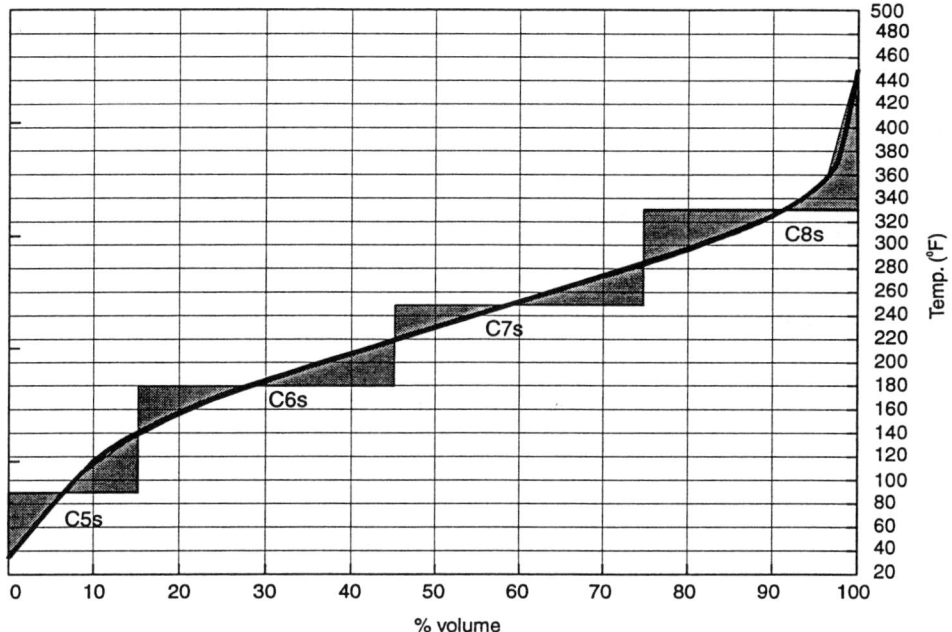

Figure 9.5. TBP curve of C_5^+ reformate (98 Ron). Feed: Kuwait 25% naphthenes

- *Step 1* Establish the distillation specification required. In the example this spec allows no more than 1 mol. % of C_5s based on total C_4s in the overhead product. It also allows no more than 1% by weight of C_4s in the bottom product.
- *Step 2* Establish the distribution of the key components. In this case it will be $nC4$ as light key and $iC5$ as heavy key.
- *Step 3* Calculate composition of the bottom product.
- *Step 4* Using the Fenske equation check that there is sufficient theoretical trays to effect this separation. Take actual trays multiplied by an efficiency factor (say, 63%) to give total theoretical trays N_M available.
- *Step 5* Theoretical trays required is calculated from the Fenske equation, again using total theoretical trays required. The minimum is estimated by dividing total theoretical by 1.5.
- *Step 6* When the viability of the split is established calculate the component balance in terms of moles, weight and volume. The reformate pseudo-component properties are used for this calculation.

An example calculation now follows.

Example Calculation

Let the C_5s to overhead be 0.9% mole of total C_4s in the feed, $= 0.009 \times (19.5 + 31.7) = 0.46$ moles/h. Then total bottoms make without C_4s = 398.5 as follows:

	Moles/h	MW	lb/h	% wt
C_4s	6.8	58	392	1.0
iC_5	19.0	72	1368	3.5
nC_5	22.4	72	1613	4.1
C_6	140.1	81	11348	29.0
C_7	130.5	102	13311	34.0
C_8	87	128	11136	28.4
Total	405.8	96.5	39168	100.0

This distribution meets spec

Check that split can be achieved in tower.

Number of actual trays   = 30 (assume 63% efficiency)
Theoretical number of trays = 18.9

$$N_M = \frac{18.9}{1.5} \qquad = 12.6$$

$$N_{M+1} \qquad\qquad = 13.6$$

Using the Fenske equation:

$$N_{M+1} = \mathrm{Log}\left[\left(\frac{\mathrm{LT\ Key}}{\mathrm{HY\ Key}}\right)_D \cdot \left(\frac{\mathrm{HY\ Key}}{\mathrm{LT\ Key}}\right)_W\right] + \mathrm{Log}\phi$$

LT Key $= nC_4$ Mean temp./press. $= 309\,°F/239$ psig
HY Key $= iC_5$ (guessed)

$$N_{M+1} = \mathrm{Log}\left[\left(\frac{0.24}{0.004}\right) \cdot \left(\frac{0.047}{0.017}\right)\right] \div \mathrm{Log}\ 1.77$$

$$\mathrm{Log}\ [60 \times 2.76] \div 0.25$$

$$= \frac{2.22}{0.25} = 9 N_M = 8$$

Required theoretical trays $= 8 \times 1.5 = 12$

Therefore tower is adequate to meet this split.

Material balance is as follows:

	Feed					Overheads			Stabilized platformate		
	Mole/h	MW	lb/h	lb/gal	gal/h	Mole/h	lb/h	gal/h	Mole/h	lb/h	gal/h
H_2	1.3	2	3	—		1.3	3	—			
C_1	4.0	16	64	—		4.0	64	—			
C_2	14.8	30	444	2.96	150	14.8	444	150			
C_3	38.5	44	1694	4.22	401	38.5	1694	401			
iC_4	19.5	58	1131	4.68	242	19.5	1131	242			
nC_4	31.7	58	1839	4.86	378	24.9	1444	297	6.8	395	81
iC_5	19.4	72	1397	5.20	269	0.4	29	6	19.0	1368	263
nC_5	22.4	72	1613	5.25	307				22.4	1613	307
C_6	140.1	81	11348	6.83	1661				140.1	11348	1661
C_7	130.5	102	13311	6.84	1946				130.5	13311	1946
C_8	87.0	128	11136	6.94	1605				87.0	11136	1605
Total	509.2	86.4	43980	6.32	6954	103.4	4809	1096	405.8	39171	5863

$$\text{Mol. wt} = 46.5 \qquad \text{Mol wt} = 96.5$$
$$\text{lb/gal} = 4.39 \qquad \text{lb/gal} = 6.68$$
$$\text{°API} = 136.7 \qquad \text{°API} = 44.9$$

9.7 Calculating Number of Theoretical Trays and thus Overall Tray Efficiency for the Stabilizer

This calculation employs the use of the Underwood equation to calculate minimum reflux at infinite stages, then the Fenske equation for minimum stages or trays at total reflux and a correlation of both to give theoretical trays. (See also Section 8.4.)

The first part using the Underwood equation is a rigorous trial-and-error calculation. The following steps describe this and the other procedures to arrive at the total number of theoretical trays:

- *Step 1*   Obtain the mol. fraction composition of the stabilizer feed (see Section 9.5).
- *Step 2*   Identify one of the key components. In this case it will be either nC_4 or iC_5.
- *Step 3*   Approximate the mean conditions of temperature and pressure in the tower. This can be obtained by references to plant log or by calculating tower-top dew point and tower-bottom bubble point from product split and tower-pressure settings.
- *Step 4*   Divide the equilibrium constant (K) of each component in the feed by the K value for one or other of the key components. The equilibrium constants are taken at the mean tower conditions in each case as established in Step 3. The resulting ratios will be the relative volatilities (ϕ) for each component.
- *Step 5*   Construct the Underwood table for calculating the value of B in the equation

$$\frac{X_F \phi}{\phi - B}$$

for each component. The sum of these must equal zero for a solution to the value of B (see also Chapter 8, Section 8.4).

● *Step 6*   The columns in this table commence with the feed composition, then the relative volatilities for each component. This is followed by a column giving the multiple of mole fraction times relative volatility for each component ($X_F\phi$). The remaining columns are trial calculation for values of B that sum to zero.

● *Step 7*   When the value of B is found this is used in the equation

$$\sum \frac{X_D - \phi}{(\phi - B)} = R_{M+1}$$

where X_D is mole fraction of each component in the distillate (or overhead product) and R_M is the minimum reflux at infinite number of trays.

● *Step 8*   Calculate minimum number of trays at total reflux using the Fenske equation, this has been described in this chapter (Section 9.6) and Chapter 8.

● *Step 9*   Using the Gilliland correlation (see Section 9.4) read off a value for

$$\frac{N - N_M}{N + 1}$$

where N is the thoeretical number of trays.

● *Step 10*   Tray efficiency is found by dividing theoretical number of trays by actual number of trays.

An example calculation now follows:

Example Calculation

Calculating minimum reflux plates using the Underwood equation.

Calculating for B (see also Section 8.4)

Comp.	Mol. frac. in feed X_p	ϕ_{ave}	$X_F\phi_{ave}$	Trial 1 $B=1.1$	Trial 2 $B=1.09$	Trial 3 $B=1.094$
				$\sum \dfrac{X_F\phi_{ave}}{\phi - B} = 0$		
H_2	0.003	152	0.456	0.003	0.003	0.003
C_1	0.008	7.2	0.058	0.010	0.009	0.009
C_2	0.029	3.4	0.094	0.043	0.043	0.043
C_3	0.076	1.9	0.144	0.180	0.178	0.179
iC_4	0.038	1.17	0.045	0.643	0.563	0.592
nC_4	0.062	1.00	0.062	−0.620	−0.689	−0.660
iC_5	0.038	0.61	0.023	−0.047	−0.048	−0.048
nC_5	0.044	0.54	0.024	−0.043	−0.044	−0.043
C_6	0.275	0.19	0.052	−0.057	−0.058	−0.058
C_7	0.256	0.08	0.020	−0.020	−0.020	−0.020
C_8	0.171	0.03	0.005	−0.005	−0.005	−0.005
Total	1.000			0.087	−0.068	0.008

Trial 3 is within acceptable limits
$B = 1.094$

Use key as nC_4 at 309 °F and 254 psia

Calculate for R_{M+1}

	Mol. frac. in distillate X_D	ϕ_{ave}	$X_D - \phi_{ave}$	$R_{M+1} = \dfrac{X_D \phi_{ave}}{(\phi_{ave} - B)}$
H_2	0.013	152	1.976	0.013
C_1	0.039	7.2	0.281	0.046
C_2	0.143	3.4	0.486	0.211
C_3	0.371	1.9	0.705	0.875
iC_4	0.189	1.17	0.221	0.908
nC_4	0.241	1.00	0.241	-2.560
iC_5	0.004	0.61	0.002	-0.004
Total	1.000			1.489

$R_{M+1} = 1.489$

$R_M \quad = 0.489R = 0.489 \times 1.5 = 0.7335$

Calculating minimum trays at total reflux

Using the Fenske equation (see also Section 8.4):

$$N_{M+1} = \text{Log} \left[\left(\frac{\text{LT Key}}{\text{HY Key}} \right)_D \cdot \left(\frac{\text{HY Key}}{\text{LT Key}} \right)_W \right] \div \text{Log} \phi$$

$$N_{M+1} = \text{Log} \left[\left(\frac{0.241}{0.004} \right) \cdot \left(\frac{0.047}{0.017} \right) \right] + \text{Log } 1.77$$

$$N_{M+1} = 9 \text{ trays}$$

$$N_M = 8$$

Using the Gilliland correlation:

$$\frac{R - R_M}{R + 1} = \frac{0.7335 - 0.489}{1.7335} = 0.141$$

From Figure 8.2:

$$\frac{N - N_M}{N + 1} = 0.52$$

$$N - 8 = 0.52 \, (N + 1)$$

$$N - 0.52N = 9$$

$$N = 19 \text{ trays}$$

$$\text{Efficiency} = \frac{19}{30} \times 100 = 63.3\%$$

This is low due to the fact that it is a relatively easy separation to achieve. The spec could be made more stringent.

9.8 Calculating Tower Loadings

This section shows how test run data (or in their absence, predicted data) can be used to establish tower traffic for the reformer stabilizer. Most of the data here have been developed, but in real test-run conditions many of these data would be obtained through lab tests or plant readings.

It will not be possible to fully condense the overheads on a reformer stabilizer. The reason of course, is that the feed to the column comes from the unit flash drum and has therefore an equilibrium quantity of hydrogen and methane in its composition. Although these quantities are only 1% or 2%, to condense them in the column reflux drum would require an unacceptably high pressure at a normal condensing temperature. A major portion of the overhead product is therefore condensed and a vapour stream is also taken off from the drum. Some of the distillate is returned as reflux while the remainder is taken off as a distillate product.

Because the reflux stream will not have the same composition as the product stream the dew point temperature at the top of the tower will be that of both streams. Thus the reflux must be known to calculate tower-top temperature. The amount and composition is calculated using the equilibrium vaporization calculation (see Section 9.5). The tower-top temperature is then calculated using that composition and fixing the reflux rate.

The pressure of the column is a compromise between the cost of high pressure (or its feasibility if close to critical) and the loss of C_3, C_4 LPG in the vapour stream from the reflux drum. Normally the reflux drum is run at about 200 psig and at 95–100 °F.

Knowing the tower-top temperature and the relative flows, the condenser duty is calculated by heat balance. When this duty is found the internal reflux (over flow from top tray to the tray below) can be calculated by a heat balance over the top of the tower. The quantity of the internal reflux and total product plus internal reflux as vapour gives the liquid and vapour traffic in this section of the tower. Note the composition of the internal reflux is the liquid composition in equilibrium with total product plus external reflux: in other words, the X column from the dew point calculation.

The loading at the bottom of the column is developed by first establishing the bottom of the tower conditions. Normally, this is by plant readings, otherwise it is calculated as the bubble point of the bottom product at the predicted pressure. To continue with calculating loads for the bottom of the tower the heat input with the feed has to be established. This is done by calculating the equilibrium vaporization of the feed at preheater outlet conditions. The amount of vaporization is determined by this calculation. Allotting enthalpy to the two phases results in fixing the total heat input with the feed.

An overall heat balance across the tower fixes the reboiler duty. When this has been established a heat balance across the bottom of the tower (with the unknown as the vapour from the reboiler) gives the vapour traffic to the bottom tray. The liquid traffic across the tray is the reboiler feed and is the sum of the vapour calculated and the bottom product. Again the composition of the vapour leaving the reboiler is that in equilibrium with the bottom product, that is, the Y column from the bubble point calculation.

Having now established the liquid and vapour traffic in the two critical sections the evaluation of the loadings can be completed. This has been described in Sections 6.14 and 8.8 for valve and sieve trays. An example calculation to predict vapour and liquid traffic now follows.

Example Calculation

Tower-top conditions

Because of the high hydrogen and methane content it will not be possible to operate at total overhead condensing.

Set reflux drum temperature at 95 °F

Let tower reflux drum pressure be 200 psig = 215 psia

Calculate equilibrium separation in reflux drum

Comp.	Dist. moles/h	K 95 °F 2/5	1st trial $V/L=1.0$ V/LK	$L=\dfrac{D}{1+(V/L)K}$	2nd trial $V/L=0.5$ V/LK	$L=\dfrac{D}{1+(V/L)K}$	3rd Trial $V/L=0.27$ V/LK	$L=\dfrac{D}{1+(V/L)K}$	Vapour Moles/h	lb/h
H_2	1.3	230	230	0.006	115	0.011	62	0.021	1.28	2.56
C_1	4.0	13.1	13.1	0.284	6.55	0.53	3.54	0.881	3.119	49.90
C_2	14.8	2.9	2.4	3.795	1.45	6.04	0.783	8.301	6.499	195.0
C_3	38.5	0.9	0.9	20.263	0.45	26.55	0.243	30.973	7.527	331.2
iC_4	19.5	0.39	0.39	14.029	0.195	16.32	0.105	17.647	1.853	107.5
nC_4	24.9	0.30	0.30	19.154	0.15	21.85	0.081	23.034	1.866	108.2
iC_5	0.4	0.13	0.13	0.354	0.065	0.38	0.035	0.386	0.014	1.0
Total	103.4			57.885		71.48		81.243	22.158	795.4
				$VL=0.786$		$V/L=0.447$		$VL=0.272$		35.9 mol. wt

Set external reflux at 1.8 : 1

Total moles/h reflux = 1.8 × 103.4 (total overhead product)
= 186 miles/h

To calculate composition of overhead vapours and its dew point

Tower-top pressure   = 220 psia

Comp.	Moles/h product	Moles/h reflux	Total overhead	Mole fract. Y	1st trial at 200 °F		2nd trial at 163 °F		MW	Weight factor	lb/gal	Vol. factor
					K	$X = Y/K$	K	$X = Y/K$				
H_2	1.3	0.05	1.35	0.005	210	NEG	—	—				—
C_1	4.0	2.02	6.02	0.021	13	0.002	11.5	0.002	16	0.03	—	—
C_2	14.8	19.00	33.80	0.117	5.2	0.023	4.4	0.027	30	0.81	2.96	0.27
C_3	38.5	70.92	109.42	0.378	2.1	0.180	1.7	0.222	44	9.77	4.22	2.32
iC_4	19.5	40.40	59.90	0.207	1.1	0.188	0.84	0.246	58	14.27	4.68	3.05
nC_4	24.9	52.73	77.63	0.268	0.86	0.312	0.613	0.437	58	25.35	4.86	5.22
iC_5	0.4	0.88	1.28	0.004	0.5	0.008	0.34	0.012	72	0.86	5.2	0.17
Total	103.4	186.00	289.4	1.000		0.713		0.946	54	51.09	4.63	11.03

2nd trial 0.86×0.713 Actual temp. = 158 °F
$= 0.613$
K at 163 °F

Tower-top conditions are   220 psia
                                 158 °F

To calculate vapour and liquid load on top tray

(a) Calculate condenser duty

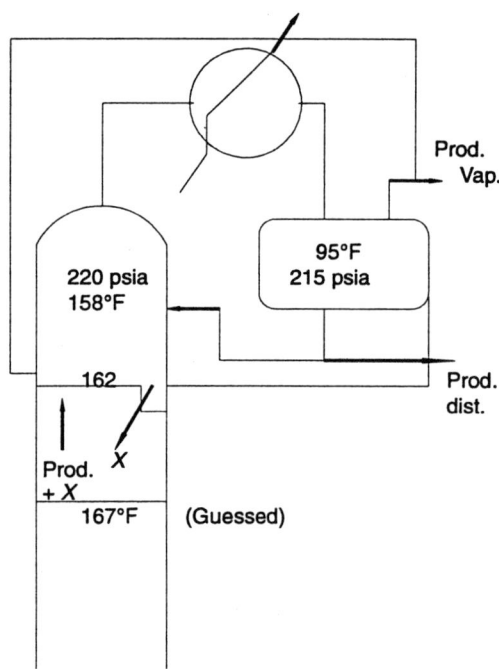

	V or L	°API or (MW)	°F	lb/h	Enthalpy	
					BTU/ lb	BTU/ h × 10⁶
In						
Prod. vapour	V	(46)	158	4809	330	1.587
Reflux vapour	V	(48)	158	9188	332	3.050
Total in				13 997		4.637
Out						
Prod. vapour	V	(36)	95	795	320	0.254
Prod. dist.	L	(48)	95	4014	170	0.682
Liquid reflux	L	(48)	95	9188	170	1.562
Condenser				By difference		2.139
Total out				13 997		4.637

All enthalpies from Maxwell, *Data Book on Hydrocarbons*

(b) Calculate vapour/liquid traffic

	V or L	°API or (MW)	°F	lb/h	Enthalpy BTU/lb	Enthalpy BTU/h × 10⁶
In						
Vapour ref. ex tray 2	V	(54)	167	X	332	$332X$
Vapour prod. ex tray 2	V	(46)	167	4809	335	1.611
Total in				$4809 + X$		$1.611 + 332\,X$
Out						
Prod. vapour	V	(36)	95	795	320	0.254
Prod. dist.	L	(48)	95	4014	170	0.682
Internal reflux	L	(54)	162	X	200	$200\,X$
Condenser duty						2.139
Total				$4809 + X$		$3.075 + 200X$

$$X = \frac{1\,464\,000}{132} = 11\,091 \text{ lb/h}$$

Traffic over the tray

Vapour to tray			Liquid from tray		
lb/h	=	15 900	Gal/h at 60	=	39.9
moles/h	=	309.9	Hot gal/h	=	51.0
ACFS	=	2.61	Hot CFS	=	0.113
lb/ft³ ρv	=	1.69	lb/h	=	11 091
Temp (°F)	=	167	lb/ft³ ρ_L	=	27.3 (hot)
Pressure (psia)	=	220	Temp (°F)	=	162

Tray analysis giving percentage flood is carried out following the steps given in Section 8.8

Calculate bottoms temperature

$$\text{Pressure at bottom of tower} = 220 + (30 \times 0.25)$$
$$= 227 \text{ psia}$$

	X_w	1st trial = 400 °F K	1st trial = 400 °F $Y = XK$	2nd trial = 435 °F K	2nd trial = 435 °F $Y = XK$	MW	Weight factor	lb/gal	Vol. factor
nC_4	0.017	2.7	0.046	3.1	0.053	58	3.1	4.86	0.64
iC_5	0.047	1.9	0.089	2.2	0.103	72	7.4	5.20	1.42
nC_5	0.055	1.7	0.094	1.9	0.105	72	7.6	5.25	1.45
C_6	0.345	0.96	0.331	1.3	0.449	81	36.4	6.83	5.33
C_7	0.322	0.44	0.142	0.59	0.190	102	19.4	6.84	2.84
C_8	0.214	0.17	0.036	0.25	0.054	128	6.9	6.94	0.99
Total	1.000		0.738		0.954	84.7	80.8	6.38	12.67

°API 53.1
Actual temp. = 440 °F

Tower-bottom condition:

$$\text{Temp.} \quad = 440\,°F$$
$$\text{Pressure} = 227 \text{ psia}$$

Feed enthalpy

Plant reading gave feed at 300 °F and at 230 psia (at heat exchanger outlet). Carry out flash vaporization at these conditions

This approximated the following results:

Vapour 112 moles/h   = 5544 lb/h (130 °API)
Liquid 397.2 moles/h = 38 436 lb/h (46 °API)
Enthalpy   = 7.567×10^6 BTU/h

Overall heat balance to establish reboiler duty

	V or L	°API or (MW)	°F	lb/h	Enthalpy	
					BTU/lb	BTU/h $\times 10^6$
In						
With feed	$V+L$	—	300	43 980	—	7.567
With reboiler			By difference			10.001
Total in				43 980		17.568
Out						
Bottom product	L	(96.5)	440	39 171	370	14.493
Overhead distillate	L	(48)	95	4014	170	0.682
Overhead vapour	L	(36)	95	795	320	0.254
Condenser duty						2.139
Total out				43 980		17.568

All enthalpies from Maxwell

Calculate vapour and liquid loading to bottom tray

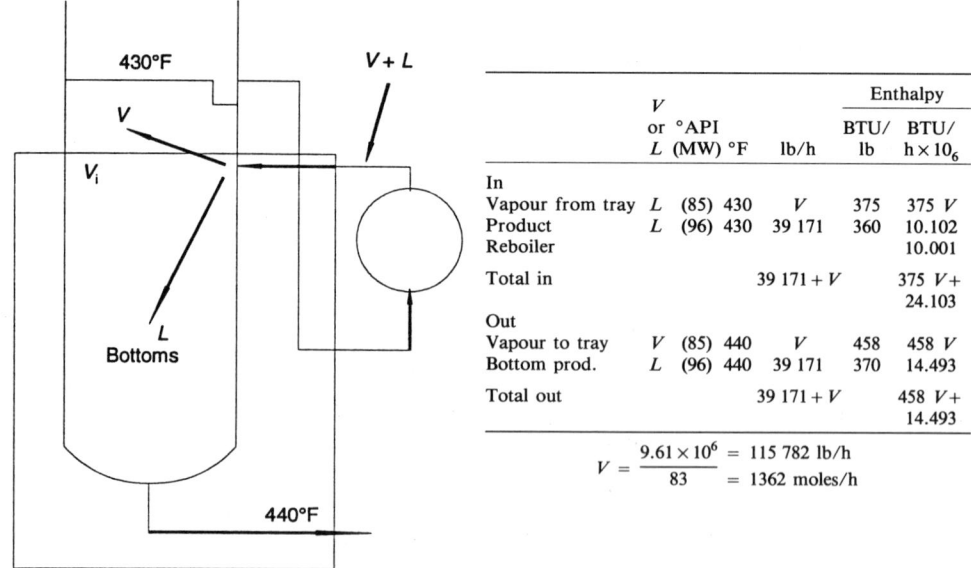

	V or L	°API (MW)	°F	lb/h	Enthalpy	
					BTU/lb	BTU/h $\times 10_6$
In						
Vapour from tray	L	(85)	430	V	375	375 V
Product	L	(96)	430	39 171	360	10.102
Reboiler						10.001
Total in				39 171 + V		375 V + 24.103
Out						
Vapour to tray	V	(85)	440	V	458	458 V
Bottom prod.	L	(96)	440	39 171	370	14.493
Total out				39 171 + V		458 V + 14.493

$$V = \frac{9.61 \times 10^6}{83} = 115\,782 \text{ lb/h} = 1362 \text{ moles/h}$$

Note: ratio of stripping to bottom product is 3.3—this is more than adequate. Feed temp. could be increased for better balance.

Vapour to tray			Liquid from tray		
lb/h	=	115 782	gal/min	=	400 at 60
moles/h	=	1 362	gal/min	=	507 at 430 °F
ACFS	=	16	CFS	=	1.126
lb/ft$_3$ ρ_v	=	2.0	lb/h	=	154 953
Temp. (°F)	=	440	lb/ft^3 ρ_L	=	38.2
Pressure (psia)	=	227			

9.9 Evaluation of Heat Exchanger Performance in Catalytic Reformers

The evaluation of heat exchanger in this section compares heat transfer coefficients in the case of shell and tube exchangers and horsepower requirements in the case of air coolers. Both these have been described in Sections 6.15 and 8.9, respectively.

In catalytic reforming the effluent from the reactors will initially be in the vapour phase. The effluent is cooled and partially condensed before entering the flash drum. In evaluating the performance of heat exchangers in this train the condition of the stream (vapour or liquid or mixed phase) must be determined at the inlet and outlet of each exchange unit. Plant data give the inlet and outlet pressure and temperature conditions. Knowing these and from lab data, the following steps are used to develop data for the equipment evaluation:

- *Step 1* Establish the fluid phase by calculating the dew point of the effluent stream. The pressure condition is guessed for that exchanger where condensing begins to occur.
- *Step 2* Almost invariably, the first exchanger is all vapour phase. Calculate its duty, and from temperature readings, and equipment data sheet for surface area calculate the overall heat transfer coefficient. Compare with those given in Table 6.2.
- *Step 3* Calculate the conditions (vapour and liquid) of the effluent stream at the outlet of the second exchanger. That is, calculate equilibrium vaporization at outlet temperature and pressure.
- *Step 4* Using the liquid vapour data from Step 3, calculate the exit enthalpy in BTU/h. Knowing the inlet enthalpy from Step 2, establish the exchanger duty.
- *Step 5* If the exchanger is shell and tube type, repeat Step 2 for this unit. If the exchanger is an air cooler, calculate the required absorbed power for the unit. (As described in the following Steps 6–8.)
- *Step 6* From plant data fix the air temperature rise across the unit. Calculate the air flow by the equation

$$\frac{\text{Duty}}{\text{SPHT} \times \text{temp. rise}}$$

(See also Section 8.9.)

● *Step 7*   Fix the static pressure drop on the air side (between 0.5 and 0.7 in. water gauge). Calculate the fan brake horsepower by the expression

$$\frac{ACFM/FAN \times \Delta P.}{6370 \times efficiency\ (say,\ 0.7)}$$

● *Step 8*   Estimate efficiency of the motor and determine its horsepower. This, multiplied by 0.746, gives the absorbed power in kilowatts. Compare with plant data and repeat Step 7 to determine fan efficiency.
● *Step 9*   Establish the vapour liquid condition of the effluent leaving the last condenser. This will be the condition at the flash drum. Allocate enthalpy values to the vapour and liquid phase.
● *Step 10*   Subtract the enthalpy of the stream leaving the last condenser from the exit enthalpy of the previous exchanger. This is the duty of the final condenser.
● *Step 11*   Repeat Step 2 for this exchanger if it is a shell and tube. Repeat Steps 6–8 if it is an air condenser.

An example calculation to establish vapour and liquid composition and enthalpy now follows:

Example Calculation

From plant data estimate pressure profile over effuent side of heat exchangers

Effluent flows tube-side through first set exchanger.

Plant data gave	Inlet 230 psig
	Outlet 215 psig
Air cooler gave	Inlet 215 psig
	Outlet 200 psig
Water trim cooler	Inlet 200 psig
	Outlet 190 psig (separator pressure)

Temperature into first exchanger 912 °F
Temperature out first exchanger 305 °F
Temperature out of air cooler 140 °F
Temperature out of water trim 95 °F

Calculate condition and enthalpy at outlet of first exchanger

Dew point at 230 psia

	Mol frac.	1st trial at 285 °F	
		K	$X = Y/K$
H_2	0.628	370	0.002
C_1	0.084	13	0.006
C_2	0.083	5.9	0.014
iC_4	0.018	2.0	0.009
nC_4	0.025	1.7	0.015
iC_5	0.007	1.0	0.007
nC_5	0.007	0.92	0.008
C_6	0.030	0.28	0.107
C_7	0.026	0.12	0.217
C_8	0.018	0.035	0.514
Total	1.000		0.921

Dew point conditions = 230 psia and 278 °F

Calculate feed conditions at outlet of air condenser

Pressure at outlet   = 215 psia
Temperature at outlet = 140 °F

Calculate vaporization

Comp.	Moles/h	K	1st trial $V/L = 20$		2nd trial $V/L = 50$		Final $V/L = 11$		Vapour		Liquid
			V/LK	$L = \dfrac{X}{1+(V/L)K}$	V/LK	$L = \dfrac{X}{1+(V/L)K}$	V/LK	$L = \dfrac{X}{1+(V/L)K}$	Moles/h	lb/h	lb/h
H_2	3219	300	—	NIL	—	—	—	—	3219	6438	—
C_1	432	12	240	1.79	600	0.72	132	3.25	428.7	6859	52
C_2	426	3.9	78	5.39	195	2.17	42.9	9.70	416.3	12 489	291
C_3	378	1.7	34	10.80	85	4.40	18.7	19.19	358.8	15 787	844
iC_4	93	0.66	13.2	6.55	33	2.74	7.26	11.26	81.7	4739	653
nC_4	126	0.50	10	11.45	25	4.85	5.50	19.38	106.6	6183	1124
iC_5	35	0.25	5	5.83	12.5	2.59	2.75	9.33	25.7	1850	672
nC_5	36	0.20	4	7.20	10	3.27	2.20	11.25	24.7	1778	810
C_6	159	0.034	0.68	94.64	1.7	58.89	0.374	115.72	43.3	3507	9373
C_7	134	0.009	0.18	113.56	0.45	92.41	0.099	121.93	12.1	1234	12 437
C_8	87	0.002	0.04	83.65	0.10	79.09	0.022	85.13	1.9	243	10 897
Total	5125			340.86		251.13		406.14	4718.8	61 107	37 153
				$V/L = 14.0$		$V/L = 19.4$		$V/L = 11.6$	Mol.wt liquid = 91.5		
									Mol.wt vap = 12.9		

Close enough.

Calculate duty of condenser

Heat in:

Weight 98 260 lb/h at 305 °F all vapour
Mol. wt = 19.17

Enthalpy 470 BTU/lb (from Maxwell)

Total heat in   = 46.182×10^6 BTU/h

Heat out:

In liquid: 37 153 lb/h 91.5 MW at 140 °F
                      =   168 BTU/lb   = 6.242×10^6 BTU/h

In vapour   61 107 lb/h at 12.9 MW   =   370 BTU/lb
                                                      = 22.610×10^6 BTU/h

Duty of exchange= 17.33×10^6 BTU/h

Evaluation and analysis of the air condenser may now follow the steps given in Section 8.9.

First heat exchanger

The first cooler exchanger exchanges hot effluent vapour with cold hydrogen recycle and cools the effluent from 912 °F to 305 °F. No condensation takes place. The duty of this exchanger in single phase is:

98 260 lb/h at 912 °F   =   960 BTU/lb × 98 260
                                 = 94.330×10^6 BTU/h
              at 305 °F   = 46.182×10^6 BTU/h
                    Duty   = 48.148×10^6 BTU/h

Evaluation and analysis of this exchanger is carried out to compare its overall heat transfer coefficient with accepted norms. See Section 6.15.

Final water/effluent condenser

The enthalpy and duty of this exchanger are calculated using the phase calculation (equilibrium flash) completed for Section 9.5 and that carried out above for all condenser outlet conditions. Plant and equipment data are used to determine overall heat transfer coefficient and this is compared with accepted norm in a manner described in Section 6.15.

9.10   Predicting the Performance of Reactor Interstage Fired Heaters

Chemical reaction in catalytic reforming includes the dehydrogenation of cycloparaffins (naphthenes), isomerization of straight paraffin chains, hydrocyclization (paraffins to cycloparaffins) and some hydrocracking. These chemical reactions occur simultaneously but at varying rates. Most of the reactions are exothermic, that is, they give out heat. Dehydrogenation of paraffins, however, occurs very rapidly in the process. Most of the conversion of cycloparaffins to aromatics (the dehydrogenation of cycloparaffins) occurs in the first reactor. This reaction is endothermic—that is, it takes in heat to react. In catalytic reforming therefore there is always a significant temperature drop over the first reactor and this is more pronounced at the beginning of the run, when the catalyst is fresh and very active.

This section describes a method of predicting this temperature drop and calculating the duty of the intermediate heater which restores the system temperature. The following steps are used in developing this prediction:

- *Step 1* Estimate the naphthene content of the feed. This method is described in Section 9.3.
- *Step 2* Assume that 90% of the naphthenes will be converted to aromatics in the first reactor.
- *Step 3* From the chemical reaction equation 1 mole naphthene = 1 mole aromatics + 3 moles H_2 calculate the moles hydrogen formed. Use an average lb/gal figure of 6.44 and mol. weight of 98 for the naphthenes.
- *Step 4* From the heat of reaction of the dehydrogenation, which is 30 500 BTU/mole of hydrogen formed, calculate total heat loss as BTU/h.
- *Step 5* Calculate the total heat in feed and recycle gas at the reactor inlet. Take inlet temperature as 890 °F. The recycle gas is taken as 2700 SCF/BbL. (Actual plant data should be used where possible.)
- *Step 6* Ignoring the formation of aromatics and the aromatic content of the effluent, subtract the heat of reaction from the total heat calculated in Step 5. This is the enthalpy in the effluent in BTU/h.
- *Step 7* Divide that portion of the total enthalpy calculated in Step 6 which is attributed to the hydrocarbons by the pounds per hour of hydrocarbons. Assume that the ratio of the outlet enthalpy of hydrocarbons to hydrogen will be the same as the inlet ratio.
- *Step 8* Using the BTU/lb of hydrocarbon enthalpy calculated in Step 7 refer to the enthalpy charts and read off the corresponding temperature. This is the outlet temperature from the reactor. Check against plant data if this is possible.
- *Step 9* The heat of reaction (in BTU/h) calculated in Step 5 is used to fix the heater duty of the interstage heater.
- *Step 10* Using the rate of fuel fired (from plant readings) and the fuel's API gravity read off the heat of combustion (BTU/lb) from heat of combustion graph in Figure A1.9. Multiply this by the rate to give total heat fired.
- *Step 11* Divide the duty of the heater by the enthalpy BTU/h fired calculated from Step 10. This multiplied by 100 is the thermal efficiency of the heater (see also Section 6.16).

An example calculation now follows:

Example Calculation

Most of the reforming action in the first reactor is the dehydrogenation of cycloparaffins (naphthenes) to aromatics. *Note*: This is *not* the only reaction that takes place but it is the major one.

Assume 90 mol. % of the naphthenes are converted in this reactor. The naphthenes in C_6 to 380 °F range will have an average mol. wt of 98 (methyl cyclohexane) and 6.44 lb/gal.

Total naphthene content (Section 9.3) is 20% vol. fraction. From Section 8.11 feed rate to the platformer will be depentanized splitter bottoms

$$11\ 775.1 - (1604.5) \qquad = \quad 10\ 170.6 \text{ gal/h}$$

$$\text{Total naphthenes} \qquad = \quad 2034 \text{ gal/h}$$

$$\text{Reacted naphthenes} \qquad = \quad 1831 \text{ gal/h}$$

$$= \quad 11\ 792 \text{ lb/h}$$

$$= \quad 120 \text{ moles/h}$$

Now

Moles naphthenes $\rightarrow$ moles aromatics $+ 3H_2$

$\therefore$ moles/h of hydrogen produced $= 360$.

The dehydrogenation reaction is endothermic and the heat of reaction is 30 500 BTU/mole H_2.

Total heat loss due to dehydrogenation reaction $= 30\ 500 \times 360 = 10.98 \times 10^6$ BTU/h.

Total flow of feed and recycle gas to the first reactor.

Feed $= 60\ 957$ lb/h $= 5811$ BPSD.

$$
\begin{array}{lll}
\text{Recycle gas say 2700 SCF/BbL} & = & 1729 \text{ moles/h} \\
\qquad\qquad\text{MW} & = & \text{say } 18.6 \\
\qquad\qquad\text{lb/h} & = & 32\ 159
\end{array}
$$

Inlet temperature to first reactor 890 °F all vapour.

$$
\begin{array}{lll}
\text{Head in with feed (65 °API)} & = & 644 \text{ BTU/lb} \times 60\ 957 \\
 & = & 39.256 \text{ BTU/h} \times 10^6 \\
\text{Heat in with gas (18 MW)} & = & 32\ 159 \times 650 \text{ BTU/lb} \\
 & = & 20.903 \text{ BTU/h} \times 10^6 \\
\\
\qquad\text{Total heat in} & = & 60.159 \\
\\
\qquad\text{Total heat out} & = & 60.159 - 10.980 \\
 & = & 49.170 \text{ BTU/h} \times 10^6
\end{array}
$$

Assume enthalpy ratios oil/gas remain the same.

Then heat out with oil vapour	$= 32.091 \text{ BTU/h} \times 10^6$
and heat out with H^2	$= 17.088 \text{ BTU/h} \times 10^6$

Temperature out approximately

$$\frac{32.091 \times 10^6}{60\,957}$$

$$= 526 \text{ BTU/lb} = 730\,°\text{F}$$

ΔT across reactor $= 120\,°\text{F}$, which is reasonable.

This can be checked against plant readings. The duty of the first inter-reactor heater will be to replace this endothermic heat. This is

$$10.98 \times 10^6 \text{ BTU/h}$$

From plant data fuel oil fired $= 853 \text{ lb/h } (22.3\,°\text{API})$

Heat of combustion
   (low heating value) $= 17\,860 \text{ BTU/lb}$

Heat fired $= 17\,860 \times 853$

$$= 15.235 \text{ BTU/h} \times 10^6$$

Heater efficiency $= \dfrac{10.98}{15.235} \times 100$

$$= 72\%$$

which is satisfactory.

10 THE HYDROTREATING PROCESSES

10.1 Process Description—the Hydrotreating Plants

NAPHTHA HYDRODESULPHURIZATION (HYDROFINING)

This uses the hydrogen stream produced by the catalytic reformer on a once-through basis. Heavy naphtha feed to the cat reformer is fed to the naphtha hydrofiner from storage. The feed stream and the hydrogen gas stream are preheated by exchange with the hot reactor effluent stream. The feed then enters the fired heater, which brings it up to the reactor temperatures (about 450 °F) and leaves the heater to enter the reactor which operates at about 400–450 psig. Sulphur is removed from the hydrocarbon as hydrogen sulphide in this reactor and the reactor effluent is cooled to about 100 °F by heat exchange with feed. The cooled effluent is collected in a flash drum where the light hydrogen-rich gas is flashed off. This gas enters the suction side of the booster compressor which delivers it to other hydrotreaters and hydrogen users. The liquid phase from the drum is pumped to a reboiled stabilizer. The overhead

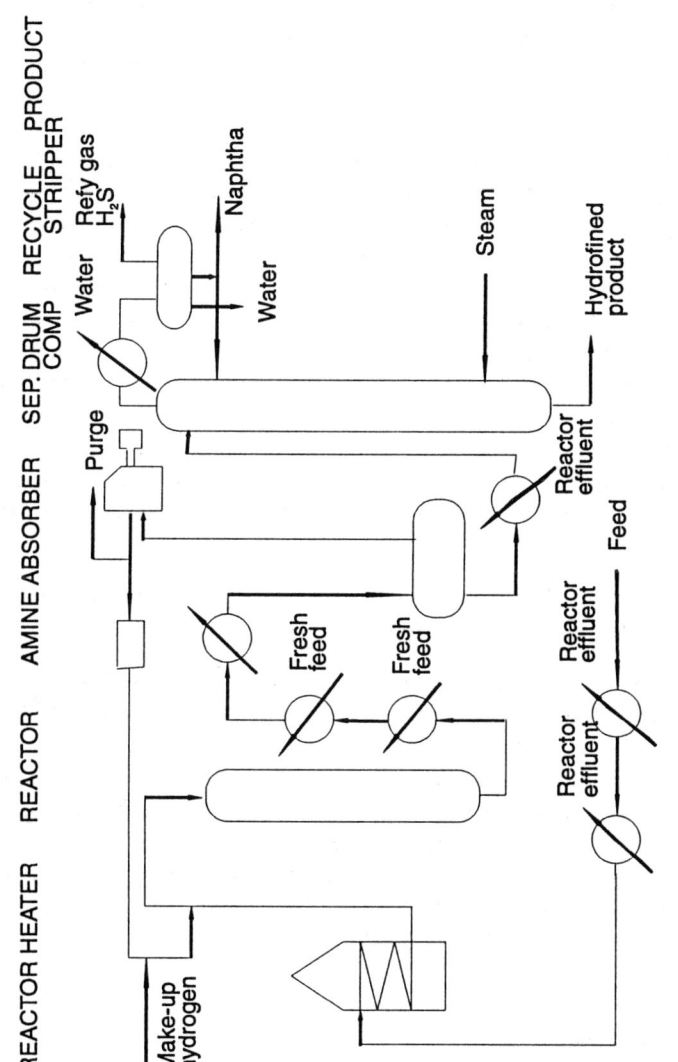

Figure 10.1. Hydrofining unit (typical middle distillate)

vapour stream from the stabilizer is routed to fuel while the bottom product, cat reformer feed, is pumped to the cat reformer.

GAS OIL HYDROFINING

This process uses a recycled hydrogen stream to desulphurize a gas oil feed. Figure 10.1 shows the gas oil feed entering the unit to be preheated with hot effluent stream before entering a fired heater. Its temperature is increased to the reactor temperature of about 750 °F in this heater. A hydrogen-rich stream is introduced at the coil outlet prior to the mixed streams entering the reactor. The reactor contains a bed of cobalt molybdenum on alumina catalyst and desulphurization takes place over the catalyst with 70–75% of the total sulphur in the oil being converted to H_2S.

The reactor effluent is cooled by the cold feed stream, water or air. This cooled effluent enters a flash drum where the gas phase and liquid phase are separated. The gas phase rich in H_2S and hydrogen enters the recycle compressor. The gas stream then enters an amine contactor where the H_2S is absorbed into the amine and removed from the system. Although Figure 10.1 shows a purge stream before the amine absorber, in most cases the purge is downstream after the amine clean-up. The purged gas is replaced by fresh hydrogen-rich make-up stream, thus maintaining the purity of the recycle gas.

The liquid phase leaving the flash drum is preheated before entering a stream stripping column where the light ends created in the process are removed as overhead products. The bottom product leaves the tower to be cooled and stored.

10.2   Predicting Reactor Conditions and Start of Run, End of Run and Degree of Desulphurization

Reactor conditions required for the proper operation of catalytic units are dependent on the type of catalyst used. In hydrotreating therefore reactor side conditions and operation are proprietary to the licensor of the process. To fulfil obligations the licensor imparts correlations that satisfy the process operation to the operator company. The company in turn is obligated and bound by secrecy agreement not to divulge these data to anyone outside the company.

Such correlations for the hydrotreaters are available within the user-company's organization. They are usually self-explanatory and simple to use. The variables for hydrotreater operation are normally:

● Reactor inlet temperature
● Reactor pressure
● Recycle rate required
● Purity of the inlet hydrogen stream
● Space velocity

These change with the degree of desulphurization required, feed cut point and quality, and the catalyst age.

There are typical conditions for the reactor side which are in the public domain:

Straight-run naphtha desulphurization

Reactor inlet temperature	650 °F
Reactor pressure	500–600 psig
Space velocity v/v/h	4–6
Hydrogen purity at inlet	75% mole (minimum)
Recycle rate scf/bbl	500–550

Desulphurization to <5 ppm total sulphur.

Diesel desulphurization

Reactor temperature	650 °F
Reactor pressure	750–1000 psig
Space velocity v/v/h	1.5–2
Recycle rate scf/bbl	600–700

Desulphurization: remove 85% sulphur in feed.

Hydrogen consumption in these units can be calculated and this depends to a large extent only on feed quality, that is, the amount of sulphur (and nitrogen in some cases) and the presence of olefinic material.

10.3 Predicting Hydrogen Consumption in Naphtha Hydrotreating

This section describes a method of predicting the hydrogen consumed in hydrotreating straight-run naphtha. Emphasis is based on this naphtha, and treating cracked naphtha and gas oils needs approximately three to five times the amount of hydrogen than straight-run.

The method for predicting hydrogen consumption has the following steps:

- *Step 1* From the naphtha TBP and the assay, estimate the weight percentage sulphur in the naphtha feed (see the Introduction to this book).
- *Step 2* Establish the throughput of the naphtha and assume that the feed will be completely desulphurized in the process.
- *Step 3* Calculate the moles of sulphur removed by the chemical equation

$$H_2 + S = H_2S$$

Fix the moles of hydrogen required to satisfy the reaction given above. That is, moles hydrogen equals moles sulphur removed.
- *Step 4* Estimate the amount of hydrogen required to saturate the hydrocarbon chain or ring after the sulphur molecule is removed. In straight-run naphtha fractions the sulphur compounds are relatively simple in structure. A figure of two or three times the hydrogen used for sulphur removal is reasonable for resaturation. Remember, this only applies to straight-run feeds.
- *Step 5* The remaining consumption of hydrogen is to replace the hydrogen lost from the system in solution with the liquid product leaving. This can be very significant quantity. Calculate using Steps 6 to 10.

● *Step 6* Establish the component analysis of the make-up gas. This is usually reformer offgas. Calculate the amount in moles/h of each component that satisfies the chemical reaction quantity calculated in Steps 3 and 4.

● *Step 7* Let x moles/h be the hydrogen that leaves in solution with the product. Calculate in terms of x the proportion of the other components related to the hydrogen that also leave in solution with the product. This is each component moles divided by the hydrogen component and multiplied by x.

● *Step 8* Add the C_1–C_5 portion of the make-up gas to the x components calculated in Step 7.

● *Step 9* The quantity calculated in Step 8 plus the moles of naphtha product is the liquid phase that leaves the separator (i.e. the unstabilized product). By definition, this is in equilibrium with the gas phase that leaves the separator drum. Calculate, using the liquid phase composition, its bubble point at the separator drum conditions.

● *Step 10* Again by definition, the Y factor calculated in Step 9 (i.e. $Y = Kx$) is the composition of the vapour phase leaving the separator drum. As $Y = Kx$ then equate and solve for x as moles hydrogen leaving in solution with the hydrotreated product.

● *Step 11* The gas phase composition is calculated by substituting the calculated value for x in Step 10. If the unit has a recycle gas stream this is also the composition of the recycle gas.

● *Step 12* The total hydrogen and hydrogen gas stream make-up to the unit can now be completed with the addition of the moles in solution calculated in Step 10.

An example calculation now follows:

Example Calculation

Hydrotreating the depentanized naphtha from Section 8.11:

From assay approx. sulphur content   = 0.05% wt

Assume 100% desulphurization.

lb/h of feed ex splitter   = 60 957

lb/h of sulphur   = 30.5

$S + H_2 \rightarrow H_2S$ then moles hydrogen required to remove sulphur $= \dfrac{30.5}{32} = 0.953$ moles H_2/h

Say, 1.0

Hydrogen is also required to saturate after sulphur removal and some light ends are produced. Let total amount of sulphur for this requirement be:

For saturation   — 3 moles/h
For related reactions   — 1 moles/h

Note: If olefins were present saturation could be four to five times higher.

Total hydrogen consumed in chemical reaction—4 moles/h

Make-up gas to satisfy this need $\qquad = \dfrac{4}{0.697}$

(See Section 9.5) $\qquad\qquad\qquad\qquad = 5.74$ moles/h

Make-up gas will also be required to make up for hydrogen in liquid solution leaving the system. This is calculated as follows:

Make up gas is:

	Mol. frac.	Moles/h
H_2	0.697	4.0
C_1	0.093	0.534
C_2	0.089	0.511
C_3	0.074	0.425
iC_4	0.016	0.092
nC_4	0.020	0.115
C_5^+	0.011	0.063
	1.00	5.74

Let x moles/h be the hydrogen in solution in the liquid leaving the separator. Then total make-up gas can be written as follows:

	Consumed in reaction	Loss in solution	Total make-up Moles/h
H_2	4.0	X	$4.0 + X$
C_1	0.534	$0.534/4\ X$	$0.534 + 0.134\ X$
C_2	0.511	$0.511/4\ X$	$0.511 + 0.128\ X$
C_3	0.425	$0.425/4\ X$	$0.425 + 0.106\ X$
iC_4	0.092	$0.092/4\ X$	$0.092 + 0.023\ X$
nC_4	0.115	$0.115/4\ X$	$0.115 + 0.029\ X$
C_5^+	0.063	$0.063/4\ X$	$0.063 + 0.016\ X$
	5.74	$1.436\ X$	$5.74 + 1.436\ X$

The four moles hydrogen has been converted. Then to maintain material balance the liquid in the separator is given below. This is in equilibrium with the separator gas. A bubble point calculation then gives the composition of the separator gas in terms of x thus:

Comp.	Separator liquid moles/h			365 psig K 100 °F*		Separator gas moles/h		
H$_2$			X	55				55 X
C$_1$	0.534	+	0.134 X	7.5	4.0	+		1.01 X
C$_2$	0.511	+	0.128 X	1.8	0.92	+		0.23 X
C$_3$	0.425	+	0.106 X	0.62	0.26	+		0.066 X
iC$_4$	0.092	+	0.023 X	0.30	0.028	+		0.007 X
nC$_4$	0.115	+	0.029 X	0.24	0.028	+		0.007 X
+C$_5$s	0.063	+	0.016 X	0.09	0.006	+		0.001 X
Oil Feed			558.9					Nil
Total	560.64	+	1.436 X		5.242	+		56.321 X

*Separator operates at 350 psig and 100 °F.

$$560.64 + 1.436x = 5.242 + 56.321x$$

$$x = 10.12 \text{ moles/h}$$

Total make-up gas $= 10.12 + 4 = 14.12$ moles/h of hydrogen

$$= \frac{14.12}{0.697} = 20.26 \text{ mole/h of gas}$$

$= 32$ SCF/Bbl of fresh feed

(bbls fresh feed $= 5811$ BPSD).

Recycle gas is 96.8 moles % H$_2$ (no purging required).

10.4 Predicting Hydrogen Consumption in Diesel Hydrotreating Where a Gas Purge is Required

Diesel hydrotreating is in many ways similar to naphtha hydrotreating. However, because of a more complex molecular structure in the diesel fraction and more complex sulphur compounds, some additional consideration must be made in this case. For example there will be a higher quantity of sulphur and this will contain disulphides and thiophenes which are complex ring compounds of sulphur. More light ends are made in the process when these compounds are broken to release the sulphur. Consequently, the process will require more hydrogen to satisfy the chemical reactions than was the case for naphtha treating. This and a requirement of most middle distillate processes for a high concentration of hydrogen at the reactor outlet (to prevent coking and rapid deactivation of the catalyst) makes a recycle gas system necessary. Also, because of the quantity of sulphur released as H$_2$S there is every probability that the recycle gas will require amine treating to remove this H$_2$S and thus retain its purity. Very often some of the recycle gas will also require to be purged off and replaced with fresh platformer (or high-hydrogen content gas). This will certainly be so if the light ends make is high and the subsequent purity of the recycle gas diminished.

In this section consideration has been given to amine treating and purging the recycle gas. The purging is an added item to be satisfied by the make-up gas consumption. The method of predicting hydrogen consumption and recycle gas purity is given by the following steps:

- *Step 1* From the feed TBP and assay calculate its sulphur content (see the introductory chapter of this book).
- *Step 2* Set the degree of desulphurization required. This will depend on catalyst and reactor conditions. Most well-designed modern diesel hydrotreaters can desulphurize heavy gas oil to remove at least 85% weight of its sulphur content.
- *Step 3* Establish the feed throughput. This will depend on the space velocity required for Step 2.
- *Step 4* Estimate the light ends produced in the process. This can be done by referring to plant records or, as a rule of thumb, this can be in the form of C_5^+ naphtha at about 6–10% by vol. of feed. Some C_5s and lighter are also formed but these are usually in small quantities.
- *Step 5* As in Section 10.3, calculate the hydrogen required to remove the sulphur molecules. Again add twice this quantity to saturate the compounds that contained the sulphur.
- *Step 6* The light ends formed through the minor cracking to release thiophenes and disulphides will need to be saturated. This consumes hydrogen. Approximately 2 moles of hydrogen will be required for this purpose per mole of light ends formed. Using past lab tests on the hydrotreater naphtha, develop the TBP and split to pseudo-components. Allocate mole weights and gravities to these components to arrive at a number of moles/h for the light ends. Calculate the hydrogen consumption.
- *Step 7* The remaining hydrogen requirements will be to replace hydrogen lost in solution with the liquid product and, of course, that lost in the purge stream. Commence with the calculation to determine solution loss as follows.
- *Step 8* Establish the component analysis of the make-up gas. This is usually reformer offgas or if the naphtha hydrotreater is operating on a once-through gas basis it will be the offgas from that unit.
- *Step 9* Calculate the amount in moles/h of each component associated in the make-up gas with the hydrogen required for the chemical reactions calculated in Steps 5 and 6.
- *Step 10* Let x moles/h be the hydrogen that leaves in solution with the liquid product from the separator and the purge gas. Calculate in terms of x the proportion of the other components in the make-up gas associated with the hydrogen.
- *Step 11* Add the C_1 to C_5 portion of the make-up gas to the x components calculated in Step 10. Add also the C_5 naphtha components which were made in the process and, of course, a guess at the number of moles of H_2S that will be in the liquid phase of the separation drum. To do this, look at the K (equilibrium constant) for H_2S at drum conditions. Use this to estimate its proportion in liquid. For example, if $K = 1$ then the split will be close to 50% in liquid and 50% in vapour.
- *Step 12* Set the amount of purge in terms of its proportion to the liquid product (that is, set the V/L for flash vaporization). This will be such as to provide a recycle

gas hydrogen content of above 63% mole after H_2S removal. This figure is trial and error. Start with $V/L = 0.1$.

- *Step 13*   Carry out a flash calculation in terms of x and using V/L set in Step 12. Solve for x. The vapour stream from this calculation is the purge gas in terms of moles/h and composition. It is also the composition of the recycle gas.
- *Step 14*   Complete the calculation for hydrogen consumption and make-up gas using the value for x above.

An example calculation now follows.

Example Calculation

Predicting hydrogen consumption in diesel hydrotreating (including purge)

Feed to diesel hydrotreater in this case will be heavy gas oil from Section 6.4 material balance.

Gas oil cut	= 610 °F → 690 °F Kuwait crude Mid-boiling point 650 °F
Unit throughput	= 5500 BPSD (blocked operation)
From assay sulphur content	= 2.1% wt

Desulphurization shall be 85%.

lb/h gas oil	= 70 078
Sulphur in feed	= 1472 lb/h
Sulphur removed	= 1251 lb/h
Hydrogen for sulphur removal	$= \dfrac{1251}{32} = 39$ moles/h

7% vol. of C_5^+ naphtha is produced in the reaction.
Say this has the following composition:

	% vol.	BPSD	gal/h	lb/h	MW	Moles/h
C_6	18.8	72.4	127	702	86	8.2
C_7	23.6	90.9	159	911	100	9.1
C_8	27.1	104.3	183	1077	114	9.4
C_9	13.2	50.8	89	535	128	4.2
C_{10}^+	17.3	66.6	117	715	142	5.0
	100.0	385	675	3940		35.9

Approximately 2 moles of H_2 per mole will be required to saturate light ends as they are produced.

Then this hydrogen $= 35.9 \times 2 = 71.8$

Then total hydrogen make-up will be
Sulphur removal $= 39$ moles/h
Saturating after desulph. $= 78$ moles/h
Light ends $= 71.8$
Total $= 188.8$ moles/h of H_2

In addition there will be losses out of the system by H_2 in liquid solution and a purge stream. See the following diagram

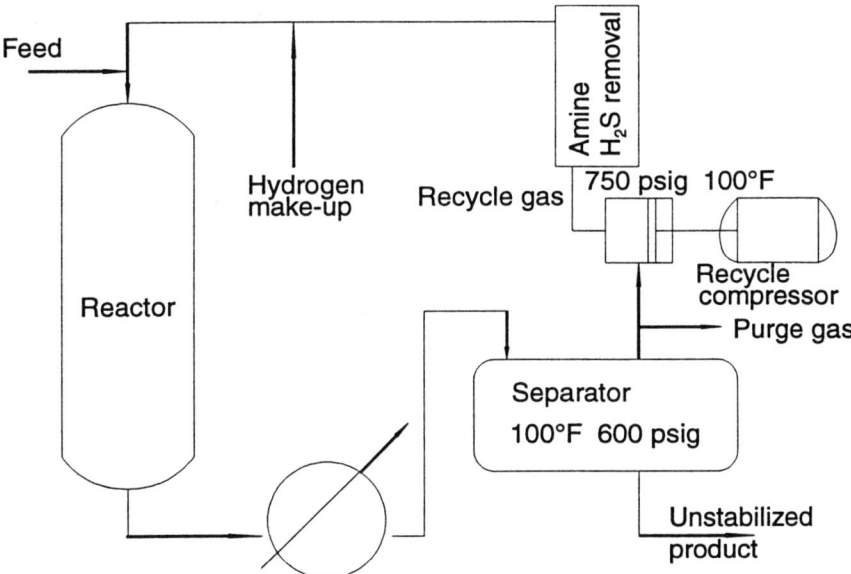

Total make-up gas for chemical reaction:

	Mole frac*	Moles/hr
H_2	0.697	188.8
C_1	0.093	25.2
C_2	0.089	24.1
C_3	0.074	20.0
iC_4	0.016	4.3
nC_4	0.020	5.4
iC_5^+	0.011	3.0
	1.000	270.8

*From reformer (see Section 9.5)

Let x moles/h be H_2 in solution lost from the system in liquid solution and purge. Total make-up gas is then:

	Consumed in reaction	Loss in solution/purge	Total make-up moles/h
H_2	188.8	X	$188.8 + \quad X$
C_1	25.2	$0.134\ X$	$25.2 + 0.133\ X$
C_2	24.1	$0.128\ X$	$24.1 + 0.128\ X$
C_3	20.0	$0.106\ X$	$20.0 + 0.106\ X$
iC_4	4.3	$0.023\ X$	$4.3 + 0.023\ X$
nC_4	5.4	$0.029\ X$	$5.4 + 0.028\ X$
$C_5{}^+$	3.0	$0.016\ X$	$3.0 + 0.016\ X$
	270.8	$1.436\ X$	$270.8 + 1.434\ X$

Set the purge to be 15% mole of liquid to stripper. Thus $V/L = 0.15$ (first trial).

Calculate flash for effluent in terms of x at $V/L = 0.15$. Solve for x. Thus:

	x_F effluent	K 615 psia 100 °F	V/LK	$V/L = 0.15$ $L = \dfrac{x_F}{1 + (V/L)K}$	Liquid to Stripper L Moles/h	Purge Gas V Moles/h	Recycle Gas Composition
H_2	x	32	4.8	$0.172\ x$	7.15	34.42	62.85
C_1	$25.2 + 0.133\ x$	3.8	0.51	$18.05 + 0.085\ x$	19.58	11.15	20.36
C_2	$24.1 + 0.128\ x$	1.2	0.18	$20.42 + 0.108\ x$	24.91	4.51	8.23
H_2S	19.0	1.0	0.15	16.52	16.52	2.48	4.53
C_3	$20.0 + 0.106\ x$	0.5	0.075	$18.60 + 0.099\ x$	22.72	1.69	3.09
iC_4	$4.3 + 0.023\ x$	0.25	0.0375	$4.14 + 0.022\ x$	5.05	0.21	0.38
nC_4	$5.4 + 0.028\ x$	0.19	0.0285	$5.25 + 0.027\ x$	6.37	0.19	0.35
C_5	$3.0 + 0.016\ x$	0.085	0.0128	$2.96 + 0.016\ x$	3.63	0.04	0.07
C_6	8.2	0.044	0.0066	8.15	8.15	0.05	0.09
C_7	9.1	0.020	0.0030	9.07	9.07	0.03	0.05
Oil	242	—		242	242	Nil	—
Total	$360.3 + 1.434\ x$			$343.16 + 0.529\ x$	365.15	54.77	100.00

$$\frac{(360.3 + 1.434x) - (343.16 + 0.529x)}{(343.16 + 0.529x)} = 0.15$$

$$17.14 + 0.905x = 51.474 + 0.0794x$$

$$x = \frac{34.334}{0.826} = 41.57 \text{ moles/h}$$

Substituting in the above table total effluent (less recycle) = 419.92 moles/h. The recycle gas H_2 content will be 62.9 mol. % before H_2S removal. If there is an amine contactor in the system H_2 purity of the gas becomes 65.8 mol. %, which is quite good.

A similar exercise was carried out using $V/L = 0.1$. This gave a purge steam of 36.59 moles/h and the recycle gas purity was 61.2% mol. with no H_2S removal. With H_2S removal this became 64.2% mol, which is borderline.

Total hydrogen to the plant at a purge ratio of 0.15 will be:

Sulphur removal	—	39 moles/h
Saturating	—	78 moles/h
Light ends	—	71.8 moles/h
Hydrogen in liquid	—	7.2 moles/h
Hydrogen in purge	—	34.4 moles/h
	=	230.4 moles/h

$$\text{Make-up gas stream from reformer} = \frac{230.4}{0.697}$$

$$= \quad 330.5 \text{ moles/h}$$

or 545 SCF/Bbl of fresh feed

10.5 Predicting the Feed to Naphtha Hydrotreater Stabilizer and to the Diesel Hydrotreater Stripper

The purpose of this section is to prepare the data for naphtha stabilizer feed and diesel stripper feed that will be used in sections referring to these equipment. The feed in both cases are taken from those calculated in Sections 10.3 and 10.4. The streams are flashed at reasonably low pressures to minimize hydrogen and light ends (C_1 and C_2) as much as possible in the feed to the two recovery units. Example calculations for both now follow.

Example Calculation

Calculating stabilizer feed for naphtha hydrotreater

The liquid from the first separator is depressurized into a second separator which also acts as a feed surge drum to the naphtha stabilizer. This drum operates at 120 psig and 100 °F. Carry out vapour/liquid flash using separator liquid from Section 10.3. Thus

	Moles/h XF	K 135 psia 100 °F	1st trial $V/L=0.03$		2nd trial $V/L=0.01$		Moles vapour	Liquid	
			V/LK	$L=\dfrac{X_F}{1+(V/L)K}$	V/LK	$L=\dfrac{X_F}{1+(V/L)K}$		lb/h	gal/h
H_2	10.12	140	4.2	1.9	1.4	4.22	5.9	8	—
C_1	1.89	19	0.57	1.2	0.19	1.59	0.3	25	—
C_2	1.81	4.2	0.126	1.61	0.042	1.74	0.07	52	17.6
C_3	1.50	1.3	0.039	1.44	0.013	1.48	0.02	65	15.4
iC_4	0.32	0.58	0.017	0.31	0.058	0.318	0.002	18	3.8
nC_4	0.41	0.42	0.013	0.40	0.042	0.408	0.002	24	4.9
C_5s	0.22	0.15	0.005	0.22	0.0015	0.220	Nil	16	3.0
Naphtha	558.9	0.01	0.0003	558.9	—	558.9	Nil	61 057	10 170.3
Total	575.17			565.98		568.88	629	61 265	10 216
				$V/L=0.016$		$V/L=0.011$			

*Mid-BPT $\simeq 231$ °F

Trial 2 is acceptable.

Calculating stabilizer feed for diesel hydrotreater

The liquid from the separator is further flashed at 100 psig at 100 °F. This second flash drum acts also as a surge drum for a steam stripper.

Vapour/liquid flash is as follows (from Section 10.4):

Comp.	Liquid feed X_F	K 115 psia 100 °F	1st trial $V/L=0.4$ V/LK	$L=\dfrac{X_F}{1+(V/L)K}$	2nd trial $V/L=0.1$ V/LK	$L=\dfrac{X_F}{1+(V/L)K}$	Final $V/L=0.14$ V/LK	$L=\dfrac{X_F}{1+(V/L)K}$	Liquid Feed to stripper lb/h	gal/h
H_2	6.5	165	66	0.1	16.5	0.37	23.1	0.27	0.5	
C_1	26.1	22	8.8	2.66	2.2	8.16	3.08	6.40	102.4	
C_2	24.9	5.0	2.0	8.30	0.5	16.60	0.7	14.65	439.5	148.5
H_2S	19.0	3.7	1.48	7.66	0.37	13.87	0.52	12.50	425.0	64.6
C_3	20.7	1.5	0.60	12.94	0.15	18.00	0.21	17.11	752.8	178.4
iC_4	4.45	0.66	0.26	3.53	0.066	4.17	0.09	4.08	236.6	50.6
nC_4	5.58	0.50	0.20	4.65	0.050	5.31	0.07	5.21	302.2	62.2
C_5	3.10	0.17	0.07	2.90	0.017	3.05	0.024	3.03	218.2	41.6
C_6	8.20	0.065	0.03	7.96	0.0065	8.15	0.009	8.13	699.2	126.2
C_7	9.10	0.024	0.01	9.01	0.0024	9.08	0.003	9.07	907.0	158.0
C_8	9.40	0.009	0.004	9.36	0.0009	9.39	0.001	9.39	1070.0	182.0
C_9	4.20	0.003	0.001	4.20	—	4.20	—	4.20	537.6	89.6
C_{10}	5.0	0.0015	—	5.0	—	5.00	—	5.00	710.0	116.2
Diesel	223.4	—	—	223.4	—	223.40	—	223.40	70 078	9635.2
Total	369.63			301.67		328.75		322.44	76 479	10 853.1
				Calc. $V/L=0.23$		Calc. $V/L=0.12$		Cal. $V/L=0.146$		

Trial 3 is within acceptable limits.
Mol. wt of liquid = 237
°API of liquid = 35.8

10.6   Setting Naphtha Hydrotreater Stabilizer Conditions

This section describes a calculation method by which the operating conditions for the naphtha stabilizer can be set. The example provided commences with an overall material balance using the data from Section 10.5. The objective of the stabilizer in this case is to prepare the hydrotreated naphtha for cat reformer feed. In doing this it must remove all the H_2S generated by the hydrotreating process and all the light ends in solution. Ideally, the cat reformer feed should have as a front end the hexanes and cyclohexanes. The C_5s and cyclopentanes are not reactive in the cat reformer process and so should be removed from the feed to that unit. Third, the cat reformer feed must be dry. A reboiled stabilizer cutting at around C_6 as front end will ensure this.

The following steps are used in setting this unit's operating conditions:

● *Step 1*   Fix the split required from the fractionator. Most of the C_5s will be allowed in the distillate. The Fenske equation is then used to determine the concentration of the key components in the distillate and product. The keys in this case are nC_5 and C_6. This method has been fully described in Sections 8.2 and 9.6.
● *Step 2*   After determining the distribution of the keys, develop the component balance and overall material balance.
● *Step 3*   Although the hydrogen and methane content of the column feed are relatively small these become significant in concentration of the overhead product.

Because of this, these towers normally operate with a partial condenser. That is, there will be a vapour product and a liquid product leaving the overhead reflux drum. Now these are a balance between the bottom of the tower condition (which must not be close to critical) and the loss of C_3s and C_4s to the gas stream. Normal conditions for this drum is 95–100 °F temperature and around 150 psig pressure.

- *Step 4* Calculate a flash vaporization at the conditions selected in Step 3 of the total overhead stream. This gives the gas stream and the liquid product stream both in quantity and content. This calculation is fully described in Section 8.7.
- *Step 5* Calculate the tower-top conditions. The pressure in this part of the tower is obtained by adding the pressure drop across the condenser and overhead piping to the reflux drum pressure.
- *Step 6* Calculate the tower-top temperature by first adding the moles reflux stream components to the total overhead product moles. A dew point calculation of these combined streams at the pressure found in Step 5 is the tower-top temperature. The amount of reflux should be checked by the short cut distillation method (see Section 8.4) to give the internal reflux required. From the split required, a reflux ratio of 1.1:1 should be satisfactory. This reflux ratio is measured against total overhead product.
- *Step 7* Set the tower-bottoms pressure by adding the total tray pressure drop to the overhead pressure from Step 5. Use about 0.25 psi per tray for this type of column. These towers normally have about 20 trays.
- *Step 8* Calculate the bottoms temperature by a bubble point calculation of the bottom product at the pressure from Step 7. (This calculation is described in Section 6.6.)

An example calculation now follows.

Example Calculation

An overall material balance for the stabilizer is as follows (Section 10.5):

	Feed					Overhead (total)*			Reformer feed		
	Moles/h	MW	lb/h	lb/gal	gal/h	Moles/h	lb/h	gal/h	Moles/h	lb/h	gal/h
H_2	4.22	2	8	—		4.22	8	—			
C_1	1.59	18	25	—		1.59	25	—			
C_2	1.74	30	52	2.96	17.6	1.74	52	17.6			
H_2S	0.50	34	17	6.55	2.6	0.50	17	2.6			
C_3	1.48	44	65	4.22	15.4	1.48	65	15.4			
iC_4	0.32	58	19	4.69	4.1	0.32	19	4.1			
nC_4	0.41	58	24	4.87	4.9	0.41	24	4.9			
iC_5	1.22	72	88	5.21	16.9	1.22	88	16.9			
nC_5	4.20	72	302	5.26	57.4	3.99	297	56.4	0.21	5	1
C_6	119.2	86	10 251	5.54	1850.4	5.7	490	88.4	113.5	9761	1762.0
C_7	140.9	100	14 090	5.74	2454.7				140.9	14 090	2454.7
NB 216	31.4	102	3203	6.03	531.2				31.4	3203	531.2
NB 231	37.3	108	4028	6.06	664.7				37.3	4028	664.7
NB 252	57.8	115	5957	6.12	973.4				51.8	5957	973.4
NB 279	50.8	120	6096	6.26	973.8				50.8	6096	973.8
NB 306	47.6	130	6188	6.36	973.0				47.6	6188	973.0
NB 330	30.9	138	4264	6.43	663.1				30.9	4264	663.1
NB 352	29.3	147	4307	6.48	664.7				29.3	4307	664.7
NB 380	14.0	156	2184	6.55	333.4				14.0	2184	333.4
Total	568.88	107.5	61 168	6.00	10 201.3	21.17	1085	206.3	547.71	60 083	9995

*Calculation of split

The stablizer has 20 actual trays. Assume 70% efficiency.
Then theoretical trays = 14
Minimum no. of trays at total reflux ≈ 9.3

$$N_{M+1} = 10.3$$

LT Key = nC_5 Let 95% C_5 to distillate
HY Key = C_6 Let X moles to distillate

Using the Fenske equation

$$N_{M+1} = \mathrm{Log}\left[\left(\frac{\mathrm{LT\ Key}}{\mathrm{HY\ Key}}\right)_D \cdot \left(\frac{\mathrm{HY\ Key}}{\mathrm{LT\ Key}}\right)_W\right] \div \mathrm{Log}\phi$$

Mean pressure (say, 265 psig), mean temp. = 290 °F

$$\phi = \frac{8.0}{4.5} = 1.78$$

$$N_{M+1} = \mathrm{Log}\left[\left(\frac{3.99}{X}\right)_D \cdot \left(\frac{119.2 - X}{0.21}\right)_W\right] \div \mathrm{Log}\ 1.78$$

$$10.3 = \mathrm{Log}\left[\frac{475.6 - 3.99\ X}{0.21\ X}\right] \div \mathrm{Log}\ 1.78$$

$$10.3 \times 0.250 = \mathrm{Log}\left[\frac{475.6 - 3.99\ X}{0.21\ X}\right]$$

$$379.3\ 3 = \frac{475.6 - 3.99\ X}{0.21\ X}$$

$$79.65X + 3.99X = 475.6$$

$$X = 5.7 \text{ moles/h } C_6 \text{ in distillate}$$

To calculate reflux-drum conditions

Hydrogen mole fraction is high at 0.199, therefore the stabilizer will have a partial condenser set drum at 100 °F and 150 psig

K at 100 °F and 150 psig

	Moles/hr X_F	K 165 psia	1st trial $V/L=1.0$ V/LK	$L=\dfrac{X_F}{1+(V/L)K}$	2nd trial $V/L=0.5$ V/LK	$L=\dfrac{X_F}{1+(V/L)K}$	3rd trial $V/L=0.65$ V/LK	$L=\dfrac{X_F}{1+(V/L)K}$	Vapour Liquid (lb/h)	Moles/h	lb/h
H_2	4.22	102	102	0.041	51	0.081	66.3	0.063	—	4.16	8
C_1	1.59	4.9	4.9	0.269	2.45	0.461	3.19	0.379	6	1.21	19
C_2	1.74	3.6	3.6	0.378	1.80	0.621	2.34	0.521	16	1.22	37
H_2S	0.50	2.6	2.6	0.139	1.30	0.217	1.69	0.186	6	0.31	11
C_3	1.48	1.1	1.1	0.705	0.55	0.955	0.72	0.860	38	0.62	27
iC_4	0.32	0.5	0.5	0.213	0.25	0.256	0.33	0.241	14	0.08	5
nC_4	0.41	0.37	0.37	0.299	0.185	0.346	0.24	0.331	19	0.08	5
iC_5	1.22	0.17	0.17	1.043	0.085	1.124	0.11	1.099	79	0.12	9
nC_5	3.99	0.14	0.14	3.500	0.070	3.729	0.091	3.657	263	0.33	24
C_6	5.70	0.05	0.05	5.429	0.025	5.561	0.033	5.518	475	0.18	15
Total	21.17			12.016		13.351		12.855	916	8.31	160

Calc. $V/L=0.76$ Calc. $V/L=0.59$ Calc. $V/L=0.65$

	Moles product/h	Moles reflux/h	Total moles/h	Mole frac. Y	Trial 1 at 280 °F K	$X=Y/K$	Trial 2 at 275 °F K	$X=Y/K$	Weight factor	Vol. factor
H_2	4.22	0.11	4.33	0.098	66.0	0.001	64.0	0.002	—	—
C_1	1.59	0.68	2.27	0.051	18.0	0.003	17.5	0.003	0.05	—
C_2	1.74	0.94	2.68	0.060	7.80	0.008	7.30	0.008	0.24	0.08
H_2S	0.50	0.33	0.83	0.019	6.20	0.003	6.00	0.003	0.10	0.015
C_3	1.48	1.55	3.03	0.088	4.00	0.017	3.90	0.017	0.75	0.178
iC_4	0.32	0.43	0.75	0.017	2.50	0.007	2.30	0.007	0.41	0.087
nC_4	0.41	0.60	1.01	0.023	2.00	0.012	1.90	0.012	0.70	0.144
iC_5	1.22	1.98	3.20	0.072	1.20	0.060	1.15	0.063	4.54	0.871
nC_5	3.99	6.58	10.57	0.239	1.05	0.228	1.00	0.239	17.21	3.272
C_6	5.70	9.93	15.63	0.353	0.58	0.609	0.55	0.642	55.21	9.966
Total	21.17	23.13	44.30	1.000		0.948		0.996	79.21	14.613

Mol wt = 79.5

°API = 85.80

Tower top temp. = 275 °F
Tower top press. = 170 psia

Tower-bottom conditions

Bottoms pressure = 155 psig + 5 psi
= 160 psig = 175 psia

Calculate bottoms temperature as bubble point at 175 psia

	Mol frac. X	1st trial = 385 °F K	$Y=XK$	2nd trial = 450 °F K	$Y=XK$
nC_5	Neg				
C_6	0.207	1.25	0.259	1.9	0.393
C_7	0.258	0.84	0.217	1.2	0.310
NB 216	0.057	0.71	0.040	1.14	0.065
NB 231	0.068	0.57	0.039	1.03	0.070
NB 252	0.095	0.46	0.044	0.86	0.082
NB 279	0.093	0.32	0.030	0.63	0.059
NB 306	0.087	0.24	0.021	0.45	0.039
NB 330	0.056	0.18	0.010	0.37	0.021
NB 352	0.053	0.14	0.007	0.28	0.015
NB 380	0.026	0.09	0.002	0.20	0.005
Total	1.000		0.669		1.059

Bottoms temperature = 445 °F
Feed temperature should be 320–350 °F.

10.7   Setting the Diesel Hydrotreater Stabilizer Conditions

The purpose of the diesel hydrotreater stabilizer is to remove the light ends from the diesel product after hydrotreating. This ensures the removal of H_2S and corrects the product's flash point to meet commercial specification. Very often when a dryness factor is not important (i.e. diesel is treated elsewhere for dryness) steam stripping of the light ends is used for stabilizing. In this section, however, a reboiled stabilizer is used as the dryness of the product diesel in this case is a criterion. The calculation proceeds with the following steps:

- *Step 1*   Using the stripper feed data developed in Section 10.5, set the distribution of key components allowed in the overhead and diesel. Keys in this case are C_9 and C_{10}. The Fenske equation is used for this purpose (see Sections 8.2 and 9.6).
- *Step 2*   Using the component and material balance note if a significant amount of C_9 and C_{10} are allowed into the diesel (say, 3–5 mole % of the diesel). A flash point calculation should be carried to check that the diesel meets this specification. (See the appropriate section in the introductory chapter of this book.)
- *Step 3*   Tower pressure will be too high to permit total condensation of the overhead product because of its hydrogen and methane content. Also, the bottom of the tower will have a temperature of the diesel bubble point, which even at crude unit condition will be above 500 °F. An overhead reflux drum pressure of 5 psig would be reasonable in this case at a temperature of 100 °F.
- *Step 4*   Calculate the flash vaporization of the total overhead product at the conditions selected in Step 3. This calculation will fix the predicted gas stream and naphtha product stream from the reflux drum. The calculation is fully described in Section 8.7.
- *Step 5*   Assume an external reflux ratio. A ratio of 1.8:1 would be reasonable in this case and is the ratio of the *reflux to total overhead product* (i.e. liquid and gas products). Add the molal component of the reflux stream to the molal component of dew point calculation at the tower-top pressure. This pressure is obtained by adding the condenser and overhead piping pressure drop to that of the reflux drum. The dew point temperature is the tower-top temperature (see Section 6.6 for details of dew point calculation).
- *Step 6*   Break down the diesel TBP curve to about 5 mid-boiling point components. Calculate their weight and molal composition (See the appropriate section in the introductory chapter of this book). Add these to the C_9 and C_{10} components and calculate the mole fractions for all the components.
- *Step 7*   Set the tower-bottom pressure. This will be the tower-top pressure plus a pressure drop across the tower for the trays. Take 0.24 psi/tray for this. A column with 30 trays is reasonable for this service.
- *Step 8*   Calculate the bubble point at the pressure developed in Step 7 of the bottoms product provided by Step 6. This is the bottoms temperature.

An example calculation now follows.

Setting diesel hydrotreater stabilizer conditions

The feed to the diesel hydrotreater stabilizer will be that calculated in Section 10.5. Develop component balance:

	Feed					Overhead*			Product		
	Moles/h	MW	lb/h	lb/gal	gal/h	Moles/h	lb/h	gal/h	Moles/h	lb/h	gal/h
H_2	0.27	2	—	—	—	0.27	—	—			
C_1	6.40	16	102	—	—	6.40	102	—			
C_2	14.65	30	440	296	149	14.65	440	149			
H_2S	12.50	34	425	6.55	65	12.50	425	65			
C_3	17.11	44	753	4.22	178	17.11	753	178			
iC_4	4.08	58	237	4.69	51	4.08	237	51			
nC_4	5.21	58	302	4.87	63	5.21	302	63			
C_5s	3.03	72	218	5.24	42	3.03	218	42			
C_6	8.13	86	699	5.54	126	8.13	699	126			
C_7	9.07	100	907	5.74	158	9.07	907	158			
C_8	9.39	114	1070	5.89	182	9.39	1070	182			
C_9	4.20	128	538	6.01	90	3.99	511	85	0.21	27	5
C_{10}	5.0	142	710	6.11	116	2.69	382	63	2.31	328	53
Diesel	223.4	314	70 078	7.28	9626				223.4	70 078	9626
Total	322.44	237	76 479	7.05	10 846	96.52	6046	1162	225.92	70 433	9684

*Check relative split.

Allow 90 mole % C_{10} into bottom
5% mole C_9 into bottom
LT Key C_9 let 95% to distillate $= 3.990$
HY Key C_{10} let this be X moles to distillate
30 actual trays at, say 65% efficient $= 19.5$ theor. $= N_{M+1} = 12$
$\phi = 1.25$

$$N_{M+1} = \text{Log}\left[\left(\frac{\text{LT Key}}{\text{HY Key}}\right)_D \cdot \left(\frac{\text{HY Key}}{\text{LT Key}}\right)_W\right] \div \text{Log}\phi$$

$$12 = \text{Log}\left[\left(\frac{3.99}{X}\right) \cdot \left(\frac{4.75 - 70}{0.21}\right)\right] \div 0.0969$$

$$1.163 = \text{Log}\left[\frac{(18.95 - 3.99X)}{0.21X}\right]$$

$$14.55 = \left[\frac{18.95 - 3.99X}{0.21X}\right]$$

$$3.056X = 18.95 - 3.99X$$

$$X = \frac{18.95}{7.046} = 2.69 \text{ moles}$$

Calculate reflux drum split

Set pressure at 5 psig and temperature at 100 °F.
Carry out flash vaporization calculation.

	Moles feed X_F	20 psia K100 °F	Trial 1 $V/L = 1.0$		Trial 2 $V/L = 5.0$		Final $V/L = 1.8$		Vapour		
			V/LK	$L = \dfrac{X_F}{1+(V/L)K}$	V/LK	$L = \dfrac{X_F}{1+(V/L)K}$	V/LK	$L = \dfrac{X_F}{1+(V/L)K}$	Liquid Moles/ (lb/h) h		lb/h
H_2	0.27	960	960	Nil	—	Nil	—	—	—	0.27	0.5
C_1	6.40	140	140	0.05	700	0.01	252	0.03	0.5	6.37	101.9
C_2	14.65	27.0	27.0	0.52	135	0.11	49	0.29	8.7	14.36	430.8
H_2S	12.50	22.0	22.0	0.54	110	0.11	40	0.30	10.2	12.20	414.8
C_3	17.11	7.80	7.8	1.94	39	0.43	14	1.14	50.2	15.97	702.7
iC_4	4.08	3.20	3.2	0.97	16	0.24	5.8	0.60	34.8	3.48	201.8
nC_4	5.21	2.40	2.4	2.17	12	0.40	4.3	0.98	56.8	4.23	245.3
C_5	3.03	0.84	0.84	1.65	4.2	0.58	1.51	1.21	87.1	1.82	131.0
C_6	8.13	0.260	0.26	6.45	1.3	3.53	0.47	5.53	475.6	2.60	223.6
C_7	9.07	0.090	0.90	8.32	0.45	6.26	0.162	7.81	781.0	1.26	126.0
C_8	9.39	0.034	0.034	9.08	0.17	8.03	0.061	8.65	1008.9	0.54	61.6
C_9	3.99	0.012	0.012	3.94	0.06	3.76	0.022	3.90	499.2	0.09	11.5
C_{10}	2.69	0.005	0.005	2.68	0.25	2.62	0.009	2.67	379.1	0.02	2.8
Total	96.52			38.31		26.08		33.31	3392.1	63.21	2654.3
			Calc. $V/L = 1.5$		Calc. $V/L = 2.7$		Calc. $V/L = 1.89$				

Close enough.

Calculate tower top temperature at 10 psig = 25 psia: assume 1.8:1 external reflux

	Moles/h prod.	Moles/h reflux	Total moles/h	Mole frac. Y	Trial 1 at 280 °F K	$X = Y/K$	Liquid weight factor	Volume factor	
H_2	0.27		0.27	—	—	—	—	—	Mol. wt Liquid
C_1	6.40	0.16	6.56	0.024	140	—	—	—	= 122.8
C_2	14.65	1.51	16.16	0.060	54.00	0.001	0.03	0.01	°API = 52.6
H_2C	12.50	1.56	14.06	0.052	41.00	0.001	0.03	0.005	lb/gal = 6.4
C_3	17.11	5.95	23.06	0.085	23.00	0.004	0.18	0.043	
iC_4	4.08	3.13	7.21	0.027	14.00	0.002	0.12	0.026	
nC_4	5.21	5.11	10.32	0.038	11.00	0.003	0.17	0.035	
C_5	3.03	6.31	9.34	0.035	5.60	0.006	0.43	0.082	
C_6	8.13	28.84	36.97	0.137	2.90	0.047	4.04	0.729	
C_7	9.07	40.73	49.80	0.184	1.50	0.123	12.30	2.143	
C_8	9.39	46.16	55.55	0.206	0.80	0.258	29.41	4.993	
C_9	3.99	20.34	24.33	0.090	0.42	0.214	26.53	4.414	
C_{10}	2.69	13.94	16.63	0.062	0.16	0.388	55.10	7.569	
Total	96.52	173.74	270.26	1.000		1.045	128.34	20.049	

Trial is close enough.

Temperature will 282 °F

Calculate tower bottom temperature at 15 psig = 30 psia

From Figure 10.2

	Vol. %	lb/gal	Weight factor	Mol. wt	Moles factor	% mole
Comp. 1 BPT 592	10	7.051	70.51	258	0.273	0.118
Comp. 2 BPT 616	10	7.106	71.06	278	0.260	0.112
Comp. 3 BPT 628	10	7.141	71.41	287	0.257	0.111
Comp. 4 BPT 640	20	7.163	143.26	287	0.499	0.215
Comp. 5 BPT 668	50	7.435	371.76	361	0.029	0.4444
Total	100	7.28	728.00	314	2.318	1.000

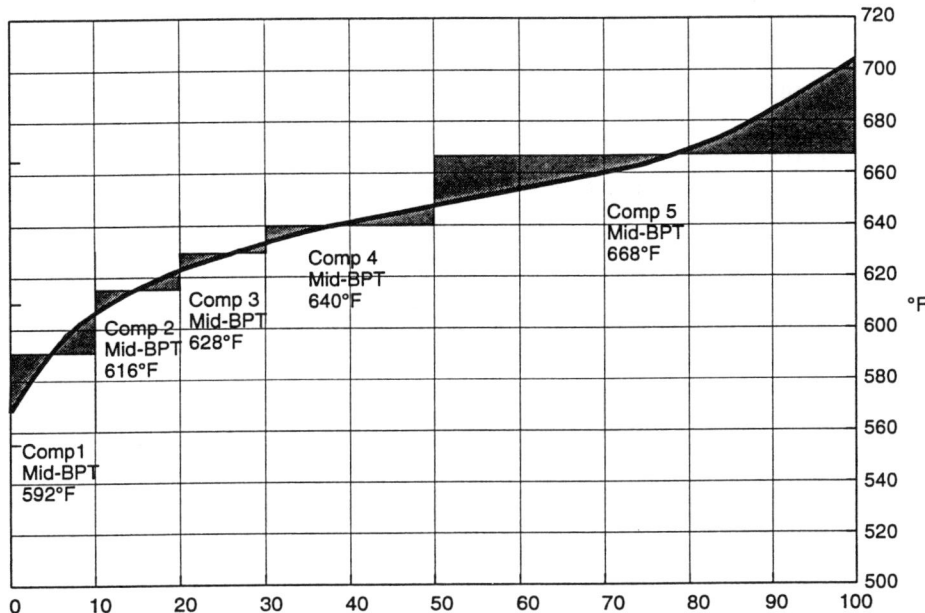

Figure 10.2. Diesel component breakdown

Calculate bubble point of bottoms at 30 psia

	Mol. frac. X	1st trial at 700 °F		Weight factor	Vol. factor	
		K	$Y = KX$			
C_9	0.001	18.7	0.19	2.43	0.40	Mol wt of
C_{10}	0.010	14.0	0.140	19.88	3.25	vapour = 278.1
						lb/gal of
BPT 592	0.116	1.5	0.174	44.89	6.37	vapour = 7.14
BPT 616	0.111	1.2	0.133	36.31	5.11	°API = 33.5
BPT 628	0.110	1.0	0.110	30.58	4.28	
BPT 640	0.213	0.93	0.198	56.83	7.93	
BPT 668	0.439	0.67	0.294	106.13	14.27	
Total	1.000		1.068	297.05	41.61	

Close enough.

Actual temperature = 705 °F

Feed to enter at about 430 °F and 27 psia

Note: At this bottoms temperature a fired heater must be used as a reboiler.

10.8 Calculating the Loadings on the Top and Bottom Trays of the Diesel Stabilizer

This calculation is made to determine the vapour and liquid loadings on the two critical trays in a diesel stripper tower. The result of this calculation may be used in the design of the tower to fix the tower diameter or, in an operating environment, to determine the column's trays performance and loadings relative to flood conditions (see also Sections 6.4 and 8.8 for details of this calculation). The following steps are used in this calculation:

- *Step 1* Using the column conditions developed in Section 10.7, calculate the overhead condenser duty by heat balance.
- *Step 2* A second heat balance over the top tray using the calculated condenser duty is made. In this balance the unknown is the overflow liquid from the top tray. The heat in and heat out values are used to equate and solve for the weight per hour of overflow.
- *Step 3* Adding the weight per hour of overhead product to the overflow calculated in Step 2 gives the vapour to the top tray from tray 2 in pounds per hour.
- *Step 4* Now the composition of the overflow liquid is that in equilibrium with the overhead material used to determine the tower top temperature. This is then the X column of the dew point calculation given in Section 10.7. Using this breakdown and the known product composition, the mol. weight, moles per hour, gravity, etc. of the vapour and liquid stream are determined. Thus the vapour loading in actual ft^3/s and its density in lb/ft^3 is fixed. The liquid load on the tray in terms of hot gallons and hot ft^3/h is also fixed together with its density at tray condition.
- *Step 5* Set a feed temperature and pressure from plant data. Calculate the flash vaporization of the feed under these conditions of temperature and pressure. This gives the amount of liquid and vapour in the mixed phase feed entering the column. Apply enthalpy values to these phases and thus obtain the total feed enthalpy in BTU/h.
- *Step 6* Using the feed enthalpy calculated in Step 5, carry out an overall heat balance over the tower. The reboiler duty in BTU/h is obtained as the difference between 'heat in' and 'heat out' in the balance.
- *Step 7* This calculated reboiler duty is used in the heat balance over the bottom of the tower. The unknown in this case is the pounds per hour of the stripping vapour leaving the reboiler. This value is obtained by equating the heat in with the heat out in the balance. Note that the feed to the reboiler is the liquid from the bottom tray, which is the stripping vapour and bottom product. The composition and properties of the stripping vapour is the vapour in equilibrium with the liquid in the bubble point calculation given in Section 10.7. The temperature of the bottom tray is taken as 15 °F below bottom temperature.
- *Step 8* From Step 7 the vapour to the bottom tray is found in terms of pounds per hour. It can also be calculated in terms of moles per hour from the bubble point calculation in Section 10.7 and in terms of hot gallons per hour or cubic feet per hour. Added to the product values, this gives the liquid loadings on the tray.

An example calculation now follows.

Example Calculation

Note: the amount of stripout vapour calculated here is more than adequate. This could be reduced to a more reasonable value (say, 1 : 1 moles stripout to bottom product) by increasing feed temperature.

Calculate the overhead condenser duty by tower top heat balance.

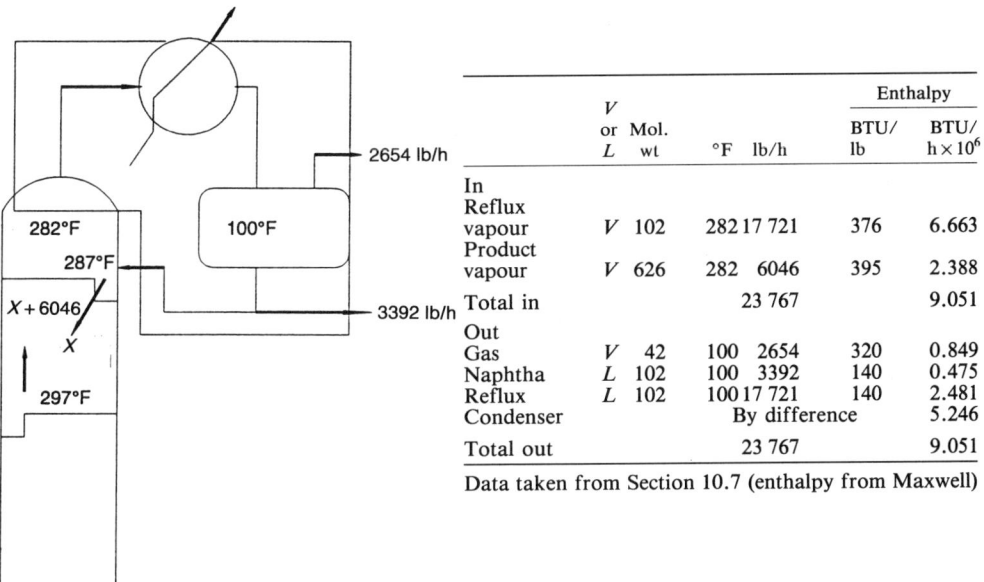

	V or L	Mol. wt	°F	lb/h	Enthalpy BTU/ lb	Enthalpy BTU/ h × 10^6
In						
Reflux vapour	V	102	282	17 721	376	6.663
Product vapour	V	626	282	6046	395	2.388
Total in				23 767		9.051
Out						
Gas	V	42	100	2654	320	0.849
Naphtha	L	102	100	3392	140	0.475
Reflux	L	102	100	17 721	140	2.481
Condenser				By difference		5.246
Total out				23 767		9.051

Data taken from Section 10.7 (enthalpy from Maxwell)

Using the calculated condenser duty determine overflow from top tray by heat balance over the tray:

	V or L	Mol. wt	°F	lb/h	Enthalpy BTU/lb	Enthalpy BTU/h × 10^6
In						
Reflux vapour ex tray 2	V	123	297	X	380	$380X$
Product vapour ex tray 2	V	63	297	6046	400	2.418
Total in				$6046 + X$		$2.418 + 380X$
Out						
Total product from drum	V + L	—	100	6046	—	1.324
Overflow ex tray 1	L	123	287	X	252	$252X$
Condenser						5.246
Total out				$6046 + X$		$6.57 + 252X$

$$(2.418 \times 10^6) + 380X = (6.57 \times 10^6) + 252X$$
$$X = \frac{4\,152\,000}{128} = 32\,438 \text{ lb/h}$$

Top tray loading will be as follows:

Vapour (per hour Basis)

	lb	Moles
Product	6046	96.52
Reflux	32 438	264.15
Total	38 484	360.67
Temp.	297 °F	
Press.	25 psia	

$$\text{ACFS} = \frac{360.67 \times 378 \times 14.7 \times 575}{3600 \times 520 \times 25}$$

$$= 32.4$$

$$\rho_v \, \text{lb/ft}^3 = 0.33$$

Liquid

lb/h	=	32 428
lbs/gal	=	6.4 at 60 °F
gal/h at 60	=	5068
gal/h at 287	=	5813 (exp. factor = 1.147)
CFH	=	58 134 × 0.1337
	=	777
ρ_L (lb/ft³)	=	41.75 at 287 °F

Calculate reboiler duty and loading over bottom tray

Enthalpy in with the feed using flash vaporization of the feed at 380 °F and 27 psia

			1st trial $V/L=0.5$		2nd trial $V/L=1.0$		3rd trial $V/L=0.3$		Liquid	
Comp.	Moles/h	K	V/LK	$L=\dfrac{X_F}{1+(V/L)K}$	V/LK	$L=\dfrac{X_F}{1+(V/L)K}$	V/LK	$L=\dfrac{X_F}{1+(V/L)K}$	lb/h	gal/h
H_2	0.27	350	175	Nil	350	Nil	105	Nil	Nil	—
C_1	6.40	125	62.5	0.1	125	0.05	37.5	0.17	2.7	—
C_2	14.65	56	28.4	0.5	56	0.25	16.8	0.82	24.6	8.3
H_3S	12.50	45	22.5	0.5	45	0.27	13.5	0.86	29.2	4.5
C_3	17.11	28	14	1.1	28	0.59	8.4	1.82	80.1	19.0
iC_4	4.08	18	9	0.4	18	0.21	5.4	0.64	37.1	7.9
nC_4	5.21	16	8	0.6	16	0.31	4.8	0.90	52.2	10.7
C_5	3.03	9.5	4.75	0.5	9.5	0.29	2.65	0.79	56.9	10.9
C_6	8.13	6.0	3.00	2.0	6.0	1.16	1.80	2.90	249.4	45.0
C_7	9.07	3.5	1.75	3.3	3.5	2.02	1.05	4.42	442.0	77.0
C_8	9.39	2.2	1.10	4.5	2.2	2.93	0.66	5.66	645.2	109.5
C_9	4.20	1.4	0.70	2.5	1.4	1.75	0.42	2.96	378.9	63.0
C_{10}	5.00	0.9	0.45	3.4	0.9	2.63	0.27	3.94	559.5	91.6
NB 592	26.36	0.03	0.015	26.0	0.03	25.59	0.009	26.12	6739.0	955.8
NB 616	25.02	0.02	0.01	25.0	0.02	24.53	0.006	24.87	6789.5	955.5
NB 628	24.80	0.01	0.005	24.7	0.01	24.55	0.003	24.73	6874.9	962.7
NB 640	48.03	—	—	48.03	—	48.03	—	48.03	13 784.6	1924.4
NB 668	99.91	—	—	99.19	—	99.19	—	99.19	35 807.6	4816.1
Total	322.44			242.32		234.35		248.82	72 553.4	10 061.9
Calculated V/L				0.33		0.376		0.30		

lb/gal liquid $= 7.21 = 31.9\,°API$

Mol. wt vapour $= \dfrac{3925.6\ \text{lb/h}}{73.62} = 53.3$

Enthalpy in feed at 380 °F

Liquid $= 72\,553.4 \times 188\ \text{BTU/lb}\qquad = 13.640 \times 10^6\ \text{BTU/h}$
Vapour $= 3925.6 \times (450-100)\ \text{BTU/lb} = \underline{1.374 \times 10^6\ \text{BTU/h}}$
$$\text{Total} = 15.014 \times 10^6\ \text{BTU/h}$$

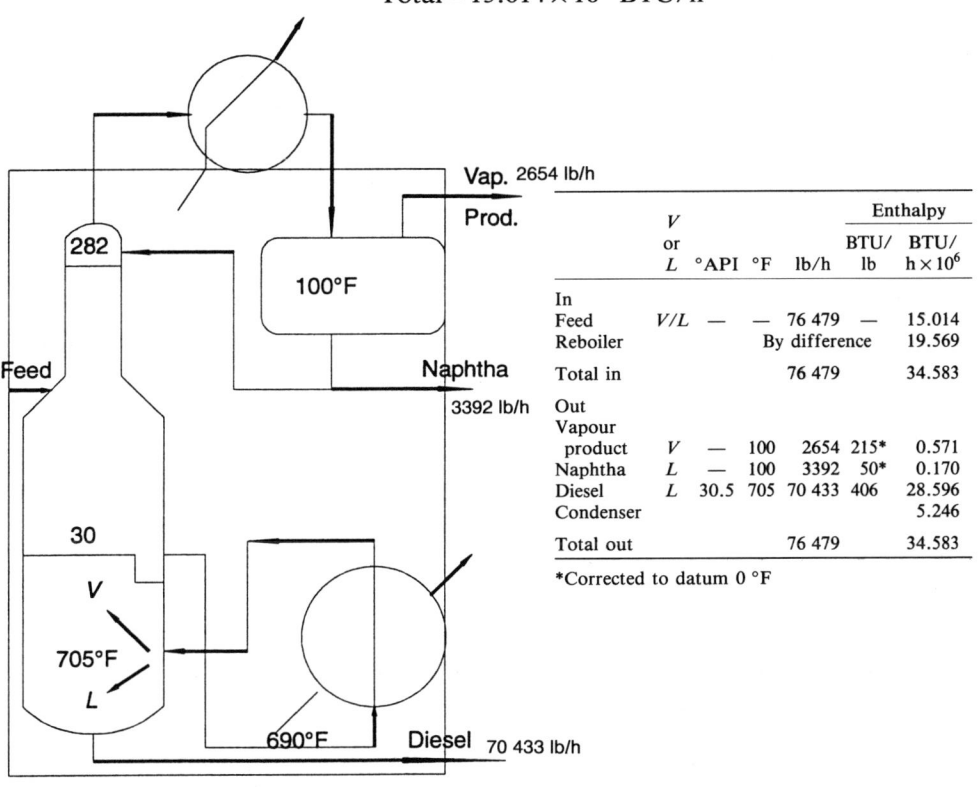

	V or L	°API	°F	lb/h	Enthalpy BTU/ lb	Enthalpy BTU/ h × 10⁶
In						
Feed	V/L	—	—	76 479	—	15.014
Reboiler			By difference			19.569
Total in				76 479		34.583
Out						
Vapour product	V	—	100	2654	215*	0.571
Naphtha	L	—	100	3392	50*	0.170
Diesel	L	30.5	705	70 433	406	28.596
Condenser						5.246
Total out				76 479		34.583

*Corrected to datum 0 °F

Calculate balance over reboiler

	V or L	°API	°F	lb/h	Enthalpy BTU/lb	Enthalpy BTU/h × 10⁶
In						
Stripping vapour	L	33.5	690	V	398	398 V
Product liquid	L	30.5	690	70 433	395	27.821
Reboiler						19.569
Total in						$47.39 + 398\ V$
Out						
Stripping vapour	V	33.5	705	V	492	492 V
Product	L	30.5	705	70 433	406	28.596
Total out						$28.596 + 492\ V$

$$V = \frac{18\,794\,000}{94} = 199\,936\ \text{lb/h}$$

Bottom tray loading

Vapour

lb/h	=	199 936 (This is more than adequate stripping vapour)
Mol. wt (Section 10.7)	=	278.1
Moles/h	=	719 (at 705 °F 30 psia)

$$\text{ACFS} = \frac{719 \times 378 \times 14.7 \times 1165}{3600 \times 520 \times 30}$$

$$= 82.9$$

ρ_v lb/ft^3 = 0.67

Liquid

Stripper liquid	Product	Total
199 936 lb/h	70 433 lb/h	270 369 lb/h
28 002 gal/h at 60 °F	9684	37 686 gal/h

lb/gal of tray liquid at 60 °F	=	7.17
Expansion factor	=	1.435
lb/gal of tray liquid at 690 °F	=	5.0
Hot gal/h	=	54 074
Hot ft^3/h	=	7230
ρ_L lb/ft^3 at 690 °F	=	37.4

10.9 Predicting the Heat Exchanger Performance in the Naphtha Hydrotreater

The naphtha hydrotreater effluent leaves the reactor in the vapour phase. It begins to condense in the first effluent feed exchanger and the condensation and cooling continues up to the flash drum. This section shows the calculation method to develop the effluent enthalpy curve for use in evaluating the heat exchanger performance or, in the case of design work, the exchanger duties and specifications. The calculation proceeds with the following steps:

- *Step 1* Establish the temperature and pressure in and out of the exchanger from plant data or in the case of a design from a layout configuration and assigned approach temperatures.
- *Step 2* Calculate the dew point of the reactor effluent. Calculate also its enthalpy in BTU/h.
- *Step 3* Calculate the flash vaporization at the exit temperature and pressure condition of each exchanger.
- *Step 4* Using the values for the vapour and liquid obtained in Step 3 calculate the total enthalpy in the stream.

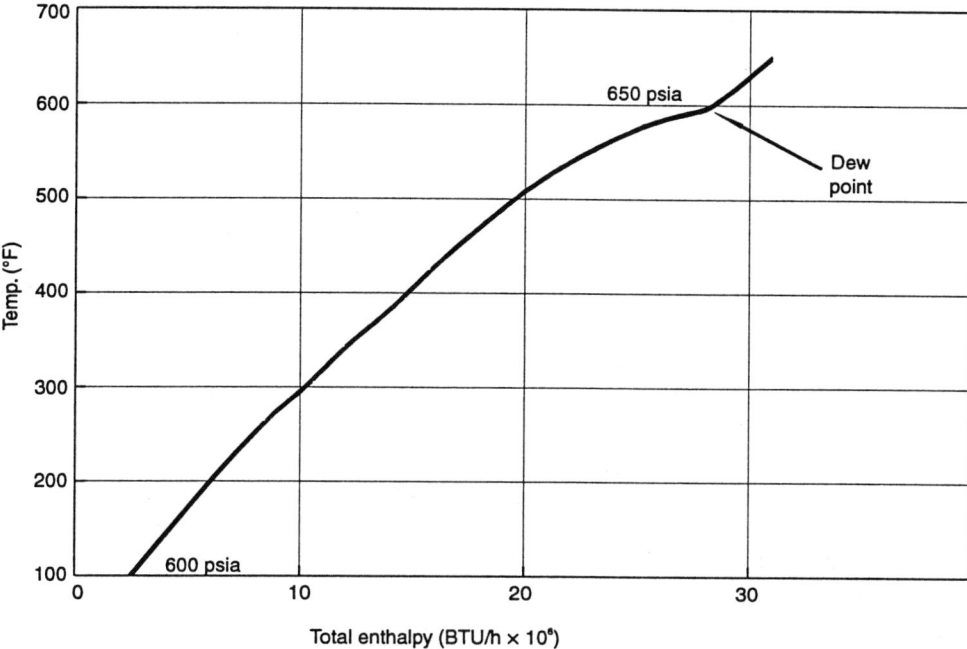

Figure 10.3. Enthalpy curve for reactor effluent

● *Step 5* Plot the enthalpy from Step 4 against temperature. An example is given in Figure 10.3.

● *Step 6* For shell and tube exchangers obtain the surface area from equipment data sheets, and the inlet/outlet temperatures from plant data. Using the temperature enthalpy curve (Step 5) calculate the overall heat transfer coefficient U by the equation

$$Q = \frac{U}{A.\Delta T_M}$$

where

Q is exchanger duty (BTU/h)

ΔT_M is log mean temperature difference (°F)

A is surface area (ft²)

Compare the result with typical U-values given in Table 6.2. In the case of a preliminary design use the values for U from Table 6.2 to calculate the surface area.

● *Step 7* Calculate the air condenser power requirement using the method given in Section 8.9. The duty for this section is read from the enthalpy curve. Compare with rated values.

An example calculation now follows.

Example Calculation

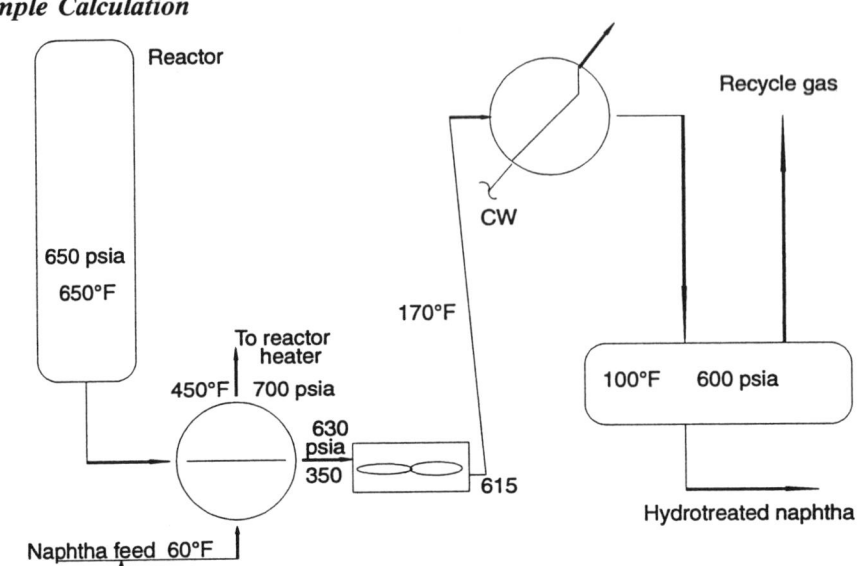

Assume total flow from the reactor is as follows:

	Moles/h	Mole frac.	lb/h	lb/gal	gal/h	
H_2	240	0.280	480	—		
C_1	38.1	0.045	610	—		
C_2	12.1	0.014	363	2.97	122	°API of effluent = 61.6
C_3	5.7	0.007	251	4.23	59	Mol. wt effluent = 74.2
iC_4	0.9	0.001	52	4.69	11	
nC_4	0.5	0.001	29	4.87	6	
C_5	1.2	0.001	86	5.23	16	
C_6	124.9	0.146	10 741	5.54	1939	
C_7	140.9	0.165	14 090	5.74	2455	
NP260	99.8	0.117	11 377	6.18	1841	
NP300	99.8	0.117	12 575	6.37	1974	
NP340	60.4	0.071	8214	6 46	1272	
NP382	29.8	0.035	4530	6.56	691	
Total	854.1	1.000	63 398	6.1	10 386	

Calculate dew point at 630 psia

		Trial 1, 550 °F		Trial 2, 600 °F	
	Mol. frac. *Y*	*K*	*X = Y/K*	*K*	*X = Y/K*
H_2	0.280	12	0.023	8.8	0.032
C_1	0.045	4.8	0.009	5.5	0.008
C_2	0.014	3.4	0.004	3.5	0.004
C_3	0.007	2.8	0.003	2.80	0.003
iC_4	0.001	2.2	Neg.	2.30	Neg.
nC_4	0.001	2.0	0.001	2.40	Neg.
C_5	0.001	1.4	0.001	2.90	Neg.
C_6	0.146	1.0	0.146	1.73	0.084
C_7	0.165	0.82	0.201	1.23	0.134
NP260	0.117	0.524	0.223	0.714	0.164
NP300	0.117	0.333	0.351	0.499	0.234
NP340	0.071	0.222	0.320	0.333	0.213
NP382	0.035	0.156	0.224	0.238	0.147
	1.000		1.498		1.023

Close enough.

Calculate flash vaporization at 350 F and 630 psia

	Moles/h XF	K 350 °F 630 psia	1st trial $V/L=0.5$		2nd trial $V/L=1.0$		Final $V/L=0.58$		Liquid	
			V/LK	$L=\dfrac{X_F}{1+(V/L)K}$	V/LK	$L=\dfrac{X_F}{1+(V/L)K}$	V/LK	$L=\dfrac{X_F}{1+(V/L)K}$	lb/h	gal/h
H_2	240	15	7.5	28.2	15	15	8.7	24.74	49.5	—
C_1	38.1	3.5	1.75	13.85	3.5	8.5	2.03	12.57	201.1	—
C_2	12.1	2.7	1.35	5.15	2.7	3.3	1.57	4.71	141.3	47.6
C_3	5.7	1.8	0.90	3.00	1.8	2.0	1.04	2.79	122.8	29.0
iC_4	0.9	1.3	0.65	0.55	1.3	0.4	0.75	0.51	24.6	6.3
nC_4	0.5	1.1	0.55	0.32	1.1	0.2	0.64	0.30	17.4	3.6
C_5	1.2	0.8	0.40	0.86	0.8	0.7	0.46	0.82	59.0	11.3
C_6	124.9	0.5	0.25	99.92	0.5	83.3	0.29	96.82	8326.5	1503.0
C_7	140.9	0.32	0.16	121.47	0.32	106.7	0.19	118.40	11 840.0	2062.7
NP260	99.8	0.079	0.04	95.96	0.079	92.49	0.046	95.41	10 876.7	1760.0
NP300	99.8	0.046	0.023	97.56	0.046	95.41	0.027	97.18	12 244.7	1922.2
NP340	60.4	0.027	0.014	59.57	0.027	58.81	0.016	59.45	8085.2	1251.6
NP382	29.8	0.016	0.008	29.56	0.016	29.33	0.009	29.53	4488.6	684.2
Total	854.1			555.97		496.14		543.23	56 482.4	9281.5

Liquid °API   =   62.6
Vapour lb/h   =   6915.6
   moles/h   =   310.9   MW = 22.24
   gal/h   =   1104.5   °API = 56.7

The flash calculation for the effluent stream is calculated in a similar manner for 610 psia and 170 °F and again at 100 °F and 600 psia. These are plotted and given in Figure 10.3. The vapour/liquid enthalpy at these temperatures are:

Temp.	Total enthalpy BTU/h $\times 10^6$
650 °F	30.81
350 °F	12.749
170 °F	5.82
100 °F	3.192

Calculating performance of the effluent/feed exchanger

From equipment data sheet the surface area of the exchanger is 1660 ft^2. Then overall heat transfer coefficient U is

$$\frac{Q}{A\Delta T_M} = U$$

$$Q = (30.81 - 12.75) \times 10^6$$
$$= 18.06 \times 10^6 \text{ BTU/h}$$
$$\Delta T_M = 242 °F$$
$$\text{Then } U = \frac{18.06 \times 10^6}{1660 \times 242}$$
$$= 45 \text{ BTU/h ft}^2 °F$$

Table 6.2 gives a value of 38 BTU/h ft^2 °F, which is satisfactory.

The air condenser is checked for power usage similar to the calculation given in Section 8.9. The final water trim cooler coefficient is checked with Table 6.2 in the same manner as the effluent/feed exchanger.

10.10   Evaluating the Performance of the Diesel Hydrotreater Reboiler

This calculation follows the steps given in Section 6.16:

- *Step 1*   Obtain details of the heater, in particular its layout and tube surface area.
- *Step 2*   From plant data obtain the duty of the heater. This is by an overall heat balance over the tower (see Section 10.8 for this calculation).
- *Step 3*   Divide this duty in BTU/h by the tube area obtained in Step 1 to obtain the heat flux. For an acceptance performance this should be within the ranges given in Section 6.16.
- *Step 4*   Obtain from plant data the quantity and characteristics of the fuel fired. Using the heat of combustion in Appendix 1, calculate the heat fired in BTU/h.
- *Step 5*   Calculate the heater's thermal efficiency from the equation:

$$\text{Efficiency} = \frac{\text{Heat absorbed} \times 100}{\text{Heat released}}$$

Check that this is within 70–80%, which is acceptable.

An example calculation now follows.

Example Calculation

Duty of the heater   = 19.569×10^6 BTU/H (see Section 10.8)

This heater is vertical single row fired from bottom (Figure 6.11 in Section 6.16). The coil area from data sheets is 1850 ft^2

Then heater flux   = $\dfrac{119.569 \times 10^6}{1850}$

=   10 578 BTU/h/ft^2

which is within the limits for this type of heater.
Plant data on fuel oil fired is 1495 lb/h.
This is 16 °API residue with a heating value of 17 580 BTU/lb (Table A1.1 in Appendix 1).

Heat released   = $1495 \times 17\,580$

=   26 282 100 BTU/h

Efficiency   = $\dfrac{\text{Heat absorbed}}{\text{Heat released}} \times 100$

= $\dfrac{19\,569\,000}{26\,282\,100} \times 100$

=   74.5%, which is satisfactory.

10.11   Estimating the Hydrogen Consumption for Olefin Saturation

This section gives a simple method for estimating the hydrogen consumption when treating cracked stocks. This method uses the correlation

$$\text{BR no.} = \frac{\%\ \text{olefins} \times 160}{\text{MW olefins}}$$

where the BR number is the bromine number of the stock. This can be obtained by lab analysis. The mol. weight of the olefins is estimated as 1.3 times the mol. weight of the paraffin with the same mid-boiling point of the cut.

The hydrogen consumption for saturation is taken at 6.5–8.0 SCF hydrogen per barrel for every unit of bromine number reduction that occurs in the process. Bromine reduction in hydrotreating is estimated as follows:

LIGHT cracked naphtha   90–100% reduction
FCCU cycle oil   12-unit reduction
Coker gas oil   40-unit reduction
Visbreaker gas oil   40–45-unit reduction

Use 6.5 SCF/BbL per bromine unit reduction for naphtha and light oil and 8.0 for the heavier feeds. An example calculation now follows:

Example Calculation

Consider a cracked naphtha 180–380 °F cut

°API = 51.0
% S = 0.89
PONA analysis given 30% olefins by vol.

Calculate bromine number from the following.

$$\text{BR no.} = \frac{\%\ \text{olefins} \times 160}{\text{MW olefins}}$$

Mol. wt of Olefins 1.3 times mol. weight paraffin with same boiling point as the cuts mid-BPT.

In this case mid-BPT is 200 °F = Heptane, mol. wt 100
Olefin mol. wt.   = 130
Then bromine no. $= \dfrac{30 \times 160}{130}$
                                          = 37

Olefin saturation is 6.5 SCF hydrogen/BbL per unit of bromine number reduction. With naphtha this is usually about 95%
Then hydrogen required $= 6.5 \times 37 \times 0.95$
                             SCF/BbL
                             = <u>228.5 SCF/BbL of feed</u>

11 FLUID CATALYTIC CRACKING

11.1 Process Description of a Fluid Catalytic Cracking Unit (FCCU)

The 'heart' of the process consists of a reactor vessel and a regenerator vessel interconnected to allow the transfer of spent catalyst from the reactor to the regenerator and of regenerated catalysts back to the reactor (Figure 11.1). The oil is cracked in the reactor section by exposing it to high temperatures and in contact with the catalyst. The heat for the oil cracking is supplied by the exothermic heat of reaction generated during the catalyst regeneration. This heat is transferred by the regenerated fluid catalyst stream itself. The oil streams (feed and recycle) are introduced into this hot catalyst stream en route to the reactor. Much of the cracking occurs in the dispersed catalysed phase along this transfer line or riser.

The final contact with the catalyst bed in the reactor completes the cracking mechanism. The vaporized cracked oil from the reactor is suitably separated from entrained catalyst particles by cyclones and routed to the recovery section of the unit. Here it is fractionated by conventional means to meet the product stream requirements. The spent catalyst is routed from the reactor to the regenerator after separation from the entrained oil. Air is introduced into the regenerator and the fluid bed of the catalyst. The air reacts with the carbon coating on the catalyst to form CO/CO_2. The hot and essentially carbon-free catalyst completes the cycle by its return to the

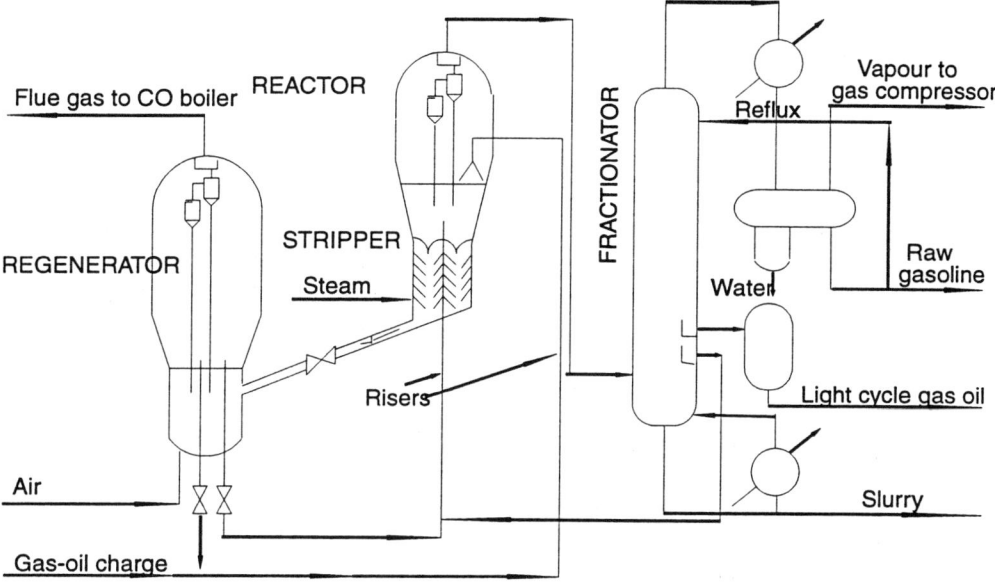

Figure 11.1. Fluid catalytic cracking

reactor. The flue gas leaving the regenerator is rich in CO. This stream is often routed to a specially designed steam generator where the CO is converted to CO_2 and the exothermal heat of reaction used for generating steam (the CO boiler).

Feedstocks to the FCCU are primarily in the heavy vacuum gas oil range. Typical boiling ranges are 640 °F (10%) to 980 °F (90%). This gas oil is limited in end point by maximum tolerable metals, although the new zeolite catalysts have demonstrated higher metals tolerance than the older silica-alumina catalyst. The process has considerable flexibility. Apart from processing the more conventional waxy distillates to produce gasoline and other fuel components, feedstocks ranging from naphtha to suitably pretreated residuum are successfully processed to meet specific product requirement.

11.2   Licensors' Correlations

The fluid catalytic cracker is usually a licensed process. Correlations and methodology are therefore proprietary to the licensor although certain data are divulged to clients under the licensor agreement. Such data are required by clients for proper operation of the unit, and may not be divulged to third parties without the licensor's express permission.

The purpose of this section is to list the type of data that would be made available to refinery personnel for operating purposes, i.e.:

(1) Method of calculating fresh catalyst addition rate
(2) Calculation of metals on equilibrium catalyst
(3) Data on 'apparent bulk density' including:

Particle density
Skeletal density
Surface area
Pore volume

(4) Catalyst evaluation by the following graphs:

Coke yield % wt	versus	conversion to $380 + /90$ LV%
C_2 yield % wt	versus	conversion to $380 + /90$ LV%
C_3 yield % wt	versus	conversion to $380 + /90$ LV%
C_4 yield & wt	versus	conversion to $380 + /90$ LV%
Adj. gasoline yield	versus	conversion to $380 + /90$ LV%
Gasoline selectivity	versus	conversion to $380 + /90$ LV%

(5) Calculation methods for:

Reactor/regenerator heat balance
Combustion heat of coke
Reactor heat calculations
Overall heat balance
UOP characterization 'K' factor calculations with data graphs

(6) Reactor temperature, pressure, catalyst/oil ratio versus yield and conversion
     graphs
(7) Residence time calculation
(8) Conversion of feed versus time graph gasoline yield % of feed versus time graph
(9) Regenerator severity—kinetics equation.

Effect of carbon level
Air distribution rate
Temperature and pressure

(10) Quality and condition of charge

These and others, including operating instructions, are required for the proper
operation of the unit. Most of the proprietary data, however, concern the
reactor/regenerator side of the process. The recovery side—that is, the equipment
required to produce the product streams from the reactor effluent—utilizes essentially
conventional techniques in their design and operating evaluation.

11.3 Prediction of Yields at Various Severities of Operation and Feed

This section gives a typical yield structure for various FCCU feedstock and severities
of operation. These were based on 100% zeolite catalyst processes and are included
here as a guide to conventional FCCU process capability. The curves are based on
processing Sassan crude. This is a light 34 API crude but containing low metals. The
feedstocks provided here are as follows:

Hydrotreated 607–948 °F gas oil (Figure 11.2)
Straight-run gas oil 503–959 °F (Figure 11.3)
Hydrocracker gas oil 750–930 °F (Figure 11.4)
Straight-run diesel 520–650 °F (Figure 11.5)
Hydrotreated diesel 520–650 °F (Figure 11.6)
Hydrocracker gas oil 380–650 °F (Figure 11.7)
Hydrocracker gas oil 650–750 °F (Figure 11.8)
Hydrotreated visbreaker gas oil 380–650 °F (Figure 11.9)
Visbreaker gas oil 380–650 °F (Figure 11.10)

11.4 Heavy Oil Cracking

Up to the late 1980s feedstocks to FCCU were limited by characteristics such as high
Condradson carbon and metals. This excluded the processing of the 'bottom of the
barrel' residues. Indeed, even the processing of vacuum gas oil feeds were limited to

Condradson carbon < 10 wt %
hydrogen content   > 11.2 wt %
metals N1 + V   < 50 ppm

During the late 1980s significant research and development breakthroughs produced
a catalytic process that can handle these heavy feeds and indeed some residues.

Feedstocks heavier than vacuum gas oil when fed to a conventional FCCU tend
to increase the production of coke and this in turn deactivates the catalyst. This is
mainly the result of:

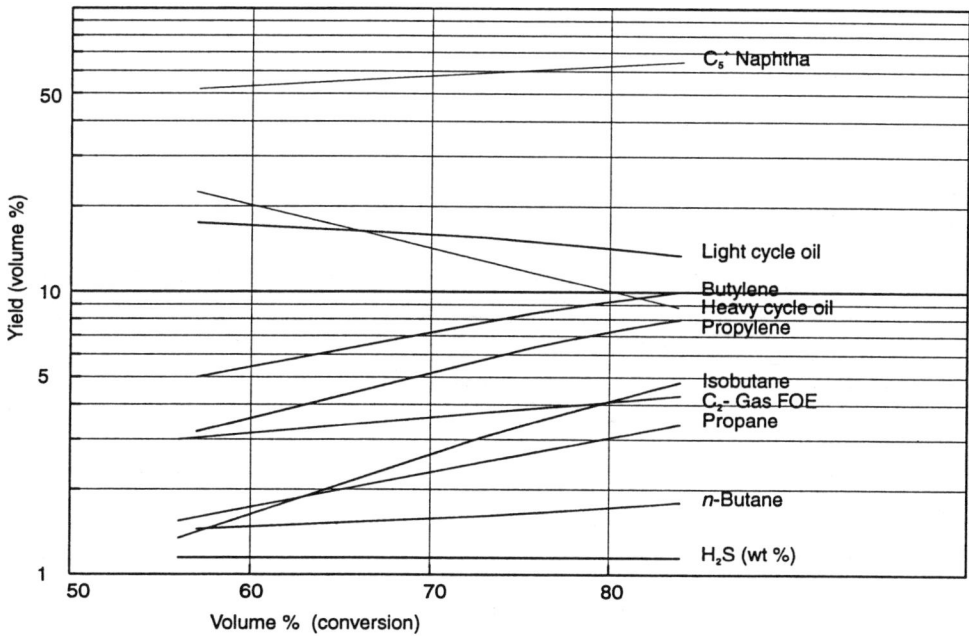

Figure 11.2. Hydrotreated gas oil feed: yield versus conversion: Sassan gas oil: 607–948 °F

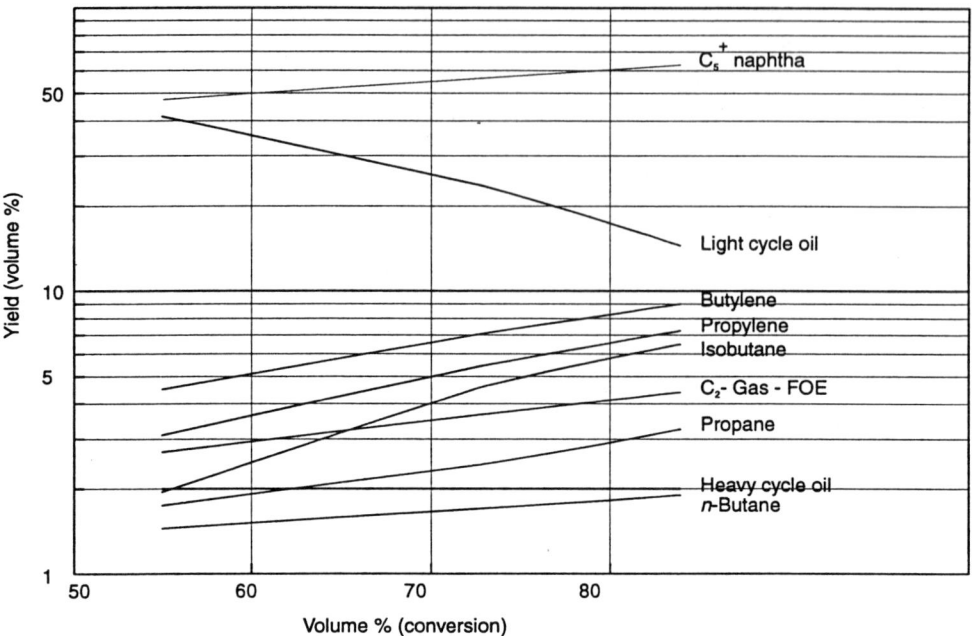

Figure 11.3. Yield versus conversion: Sassan gas oil: 503–959 °F: catalyst D + L – 40

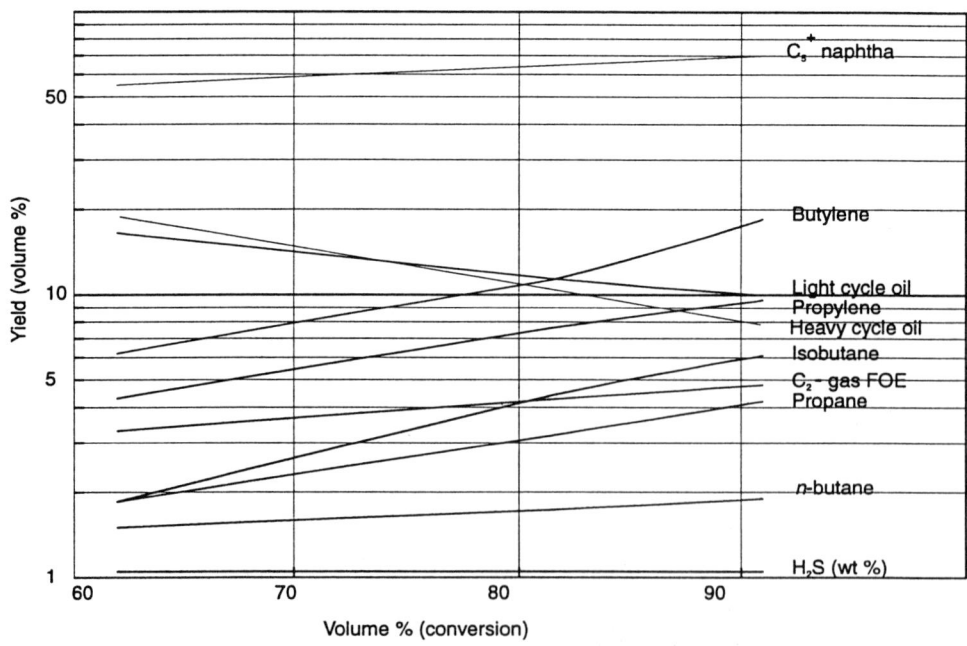

Figure 11.4. Yield versus conversion: hydrocracker gas oil: 750–930 °F

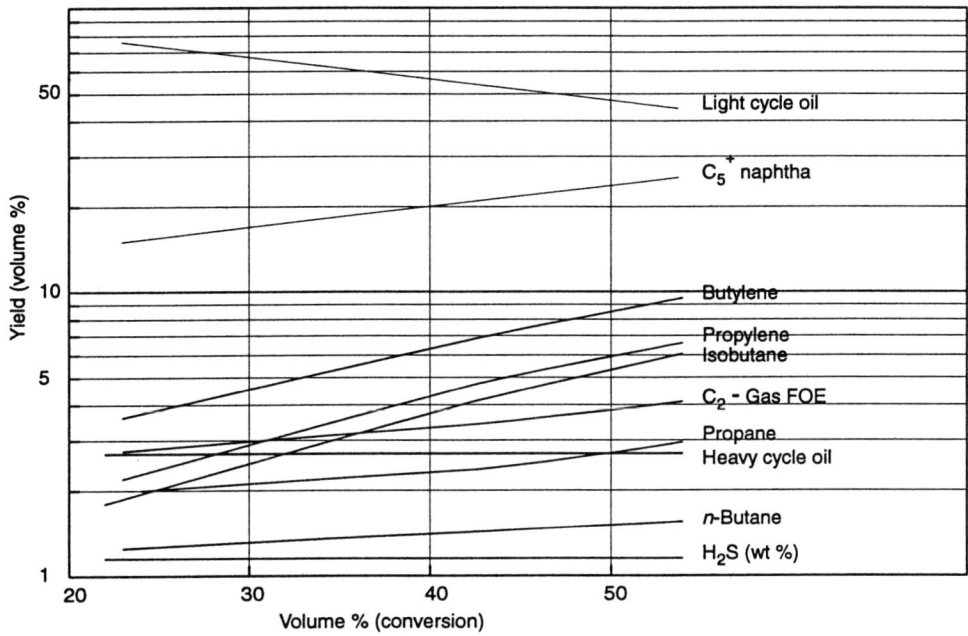

Figure 11.5. Yield versus conversion: straight-run diesel: 520–650 °F

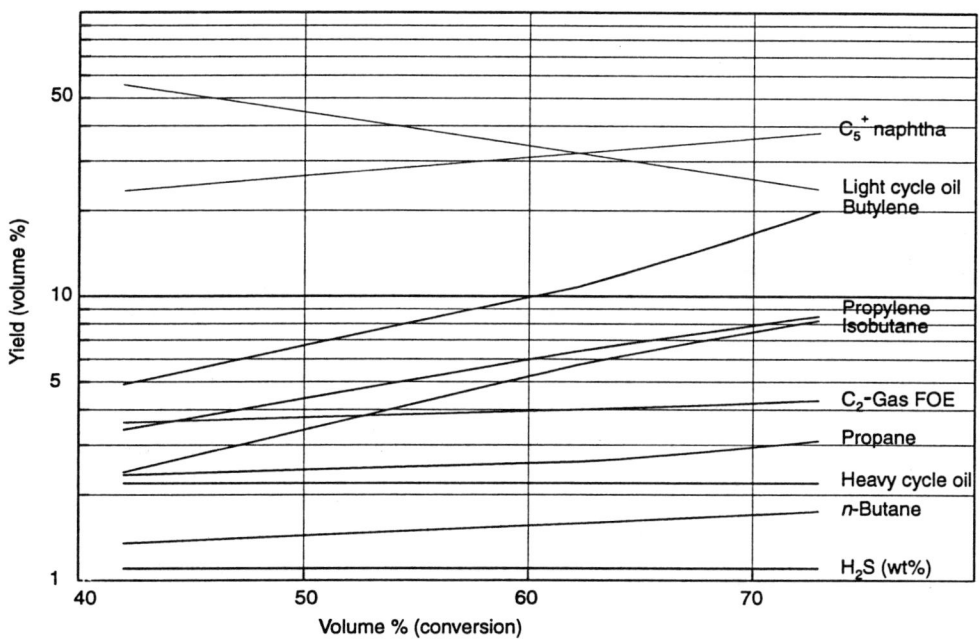

Figure 11.6. Yield versus conversion: hydrotreated gas oil feed: straight-run diesel: 520–650 °F

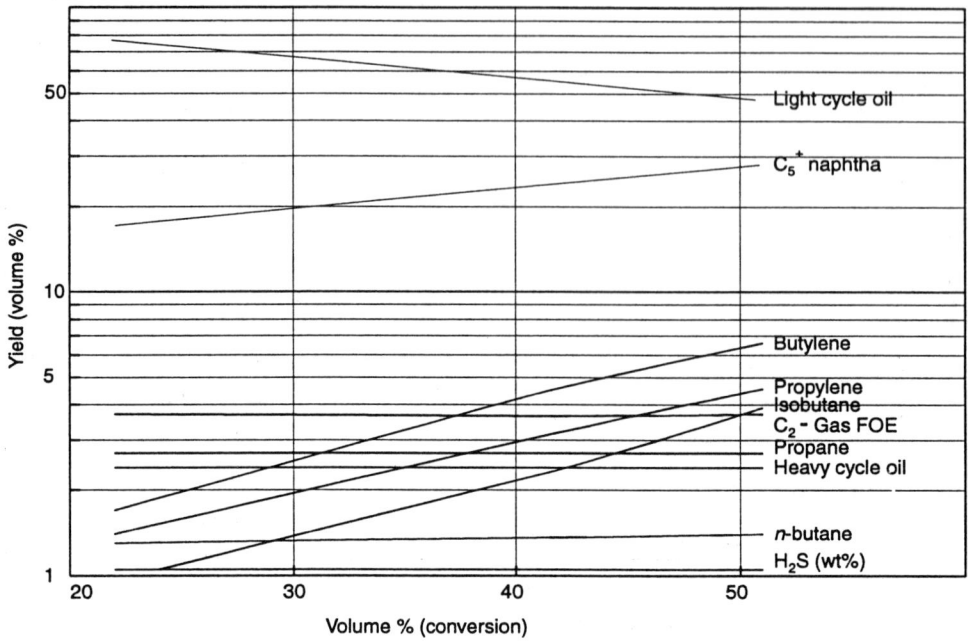

Figure 11.7. Yield versus conversion: hydrocracker gas oil: 380–650 °F

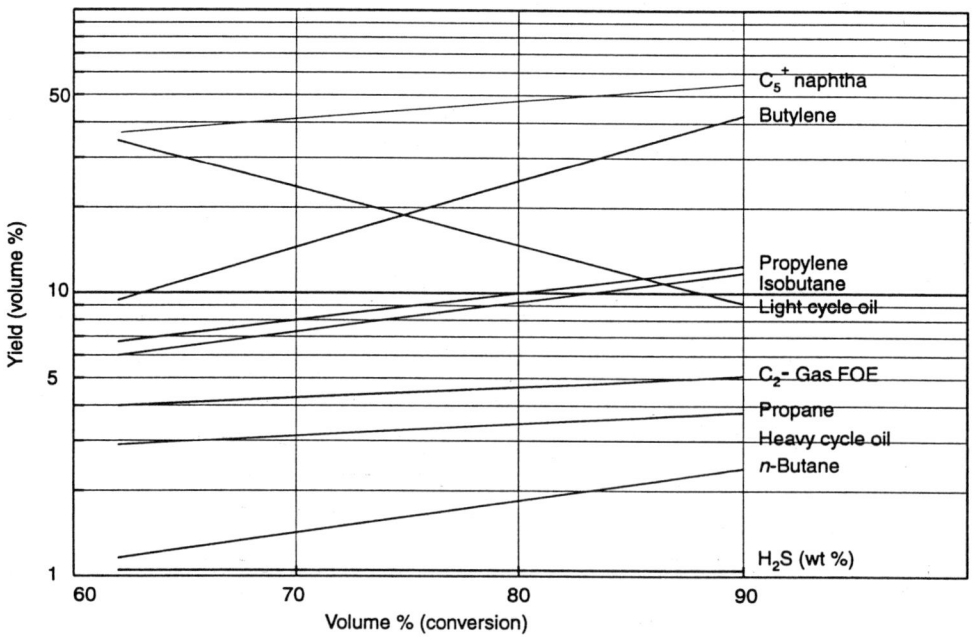

Figure 11.8. Yield versus conversion: hydrocracker gas oil: 650–750 °F

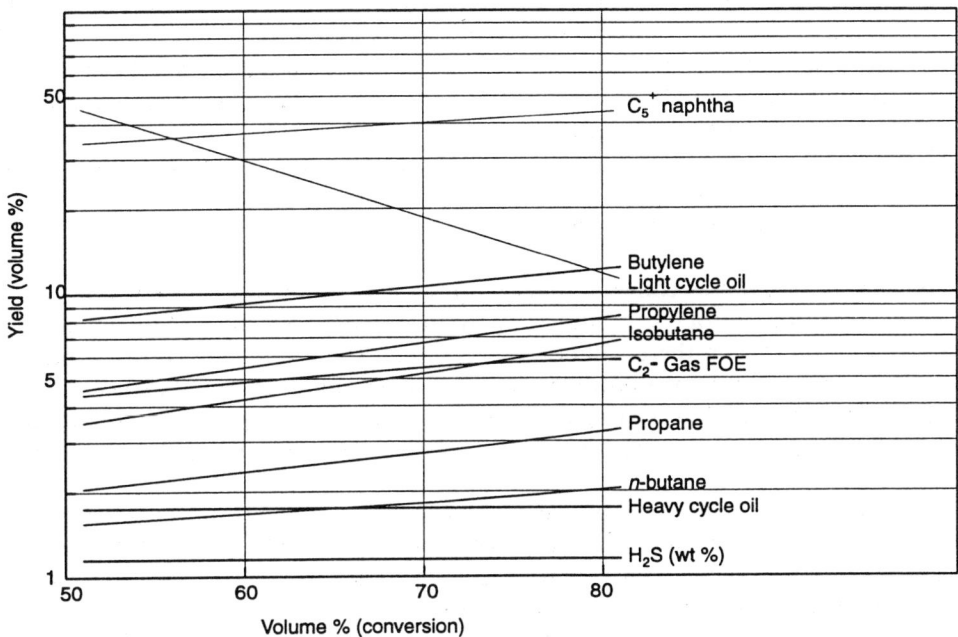

Figure 11.9. Hydrotreated gas oil feed: yield versus conversion: visbreaker gas oil: 380–650 °F

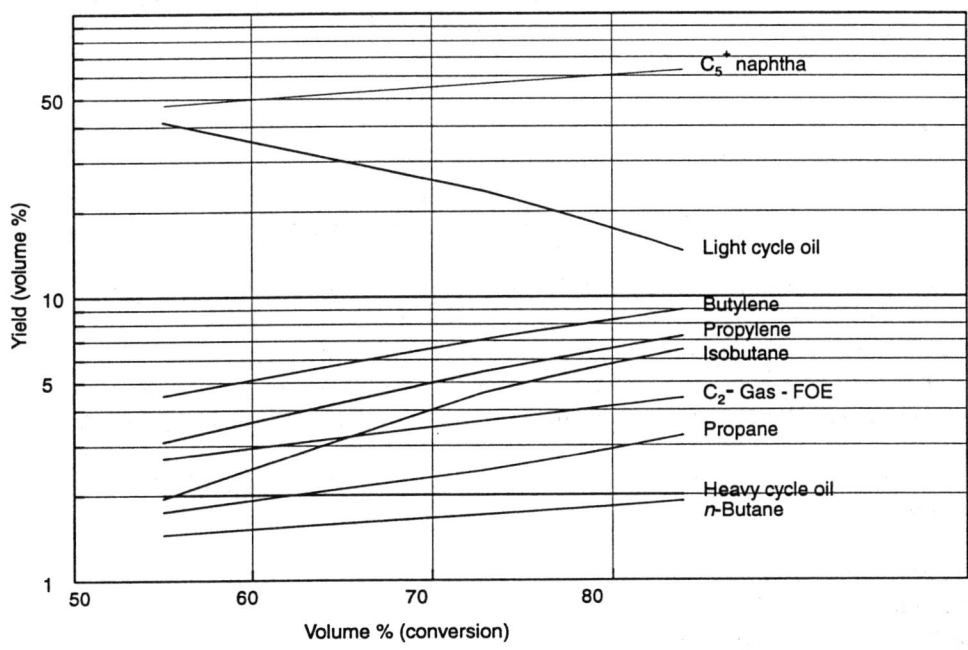

Figure 11.10. Yield versus conversion: visbreaker gas oil: 380–650 °F

● A high portion of the feed that does not vaporize. The unvaporized portion quickly cokes on the catalyst, choking its active area.
● The presence of high concentrations of polar molecules such as polycyclic aromatics and nitrogen compounds. These are absorbed into the catalyst's active area causing instant (but temporary) deactivation.
● Heavy metals contamination that poison the catalyst and affect the selectivity of the cracking process.
● High concentration of polynaphthenes that dealkylate slowly.

In FCCU the process conventional feedstock cracking temperature is controlled by the circulation of hot regen catalyst. With the heavier feedstocks, with an increase in Condradson carbon there will be a larger coke formation. This in turn produces a high regen catalyst temperature and heat load. To maintain heat balance catalyst circulation is reduced, leading to poor or unsatisfactory performance. Catalyst cooling or feed cooling is used to overcome this high catalyst heat load and to maintain proper circulation.

The extended boiling range of the feed as in the case of residues tends to cause an uneven cracking severity. The lighter molecules in the feed are instantly vaporized on contact with the hot catalyst and cracking occurs. In the case of the heavier molecules vaporization is not achieved as easily. This contributes to a higher coke deposition with a higher rate of catalyst deactivation. Ideally, the whole feed should be instantly vaporized so that a uniform cracking mechanism can commence. The mix temperature (which is defined as the theoretical equilibrium temperature between the uncracked vaporized feed and the regenerated catalyst) should be close to the feed dew point temperature. In conventional units this is about 20–30 °C above the riser outlet temperature. This can be approximated by the expression:

$$T_M = T_R + 0.1 \ \Delta H_C$$

where

$$
\begin{aligned}
T_M &= \text{the mix temperature} \\
T_R &= \text{riser outlet temperature (°F)} \\
\Delta H_C &= \text{heat of cracking (BTU/lb)}
\end{aligned}
$$

This mix temperature is also slightly dependent on the catalyst temperature.

Cracking severity is affected by polycyclic aromatics and nitrogen. This is so because these compounds tend to be absorbed into the catalyst. Raising the mix temperature by increasing the riser temperature reverses the absorption process. Unfortunately, a higher riser temperature leads to undesirable thermal cracking and production of dry gas.

The processing of heavy feedstocks therefore requires special techniques to overcome:

● Feed vaporization
● High concentration of polar molecules
● Presence of metals

Some of the techniques developed to meet heavy oil cracking processing are as follows:

- Two-stage regeneration
- Riser mixer design and mix temperature control (for rapid vaporization)
- New riser lift technology minimizing the use of steam
- Regen catalyst temperature control (catalyst cooling)
- Catalyst selection for:

  Good conversion and yield pattern
  Metal resistance
  Thermal and hydrothermal resistance
  High-gasoline RON

These are discussed in Sections 11.5 and 11.6

11.5 Two-stage Regeneration and Regen Catalyst Temperature Control

An important issue in the case of heavy oil fluid catalytic cracking is the handling of the high coke laydown and the protection of the catalyst. One technique that limits the severe conditions in regeneration of the spent catalyst is two-stage regenerator. Figure 11.11 shows the layout.

The spent catalyst from the reactor is delivered to the first regenerator. Here the catalyst undergoes a mild oxidation with a limited amount of air. Temperatures in this regenerator remain fairly low, around 1290–1380 °F. From this first regenerator

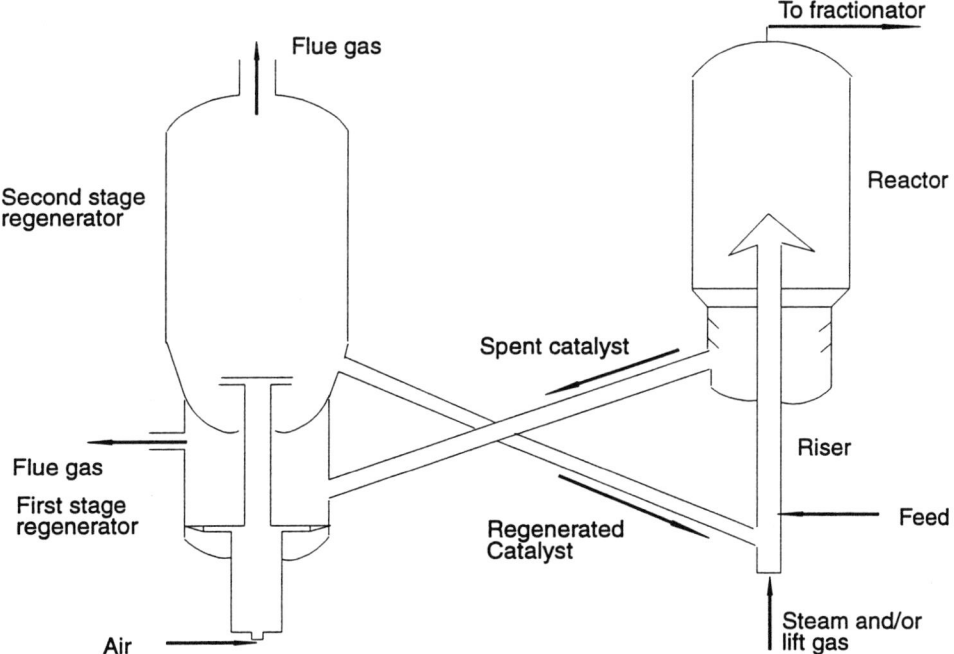

Figure 11.11. Two-stage regenerator

the catalyst is pneumatically conveyed to a second. Here excess air is used to complete the carbon burn-off and temperatures up to 1650 °F are experienced. The regenerated catalyst leaves this second regenerator to return to the reactor via the riser. The technology that applies to the two-stage regeneration process is innovative in that it achieves the burning off of the high coke without impairing the catalyst activity. In the first stage the conditions encourage the combustion of most of the hydrogen associated with the coke. A significant amount of the carbon is also burned off under mild conditions. These conditions inhibit catalyst deactivation.

All the residual coke is burned off in the second-stage regenerator with excess air and in a dry atmosphere. All the steam associated with hydrogen combustion and carry-over from the reactor has been dispensed within the first stage. The second regenerator is refractory lined and there is no temperature constraint. The catalyst is allowed to come to equilibrium. Even at high regen temperatures under these conditions lower catalyst deactivation is experienced. The two-stage regeneration technique leads to a better catalyst regeneration as well as a lower catalyst consumption. Typically the clean catalyst contains less than 0.05% wt of carbon. This is achieved with an overall lower heat of combustion (see Figures 11.12 and 11.13). Since the unit remains in heat balance coke production stays essentially the same.

It has been found that there is a specific catalyst temperature range that is desirable for a given feed and catalyst system. A unique dense phase catalyst cooling system provides a technique through which the best temperature and heat balance relationship can be maintained.

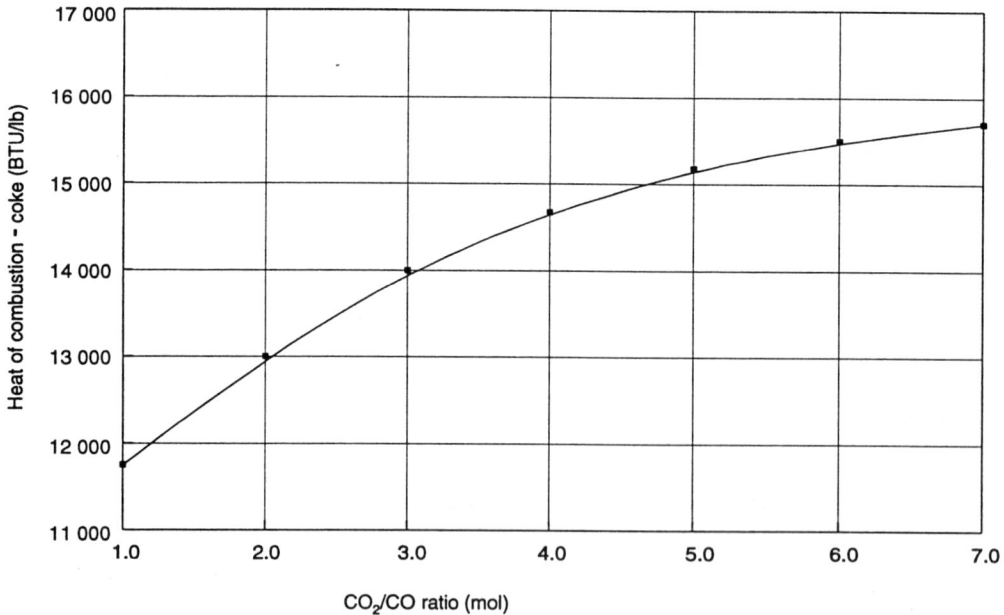

Figure 11.12. Heat of combustion versus CO_2/CO ratio in flue gas

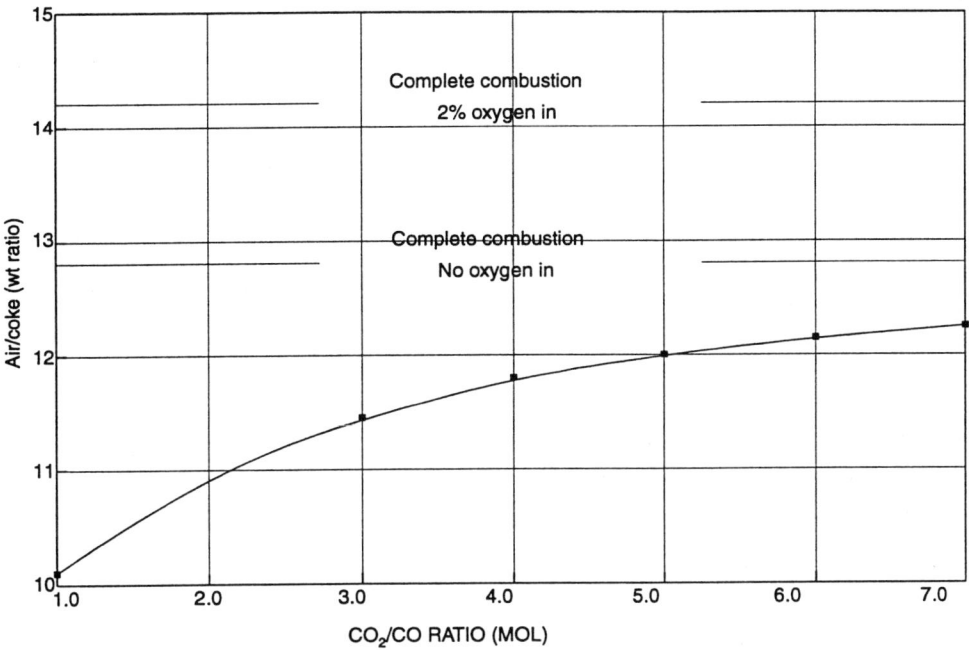

Figure 11.13. Combustion air requirement versus CO_2/CO ratio in flue gas

Consider the enthalpy requirements for a FCC reactor given in the following table:

	Per pound of feed	
	BTU	%
Feed heating/vaporizing	530.0	69.00
Stripping steam enthalpy	5.0	0.65
Feed steam for dispersion	12.7	1.65
Feed water for heat balance	18.4	2.40
Heat of reaction	200.0	26.04
Heat loss	2.0	0.26
Total	768.1	100.00

It can be seen from this table that 69% of the enthalpy contained in the heat input to the reactor is required just to heat and vaporize the feed. The remainder is essentially available for conversion. To improve conversion it would be very desirable to allow more of the heat available to be used for conversion. The only variable that can be changed to achieve this requirement is the feed inlet enthalpy, that is, through preheating the feed. Doing this, however, immediately reduces the catalyst circulation rate to maintain heat balance. This, of course, has an adverse effect on conversion. The preheating of the feed can, however, be compensated for by cooling the catalyst. Thus the catalyst circulation rate can be retained and, in many cases, increased. Indeed, by careful manipulation of the heat balance the net increase in catalyst circulation rate can be as high as 1 unit cat/oil ratio. The higher equilibrium catalyst activity possible at the lower regeneration temperature also improves the unit yield pattern. This is demonstrated in the following table:

Feedstock

		Yields:		
°API	24.5		Without catalyst cooling	With catalyst cooling
Conradson carbon	1.6			
		H_2S wt %	0.1	0.19
		C_2^- wt %	3.4	2.00
		C_3 LV%	9.9	10.34
		C_4 LV%	13.9	14.51
		C_5^+ (430 : EP) LV%	58.2	60.87
		LCO (650 : EP) LV%	17.1	15.54
		CLO LV%	8.6	8.10
		Coke wt%	5.9	6.07
		Conversion LV%	74.3	76.36

In summary, catalyst cooling will:

- Slightly increase unit coke
- Give a higher plant catalyst activity
- Be able to handle more contaminated feeds
- Improve conversion and unit yield
- Provide better operating flexibility

In residue cracking commercial experience indicates that operations at regenerated catalyst temperatures above 1350 °F result in poor yields with high gas production due to local thermal cracking of the oil on contact. Where certain operations require high regen temperatures the installation of a catalyst cooler will have a substantial economic incentive. This will be due to improved yields and catalyst consumption.

There are two types of catalyst coolers available:

- Back-mix
- Flow-through

These are shown in Figure 11.14. Both coolers are installed into the dense phase section of the regenerator.

- *The back-mix cooler*   Boiler feedwater flows tube-side in both cooler types. The catalyst in the back-mix cooler circulates around the tube bundle on the shell side. The heat transfer takes place in a dense low-velocity region so erosion is minimized. the back-mix cooler can remove approximately 50 million BTU/h.

- *The flow-through cooler*   As the name suggests, the catalyst flows once through on the shell side of this cooler. Again erosion is minimized by low-velocity operation in the dense phase. This type of cooler is more efficient than the back-mix. This unit can achieve heat removal as high as 100 million BTU/h.

11.6   Mix Temperature Control and Lift Gas Technology

The equilibrium temperature between the oil feed and the regenerated catalyst must be reached in the shortest possible time. This is required in order to ensure the rapid and homogeneous vaporization of the feed. To ensure this it is necessary to design

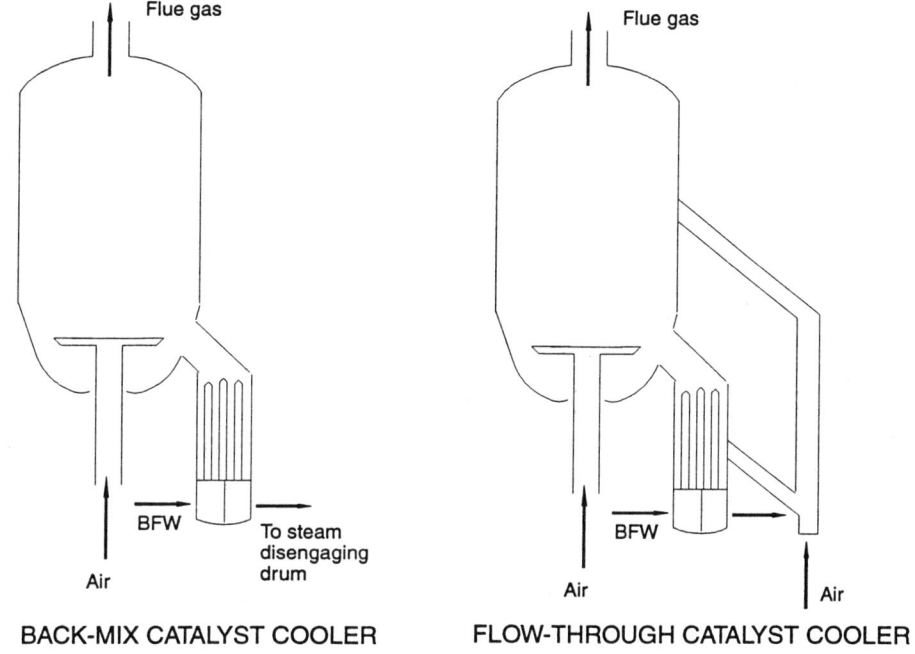

Figure 11.14. Typical catalyst coolers

and install a proper feed injection system. This system should ensure that any catalyst back-mixing is eliminated and that all the vaporized feed components are subject to the same cracking severity.

Efficient mixing of the feed finely atomized in small droplets is achieved by contact with a pre-accelerated dilute suspension of the regen catalyst. Under these conditions feed vaporization takes place almost instantaneously. This configuration is shown in the diagram below.

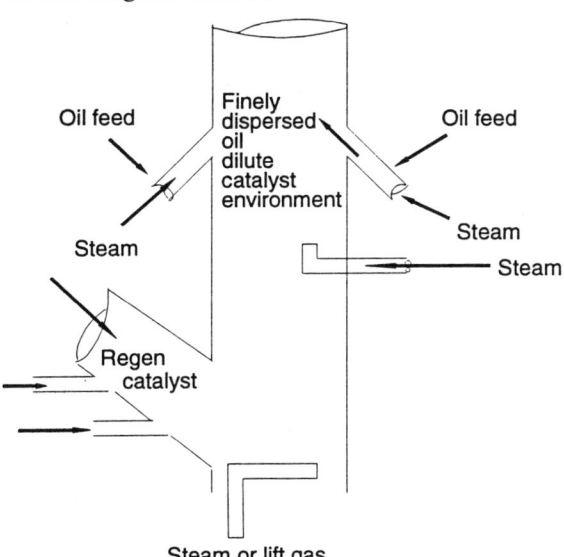

The regen catalyst stream from the regenerator is accelerated by steam or lift gas injection to move up the riser. The oil feed is introduced atomized by steam into the catalyst environment. The main motive steam into the riser is introduced below the feed inlet point. Good mixing occurs in this section with maximum contact between oil, catalyst and the steam.

In heavy oil cracking the proper selection of catalyst enables even the most bulky molecules to reach the active catalyst zone. Such zeolite catalysts have a high silica-to-alumina ratio which cracks the heavy molecules into sizes that can enter the active zone.

Efficient mixing of the catalyst and feed together with the catalyst selection ensures:

- Rapid vaporization of the oil
- Uniform cracking severity of the oil

Another problem that is met with in heavy oil cracking is the possibility of the heavier portion of the oil being below its dew point. To ensure that this is overcome the 'mix temperature' (see Section 11.4) must be set above the dew point of the feed. As stated in Section 11.4, the presence of polycyclic aromatics also affects cracking severity. Increasing the mix temperature to raise the riser temperature reverses the effect of polycyclic aromatics. In so doing, however, thermal cracking occurs, which is undesirable. To solve this problem it is necessary to be able to control the riser temperature independently to mix temperature.

Mix temperature control (MTC) is achieved by injecting a suitable heavy-cycle oil stream into the riser above the oil feed injection point. This essentially separates the riser into two reaction zones. The first is between the feed injection and the cycle oil inlet. This zone is characterized by a high mix temperature, a high catalyst-to-oil ratio and a very short contact time.

The second zone above the cycle oil inlet operates under more conventional catalytic cracking conditions. The riser temperature is maintained independently by the introduction of the regen catalyst. Thus an increase in cycle oil leads to a decrease in riser temperature which introduces more catalyst. This finally increases the mix temperature and the catalyst-to-oil ratio and decreases the regenerator temperature.

LIFT GAS TECHNOLOGY

As described earlier, it is highly desirable to achieve good catalyst/oil mixing as early and as quickly as possible. The method described to achieve this requires the pre-acceleration and dilution of the catalyst stream. Traditionally, steam is the medium used to maintain catalyst bed fluidity and movement in the riser. Steam, however, has a deleterious effect on the very hot catalyst that is met in residue cracking processes. Under these conditions steam causes hydrothermal deactivation of the catalyst.

Much work has been done in reducing the use of steam in contact with the hot catalyst. Some of the results of the work showed that if the partial pressure of steam is kept low, the hydrothermal effects are greatly reduced in the case of relatively metal-free catalysts. A more important result of the work showed that light hydrocarbons impart favourable conditioning effects to the freshly regenerated catalyst. This was even pronounced in catalysts that were heavily contaminated with metals.

Light hydrocarbon gases have been introduced in several heavy oil crackers since 1985. They have operated either with lift gas alone or mixed with steam. The limitations to the use of lift gas rests in the ability of downstream units to handle the additional gas. The following table compares the effect of lift gas in residue operation with the use of steam.

Feed: Atmospheric residue 4.3 wt % con. carb.

Product distribution	Lift gas	Steam
C_2 wt%	3.2	4.0
C_3/C_4 LV%	11.4/15.1	11.6/15.4
C_5 Gasoline LV%	56.9	55.0
LCO + slurry LV%	23.9	26.4
Total C_3 + LV%	107.3	106.4
Coke wt%	8.6	8.5
H_2 SCFB	70	89
H_2/C_1 mol.	0.74	0.85
Catalyst		
Material	<- same ->	
SA, M_2/G	91	90
Ni + V wt ppm	7100	7300

*SA = surface area of equilibrium FCC catalyst, M_2/g.

As can be seen, the use of lift gas as an alternative to steam gives:

- Lower hydrogen production
- Lower hydrogen/methane ratio
- Increase in liquid yield

11.7 Building the FCCU 'Syn' Crude Assay

The cracked products leaving the FCCU reactor represent a wide range of cuts. This reactor effluent is often referred to as a 'syn' crude because of its wide range of boiling-point material. The purpose of this section is to build up sufficient data similar to a crude assay that can be used for process-evaluation work or design of the recovery side units. Plant data should be used to define yields, lab data to establish product properties in the case of process evaluation and pilot plant or licensor data for design work.

The 'syn' crude assay should contain at least a TBP curve with an analysis of light ends, a gravity versus mid-boiling point curve and a PONA for the naphtha and sulphur content versus mid-boiling point for the 'syn' crude. The following steps are used to develop the data:

- *Step 1*   Obtain lab data for the product cuts coming off the fractionator. These should be:
Component analysis of C_4 and lighter
ASTM distillation of the distillates
API (or Sg) for the distillates
Sulphur content for the distillates
- *Step 2*   Endeavour to obtain the plant data for yields as close to that time when the lab samples were taken.
- *Step 3*   Calculate the TBP curves from the ASTM distillation (see Section 6.3).
- *Step 4*   Calculate the Me_{AVBP} (mean average boiling point) for each distillate stream (see the appropriate section in the introductory chapter of this book).
- *Step 5*   From the curve in Appendix 1 read off mol. wt and the characterization factor 'K' from the API and the AVBP.
- *Step 6*   Plot the TBP curves for each product. Break up the curves with boiling cut points. Make sure that where a cut point is used in one product it is also represented in the adjacent one.

- *Step 7*   Table the percentage volume of each fraction by the yield of the product (percentage volume of fractionator feed). Add the result for each fraction for each product.
- *Step 9*   Plot the total volume percentage from Step 8 against the cut points from Step 6. This is the 'syn' crude TBP curve.
- *Step 10*   Using the Me_{AVBP} as the product mid-boiling point plot this against lab data for gravity Figure 11.17. Do the same for the sulphur content Figure 11.18.

A worked example now follows:

Example Calculation

Lab data and yields are as follows:

C_3 and lighter	9.4% weight
C_4s	10.7% weight
H_2S	0.7% weight
C_5 to 356 °F	45.6% weight 56% vol.
Cycle oil	19.6% weight 19.4% vol.
Slurry	5.9% weight
Coke	8.1% weight
	100% weight

C_5 to 380 °F

% vol.	ASTM (°F)	TBP (°F)	Me_{AVBP} (°F)	°API	Mol. wt	K	S	
IBP	105	37						
10	130	86						$P = 40$
30	173	154						$O = 27$
50	225	226	215	55	100	11.8	0.035	$N = 10$
70	281	297						$A = 13$
90	356	376						
100	400	423						

$$VABP = \frac{t10 + 2(t50) + t90}{4} = \frac{130 + (2 \times 225) + 356}{4} = 234$$

$$\text{Slope} \frac{t90 - t10}{80} = 2.8 \text{ °F/\%, correction} = -9 \text{ °F}$$

Cycle oil

% vol.	ASTM (°F)	TBP (°F)	Me_{AVBP} (°F)	°API	Mol. wt	K	S	Visc. at 100 °F (cST)
IBP	415	370						
10	480	469						
30	510	523						
50	530	558	539	17.5	195	10.5	1.3	3.5
70	565	604						
90	680	725						
100	800	852						

$$VABP = \frac{480 + (2 \times 530) + 680}{4} = 555 \text{ °F}$$

$$\text{Slope} = \frac{680 - 480}{80} = 2.5 \text{ °F/\%}$$

$$\text{Correction} = -16$$

Slurry

% vol.	ASTM (°F)	TBP °F	Me_{AVBP}	°API	Mol. wt	K	S
IBP	500	493					
10	615	651					
30	720	774					
50	780	853	761	1.9	280	10.2	2.1
70	850	937					
90	950	1045					
100	1080	1175					

VABP = 782   Slope = 4.2 Corr. = −21

Calculating syn crude TBP curve

		Naphtha		Cycle oil		Slurry		Totals	
		% vol. cut	% vol.	% vol. cut	% vol.	% vol. cut	% vol.	% vol.	CUM % vol.
IBP To	C_5	12.0	6.7					6.7	20.6
160 °F		18.0	10.1					10.1	27.3
225		20.0	11.2					11.2	37.4
300		20.0	11.2					11.2	48.6
350		15.0	8.4					8.4	59.8
400		10.0	5.6	2.0	0.4			6.0	68.2
470		5.0	2.8	8.0	1.6			4.4	74.2
525				20.0	3.9	1.0	0.1	4.0	78.6
560				20.0	3.9	3.0	0.2	4.1	86.6
600				20.0	3.9	3.0	0.2	4.1	86.6
650				10.0	1.9	5.0	0.3	2.2	90.7
725				10.0	1.9	9.0	0.4	2.3	92.9
850				10.0	1.9	30.0	1.4	3.3	95.2
935						20.0	1.0	1.0	98.5
1050						20.0	1.0	1.0	99.5
1175						10.0	0.5	0.5	100.0
Total		100.0	56.0	100.0	19.4	100.0	5.0	80.4	—

These are plotted in Figures 11.15 and 11.16

11.8 Developing the Fractionator Material Balance

The material balance for the fractionator is calculated from the data developed in Section 11.7. The following steps are used to develop this balance:

- *Step 1* Establish the total feed to the fractionator in BPSD (or gal/h) from plant data.
- *Step 2* Calculate the C_3 and lighter and C_4 properties from the lab data, that is, the mole weight and gravity for each of the two streams. Their total yield on feed is obtained by difference in plant data.
- *Step 3* Establish the liquid products' cut points from their yield and the TBP curve from Section 11.7.
- *Step 4* Commence building the material balance from yield as percentage volume of $C_5{}^+$ products, their gravity and mol. weight.
- *Step 5* The gas streams are usually quoted in percentage weight or moles. Using the properties calculated in Step 2, work back to volume and/or weight.
- *Step 6* Complete the balance and check against plant readings for total fractionator feed.

An example calculation now follows.

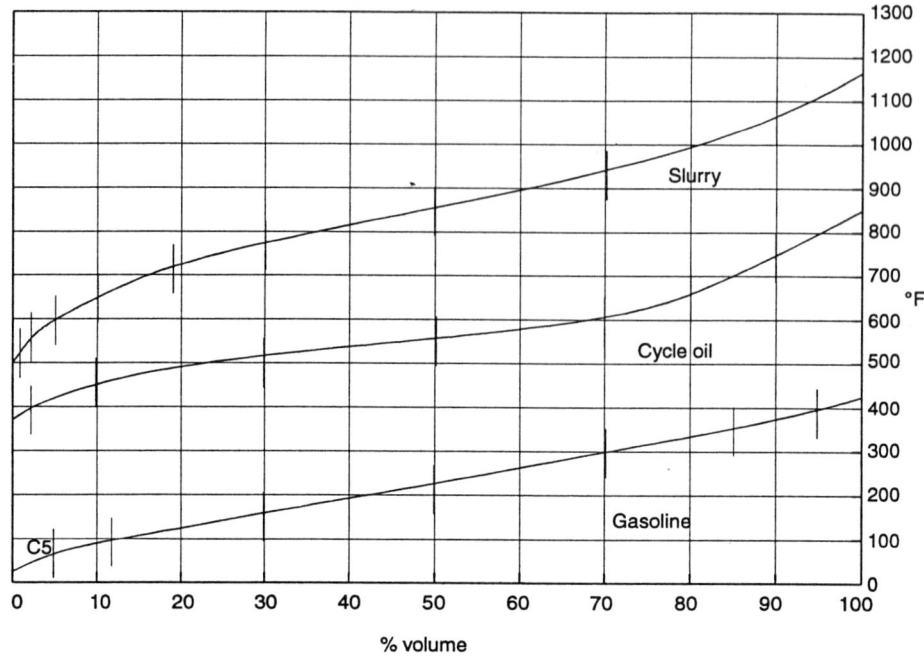

Figure 11.15. FCCU product TBP curves

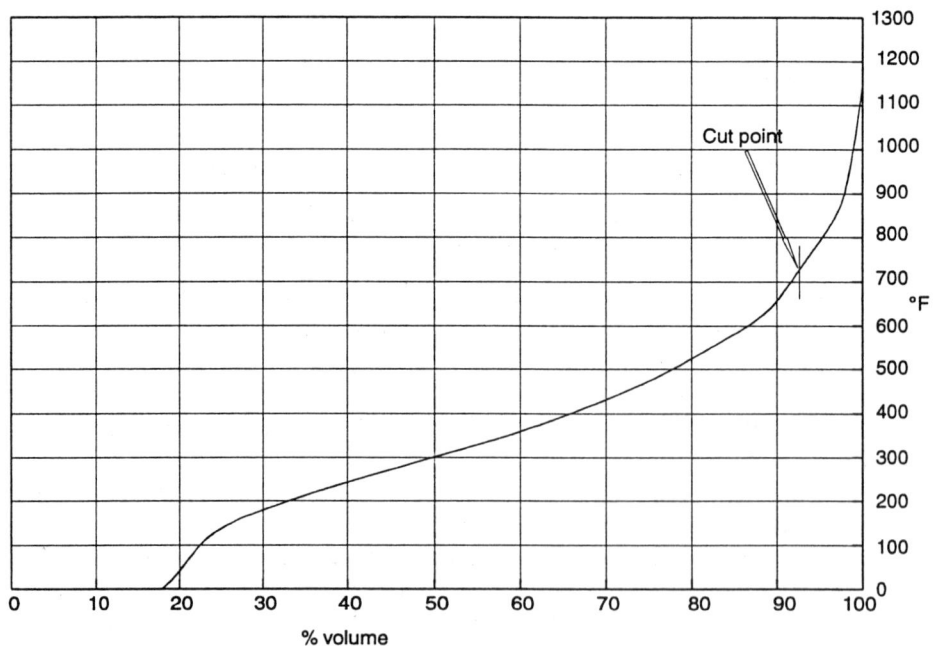

Figure 11.16. FCCU syn crude TBP curve

Example Calculation

Bottom product from the syn tower will be clarified slurry. Its cut point will be 725 °F. Feed to the unit is 21.3 °API gas oil at 25 000 BPSD conversion % vol. will be 75%.

Total feed to fractionator	=	18 750 BPSD
°API	=	42.0
Steam	=	2192 lb/h
Temperature of fractionator feed	=	921 °F

Gas analysis

C_3 and lighter	% mol.	MW	Weight	lb/gal	Vol. factor
H_2	22.2	2	40	—	
C_1	22.2	16	320	—	
C_2	18.9	30	510	2.96	172
C_3	32.3	42	1218	4.35	280
C_3	4.4	44	176	4.22	41.7
Total	100	22.6	2264	4.59	493.7

9.4 wt % on feed

C_4s	Vol. %	lb/gal	Weight factor	MW	Moles factor	
C_4	50	5.0	250	56.1	4.46	52.3
iC_4	40	4.68	187.2	58.0	3.23	37.9
nC_4	10	4.86	48.6	58.0	0.84	9.8
Total	100	4.86	485.6	57.0	8.53	

10.7 wt % on feed

Syn tower material balance

Stream	Cut points (°F)	Volume %	BPSD	gal/h	lb/gal	Weight lb/h	% wt	Mol. wt	Moles Moles/h	% mol.	°API
Feed	—	100	18 750	32 813	6.53	214 155	100	87.7	2442.5	100.0	48.5
Products											
C_3 and lighter	—	13.4	2506	4386	4.59	20 131	9.4	22.6	890.8	36.4	125.0
C_4s		14.4	2694	4715	4.86	22 915	10.7	57.0	402.0	16.5	110.5
C_5 to 350 °F	350	26.1	4894	8565	6.10	52 173	24.4	100	521.7	21.4	61.0
HY naphtha	400	12.8	2400	4200	7.00	29 400	13.7	147	200.0	8.2	36.5
LCO	725	26.5	4969	8695	8.00	69 580	32.5	195	356.7	14.6	16.0
Slurry	+725	6.8	1275	2232	8.95	19 976	9.3	280	71.3	2.9	1.0
Total		100.0	18 750	32 813	6.53	214 155	100.0	87.7	2442.5	100.0	

°API are read from Figure 11.17

11.9 Setting Main Fractionator Conditions

This section provides the method for predicting main fractionator draw-off conditions. It includes the calculation for the distillate vapours leaving the quench section, the

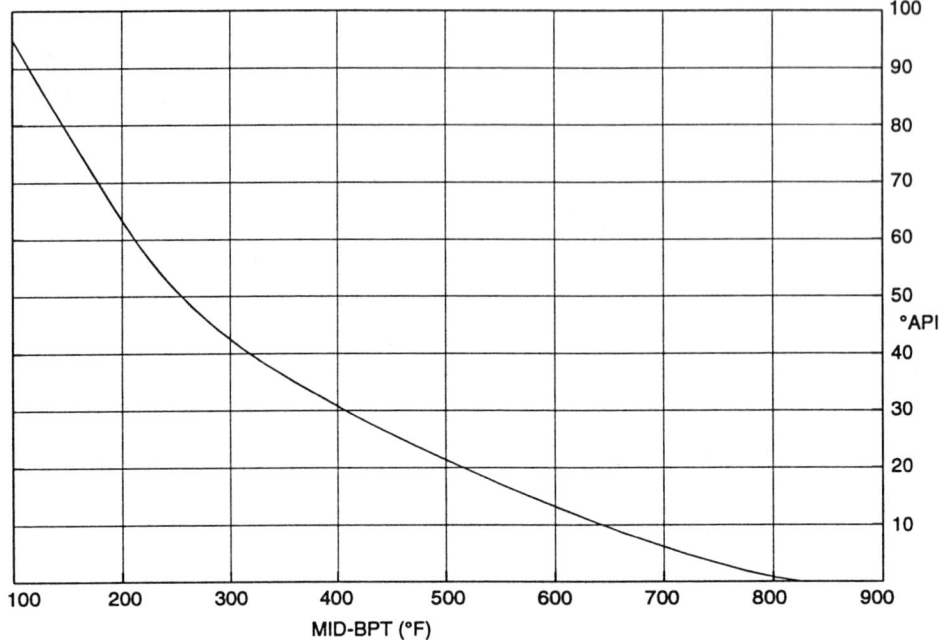

Figure 11.17. FCCU syncrude. Gravity versus mid-boiling point

side-stream draw-off temperatures and the tower-top conditions. Detailed conditions and performance of the quench will be given in a later section in this chapter. The following steps are used for this method:

- *Step 1*   From the fractionator feed TBP (see section 11.7) develop the whole feed EFV curve (Figure 11.19).
- *Step 2*   The feed to the fractionator is vapour entering the bottom of the tower. It is quenched to provide a liquid slurry stream leaving the bottom of the tower and a vapour (of gas and distillates) passing up the tower from the top of the quench section. This is equivalent to the flash zone of a crude tower (see Section 6.4). Fix the cut point on the atmospheric EFV and allow 3% as overflash.
- *Step 3*   Fix the pressure in the bottom of the tower. The reflux drum is usually between 14 and 16 psig at 100 °F. etc. Allow 5 psi for tower pressure drop.
- *Step 4*   Calculate the partial pressure of hydrocarbons directly above the quench tower. Draw the EFV curve at this partial pressure. The temperature of the vapours leaving the quench section is the cut point determined in Step 2 on this EFV curve.
- *Step 5*   Calculate the FRL (flash reference line) for all the side-stream products. This is developed from the product TBP. (*Note*: in the example a heavy cycle oil is taken off but is only used as a pumparound. Its draw-off temperature still needs to be calculated.)

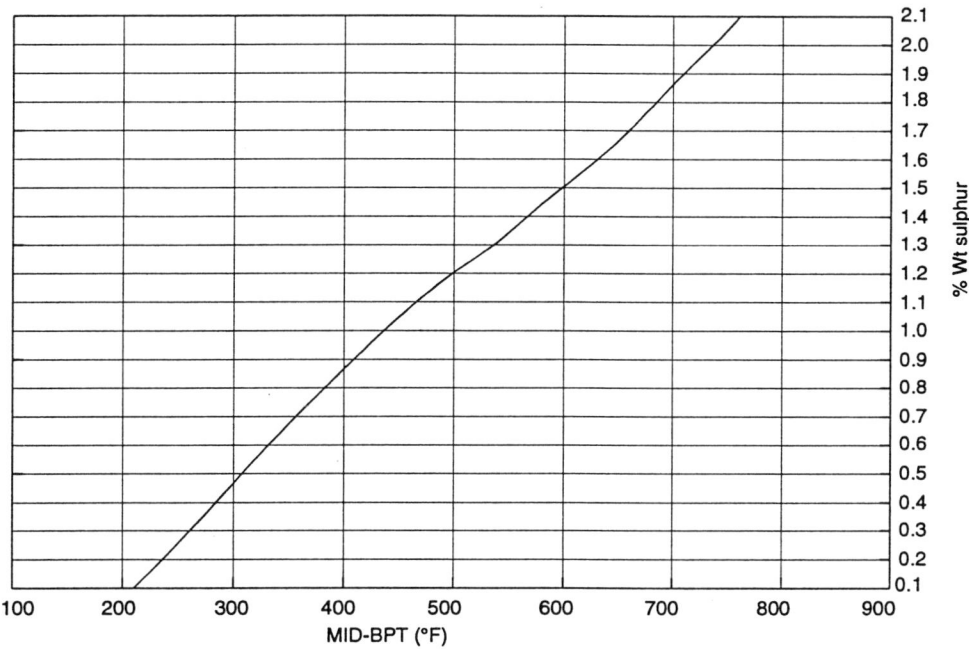

Figure 11.18. FCCU syncrude. Total sulphur versus mid-boiling point

- *Step 6*   Calculate the partial pressure of product hydrocarbon and overflow at each draw-off location. Plot a straight-line pressure profile for the total pressure at each draw-off (Figure 11.21).
- *Step 7*   Using the vapour pressure curves in Appendix 1, establish the initial point of the FRL for each draw-off at its partial pressure. This is the stream's draw-off temperature.
- *Step 8*   Develop the overhead component balance. It will be necessary to break down the overhead gasoline to PSEUDO-components (see the appropriate section in the introductory chapter). This is given in Figure 11.20.
- *Step 9*   Carry out a flash vaporization calculation at the reflux drum condition for the overhead product. (*Note:* Because of the hydrogen and methane content it will not be possible totally to condense the overheads.)
- *Step 10*   Calculate the amount of overflow required from the top tray to give a 30 °F gap between 95% gasoline and 5% ASTM naphtha. Use the method given in Section 6.7 for this.
- *Step 11*   Add the reflux component to the product components and calculate the dew point for the total at tower-top pressure. The answer will be the tower-top temperature. There will be steam present, so tower-top pressure used must be the hydrocarbon partial pressure.

An example calculation now follows.

Example Calculation

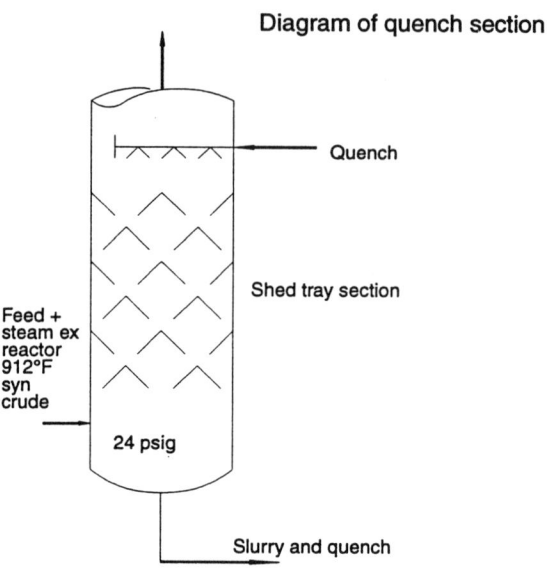

Diagram of quench section

Calculate the EFV curve for syn crude

Slope of DRL   8.5 °F/%
Slope of FRL   5.4 °F/%
▲ T_{50}°F (DRL − FRL) = 40   FRL 50%   250 + 40 = 290 °F

% vol.	▲ T (TBP) / (DRL) °F	▲ T flash/FRL / TBP/DRL	▲ T flash/FRL°F	EFV (°F)
0	− 100	0.2	− 20	20
10	0	0.4	0	85
20	+ 90	0.38	34	169
30	+ 110	0.37	41	226
40	+ 90	0.37	33	273
50	+ 50	0.37	19	309
60	+ 15	0.37	19	309
70	0	0.37	0	395
80	+ 10	0.37	4	449
90	+ 40	0.37	15	515
100	∞	0.37	∞	∞

This EFV at atmospheric pressure is drawn as Figure 11.19.

Total moles distillate = 2371.2/h

Add about 3% vol. of feed as over flash

Approx mol wt = 310
Approx lb/gal = 8.6
Moles overflash = 27.3

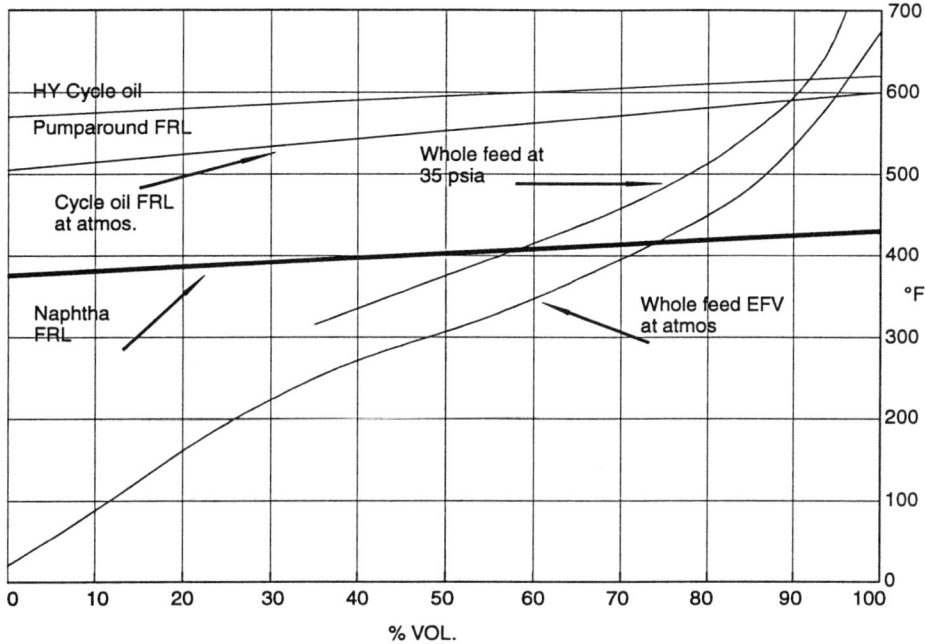

Figure 11.19. Cat cracker fractionator EFV curves

There will be about 1.5 wt% steam from the reactor

$$= \quad 3212 \ \text{lb/hr}$$
$$= \quad 178 \ \text{moles/h}$$

The pressure above the shed tray section is 23 psig

$$= \quad 38 \ \text{psia}$$

Partial pressure $\quad = \quad \dfrac{2398.5}{2576.5} \times 38$

$$= \quad 35.3 \ \text{psia}$$

Slurry cut point is 93.2% vol.

Temp. above quench is 635 °F

Calculating side-stream draw-off temperature

HCO This will not be taken off as a product but will be cooled for main tower pumparound.

Cut TBP 525–725 °F	=	13% vol.
DRL slope	=	1.53
FRL slope	=	0.45
DRL 50%	=	595 °F
FRL 50%	=	595 °F
Estimated pumparound rate	=	77 960 gal/h
	=	647 400 lb/h
	=	2600 moles/h

Total moles vapour passing through HCO

Draw-off trays 24 (top tray = tray 1) total trays = 34. Pressure = 22.5 psig
                                                              = 37.5 psia

Moles distillate product = 2371.2

Moles overflow, say 2.0 × total products

	=	4742.4 moles/h
Steam	=	178 moles/h
Partial pressure	=	$\dfrac{4742.4}{7291.6} \times 37.5$ psia
	=	24.5 psia
FRL IBP at atmos.	=	570 °F
AT 24.5 psia	=	612 °F
From Figure 6.70 correction	=	−40 °F
Actual DO Temperature	=	572 °F

LCO product draw-off (Tray 19)

Total HC vapour passing through tray	=	2371.2
Estimated overflow = 1.0 × product	=	356.7
Total	=	2727.9

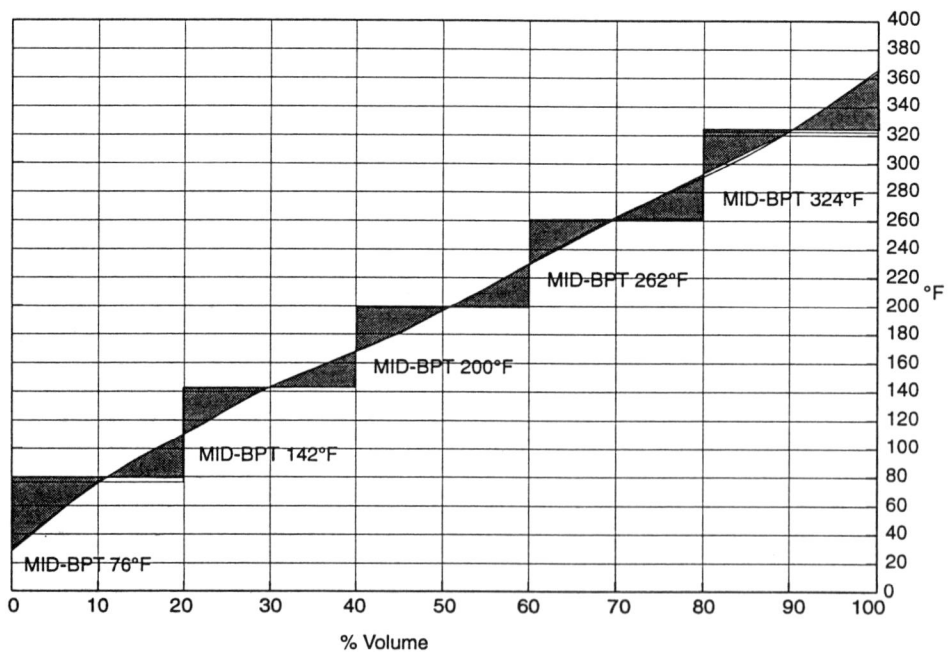

Figure 11.20. FCCU C_5^+ gasoline TBP and pseudo components

Steam   =   =   178 moles/h
Total pressure on tray   =   22 psig   =   39 psia

Partial pressure   = $\dfrac{713.4}{2905.9} \times 37$ =   9 psia

FRL at atmos.   =   508 °F

At 9.1 psia   =   475 °F
Correction factor from Figure 6.7   =   −40
DO   = $\underline{435}$

Note: In this fractionator because of the presence of light incondensible material all vapour lighter than the cut is considered inert.

Naphtha product

Draw-off tray   =   tray 8
Total HC vapour   =   2014.5 moles/h
Estimated overflow   = $\underline{\ \ 400.0\ \ }$

                        Total =   2414.5

Steam   =   178 moles (from reactor)
                                   $\underline{242}$ moles/h (from LCO stripper)
                                            (0.5 lb steam/gal/h)

                        Total =   420 moles/h

Partial pressure   = $\dfrac{600}{2834.5} \times 35$ psia

                               =   7.4 psia

FRL = 380 at atmos.   =   332 at 7.4 psia
Corrected   =   332 − 25 = 307 °F

Naphtha stripper is reboiled (i.e. no steam)

Calculate C_5^+ gasoline components (Figure 11.20)

Comp.	% vol.	Volume		°API	lb/gal	lb/h	Mol. wt	Moles/h
		BPSD	gal/h					
Mid-BPT 76	20	978.8	1717	104.5	4.98	8551	80.5	106.2
142	20	978.8	1717	80	5.56	9701	86	112.8
200	20	978.8	1717	62	6.10	10 474	98	106.9
262	20	978.8	1717	48	6.6	11 332	112	101.2
324	20	978.8	1717	37	7.06	12 115	128	94.6
Total	100	4894	8585	62	6.10	52 173	87.7	521.7

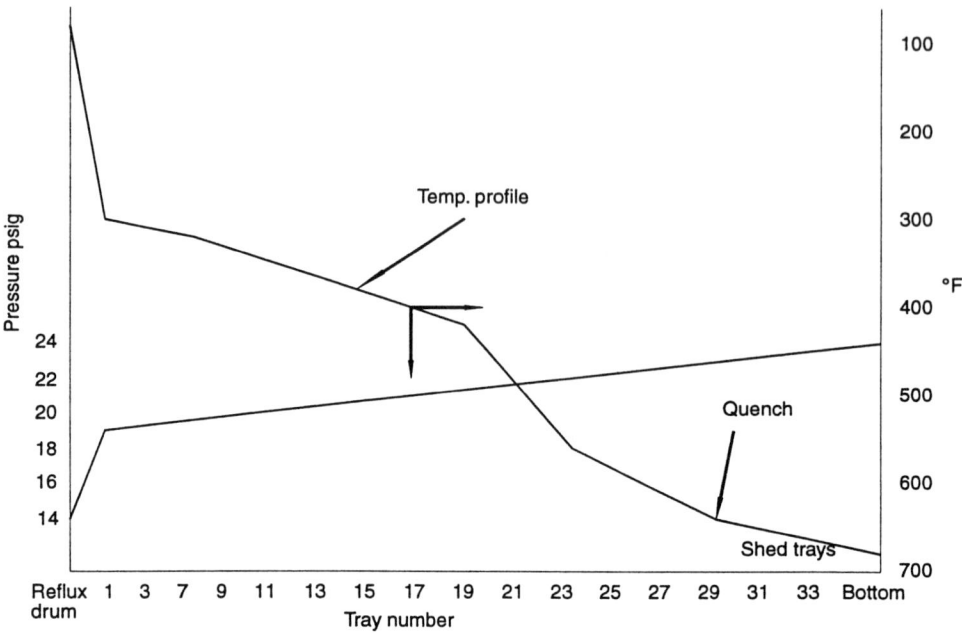

Figure 11.21. Typical syn tower profile

Composition of total overhead product (moles)

Comp.		Moles/h	Mol. frac.
H_2		197.8	0.109
C_1		197.8	0.109
C_2		168.4	0.093
C_3		287.6	0.159
C_3		39.2	0.022
C_4		210.2	0.116
iC_4		152.4	0.084
nC_4		39.4	0.022
Mid-BPT	76	106.2	0.059
	142	112.8	0.062
	200	106.9	0.059
	262	101.2	0.056
	324	94.6	0.052
		1814.5	1.000

To set tower-top temperature

Reflux drum will be operated at 14 psig and 100 °F. It will be a partial condenser.

Flash vaporization of overhead product

Comp.	Moles/h	K 29 psia 100 °F	1st trial $V/L=1.0$ V/LK	$L=\dfrac{X_F}{1+(V/L)K}$	2nd trial $V/L=5.0$ V/LK	$L=\dfrac{X_F}{1+(V/L)K}$	Final $V/L=3.0$ V/LK	$L=\dfrac{X_F}{1+(V/L)K}$	lb/h liquid	gal/h liquid
H_2	197.8	680	680	0.3	3400	Nil	2400	Nil	—	—
C_1	197.8	94	94	2.1	470	0.4	282	0.7	11	—
C_2	168.4	19	19	8.4	95	1.8	57	2.9	87	29.3
C_3	287.6	6.4	6.4	38.9	32	8.7	19.2	14.2	596	137.0
C_3^-	39.2	5.2	5.2	6.3	26	1.5	15.6	2.4	106	25.1
C_4	210.2	2.17	2.17	66.3	10.9	17.7	6.51	28.0	1568	313.6
iC_4	152.4	2.10	2.10	49.2	10.5	1.3	6.3	20.9	1212	259.0
nC_4	39.4	1.60	1.60	15.2	8.0	4.4	4.8	6.8	394	81.1
BPT 76	106.2	0.79	0.79	59.3	3.95	21.5	2.37	31.5	2536	509.2
BPT 142	112.8	0.25	0.25	90.2	1.25	50.1	0.75	64.5	5547	997.7
BPT 200	106.9	0.069	0.069	100.0	0.35	79.2	0.21	88.3	8653	1418.5
BPT 262	101.2	0.017	0.017	99.5	0.09	92.8	0.051	96.3	10 786	1634.2
BPT 324	94.6	—		94.6	—	94.6	—	94.6	12 115	1717.0
Total	1814.5			630.3		386.0		451.1	43 611	7121.5
Calculated V/L				1.88		3.7		3.0		

Calculate amount of overflow required from top tray

ASTM gap required between gasoline/naphtha 30 °F

Number of trays $\qquad = \quad 8$

Difference in 50% TBP between naphtha and gasoline

Naphtha 50% $\qquad = \quad 355$ °F
Overheads 50% $\qquad = \quad 175$ °F
Δ $\qquad = \quad 180$ °F

Referencing overhead and side-stream max. stripping steam curve (Figure 6.2)

Reflux ratio × no. tray $\qquad = \quad 12$
Reflux ratio $= \dfrac{12}{8}$ $\qquad = \quad 1.5$

Assume temp. is 250 °F

Total gal/h at 60 °F of product $\quad = \quad 17\,641$

$17\,641 \times 1.5$ $\qquad = \quad 26\,461$ gal/h
Approx lb/h $\qquad = \quad 146\,329$
Approx moles/h $\qquad = \quad 1330$ (app m.w. = 110)

which is $\dfrac{1330}{1814.5}$ $\qquad = \quad 0.7:1$

Let external reflux be $1:1$. This should be adequate.
Total moles product $\quad = \quad 1814.5$
Then moles reflux $\qquad = \quad 1814.5$

To calculate tower-top dew point temperature

Comp.	Moles product	Moles reflux	Total moles	Mole fraction
H_2	197.8	—	197.8	0.055
C_1	197.8	2.8	200.6	0.055
C_2	168.4	11.7	180.1	0.050
C_3—	287.6	57.1	344.7	0.095
C_3	39.2	9.7	48.9	0.013
C_4—	210.2	112.6	322.8	0.089
iC_4	152.4	84.1	236.5	0.065
nC_4	39.4	27.3	66.7	0.018
BPT 76	106.2	126.7	232.9	0.064
BPT 142	112.8	259.4	372.2	0.103
BPT 200	106.9	355.2	462.1	0.127
BPT 262	101.2	387.4	488.6	0.135
BPT 324	94.6	380.5	475.1	0.131
Total	1814.5	1814.5	3629.0	1.000

$$\text{Total moles steam} \quad = \quad 420$$
$$\text{Total moles HC} \quad = \quad 3629$$
$$\text{Partial pressure} \quad = \quad \frac{3629}{4049} \times 34$$
$$= \quad 30 \text{ psia}$$

	Mol. frac. Y	K 30 psia 250 °F	$X = Y/K$	K 30 psia 270 °F	$X = Y/K$	MW	Weight factor	lb/gal	Vol. factor	
		1st trial		2nd trial						
H_2	0.055	440	Neg	—	Neg	—	—	—		
C_1	0.055	115	Neg	—	Neg	—	—	—		
C_2	0.050	39	0.001	42	0.001	30	0.03	2.96	0.01	°API = 42
C_3	0.095	19	0.005	22	0.004	42	0.17	4.35	0.04	
C_3	0.013	17	0.001	18.5	0.001	44	0.04	4.22	0.01	
C_4	0.089	10	0.009	12	0.007	56	0.39	5.00	0.88	
iC_4	0.065	9.2	0.007	11.0	0.006	58	0.35	4.68	0.07	
nC_4	0.018	7.8	0.002	8.8	0.002	58	0.12	4.86	0.02	
BPT 76	0.064	6.0	0.011	7.33	0.009	80.5	0.72	4.98	0.14	
BPT 142	0.103	2.47	0.042	3.17	0.032	86	2.75	5.56	0.49	
BPT 200	0.127	1.00	0.127	1.33	0.095	98	9.31	6.10	1.53	
BPT 262	0.135	0.40	0.338	0.55	0.245	112	27.44	6.60	4.16	
BPT 324	0.131	0.15	0.873	0.21	0.624	128	79.87	7.06	11.31	
Total	1.000		1.415		1.025	118	121.19	6.79	17.86	

The second trial is close enough

$$\text{Tower-top temp.} \quad = \quad 0.21 \times 1.026$$
$$= \quad 0.215 \times 30 \quad = \quad 6.46 \text{ psia}$$
$$= \quad \underline{272 \text{ °F}}$$

Tower-top condition is 19 psig − 272 °F

$$\text{Reflux drum} \quad = \quad 14 \text{ psig}$$
$$100 \text{ °F}$$

11.10 Calculating Main Fractionator Condenser Duty, Reflux and Top Tray Loading

The method for calculating condenser and tray loading at the top of the main fractionator is similar to the stabilizers in the reformer and hydrotreater. The overhead condenser condenses some of the overhead product and allows the remainder to go forward as a vapour to the gas-processing units. There will be too much light ends to consider total condensation. This method is as follows:

- *Step 1* Using the external reflux fixed in Section 11.9, calculate the condenser duty by heat balance.
- *Step 2* Calculate the heat balance over the top tray with the overflow (internal reflux) being the unknown. The condenser duty calculated in Step 1 is used here. Solve for overflow as heat in equals heat out.
- *Step 3* Check that the internal reflux ratio calculated meets the fractionation requirement used in Section 11.9.
- *Step 4* Using the properties of the internal reflux stream (which is the liquid in equilibrium with the total overhead vapour in the dew point calculation in Section 11.9) calculate vapour flow in ACFS and hot liquid flow in CFS.

An example calculation now follows.

Example Calculation

1. Calculate heat balance over condenser

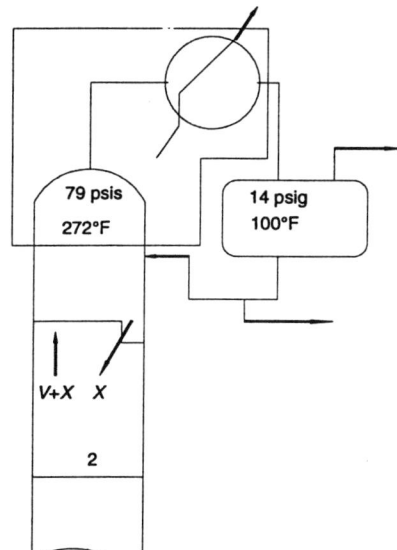

	V or L	°API	°F	lb/h	Enthalpy BTU/ lb	Enthalpy BTU/h ×10⁶
In						
Product vapour	V	87	272	95 219	288	27.423
Reflux vapour	V	80.5	272	175 420	270	47.363
Steam	V		272	7560	1227	9.276
Total in				278 199		84.062
Out						
Vapour product	V	109	100	51 608	224	11.560
Liquid product	L	60.5	100	43 611	46	2.006
Reflux product	L	60.5	100	175 420	46	8.069
Water				7560	100	0.756
Condenser			By difference			61.671
Total out				278 199		84.0672

2. Calculate lb/h of overflow (χ)

	V or L	°API	°F	lb/h	Enthalpy BTU/lb	Enthalpy BTU/h × 10⁶
In						
Vapour from tray 2	V	87	280	95 219	294	27.994
	V	42	280	X	258	258 X
Total in				95 219 + X		27.994 + 258 X
Out						
Product vapour	V	109	100	51 608	224	11.560
Product liquid	L	60.5	100	43 611	46	2.006
Overflow to tray 12	L	42	275	X	134	134 X
Condenser						61.671
Total out				95 219 + X		75.237 + 134 X

$$X = \frac{4\,724\,300}{124} = 380\,992 \text{ lb/h}$$

$$= \frac{380\,992}{118}$$

$$= 3229 \text{ moles/h}$$

which is more than adequate to meet required fractionation. (See Section 11.9)

Top tray loading—vapour flow

	lb/h	Moles/h
Product	95 219	1814.5
Reflux	380 992	3229
	476 211	5043.5

$$\text{ACFS} = \frac{5043.5 \times 378 \times 14.7 \times 740}{520 \times 33.7 \times 3600}$$

$$= 328.7 \text{ ACFS}$$

$$\rho_v = 0.402 \text{ lb/ft}^3$$

Liquid flow

$$\begin{aligned}
\text{lb/h} &= 380\,992 \\
\text{lb/gal} &= 6.79 \text{ at } 60\,°\text{F} \\
\text{Hot lb/gal} &= 6.078 \text{ at } 275\,°\text{F} \\
\text{Hot gal/mm} &= 1044.7 \\
\text{CFS} &= \frac{1044.7}{450} = 2.32 \\
\rho_v &= 45.6 \text{ lb/ft}^3
\end{aligned}$$

11.11   Calculating Quench Section Loading and Tray Performance

The purpose of the quench section at the bottom of the main fractionator is to cool the reactor effluent to a suitable flash temperature. The area between the top of the quench section and the bottom wash tray can be considered as a flash zone. The vapours rising from it are the distillate product vapours and their cut point is controlled by the temperature at the top of the quench section. In practice, everything lighter than the slurry is allowed to leave the quench section as vapour. The slurry is condensed in the quench section to leave at the bottom of the tower.

The quench is affected by the circulation of a quench stream with the same composition as the slurry. This is taken from the bottom of the tower, cooled by heat exchange with other streams, and returned to the top of quench section. It moves down this section, cooling the hot reactor effluent and condensing the slurry portion of the feed.

Because the slurry stream contains solids (entrained catalyst) the use of conventional trays in the quench section is undesirable. Downcomers would become choked and, in the case of valve trays, there would be fouling of the valves themselves. Shed baffle trays or disk and donut trays (see Figure 11.22) are normally used to enhance heat and mass transfer in this section of the tower. These are not readily prone to fouling and do provide good means of heat transfer.

The purpose of this section is to provide a calculation method to predict the performance of these trays or as a basis for their design. These will be evaluated for entrainment and for heat transfer capability. The following steps describe this calculation:

- *Step 1*   Calculate the heat balance over the quench section. The temperature of the vapour leaving the section has been calculated in Section 11.9.
- *Step 2*   The unknown in the heat balance to be completed in Step 1 is the heat-removal duty of the quench section. The inlet reactor effluent is found in plant data and so is the bottom slurry and recycle stream from the tower. The quench duty is provided by difference in this heat balance.
- *Step 3*   Using the quench duty calculated in Step 2, set the quench return temperature and calculate the recycle rate by heat balance. (*Note*: Actual plant flow and temperature readings may be available and these could eliminate the need for Steps 1 and 3.)
- *Step 4*   Using the calculated (or plant) data for recycle rate, establish the loading (vapour and liquid) on the bottom row of shed trays. This will be recycle plus bottom product for the liquid load and the reactor effluent vapour as the vapour load.
- *Step 5*   Assume that the reactor vapours to the bottom tray are at inlet conditions and that liquids leaving the bottom tray are at the temperature of the slurry stream leaving the bottom of the tower.
- *Step 6*   From the equipment data sheet calculate the horizontal free area for the bottom row. Table A1.1 should be used for this.
- *Step 7*   Calculate the liquid and vapour mass velocities in $lb/h/ft^2$ of free area. From Figure 11.23 read off the heat transfer coefficient H_o in $BTU/h/°F/ft^2$ of free area.

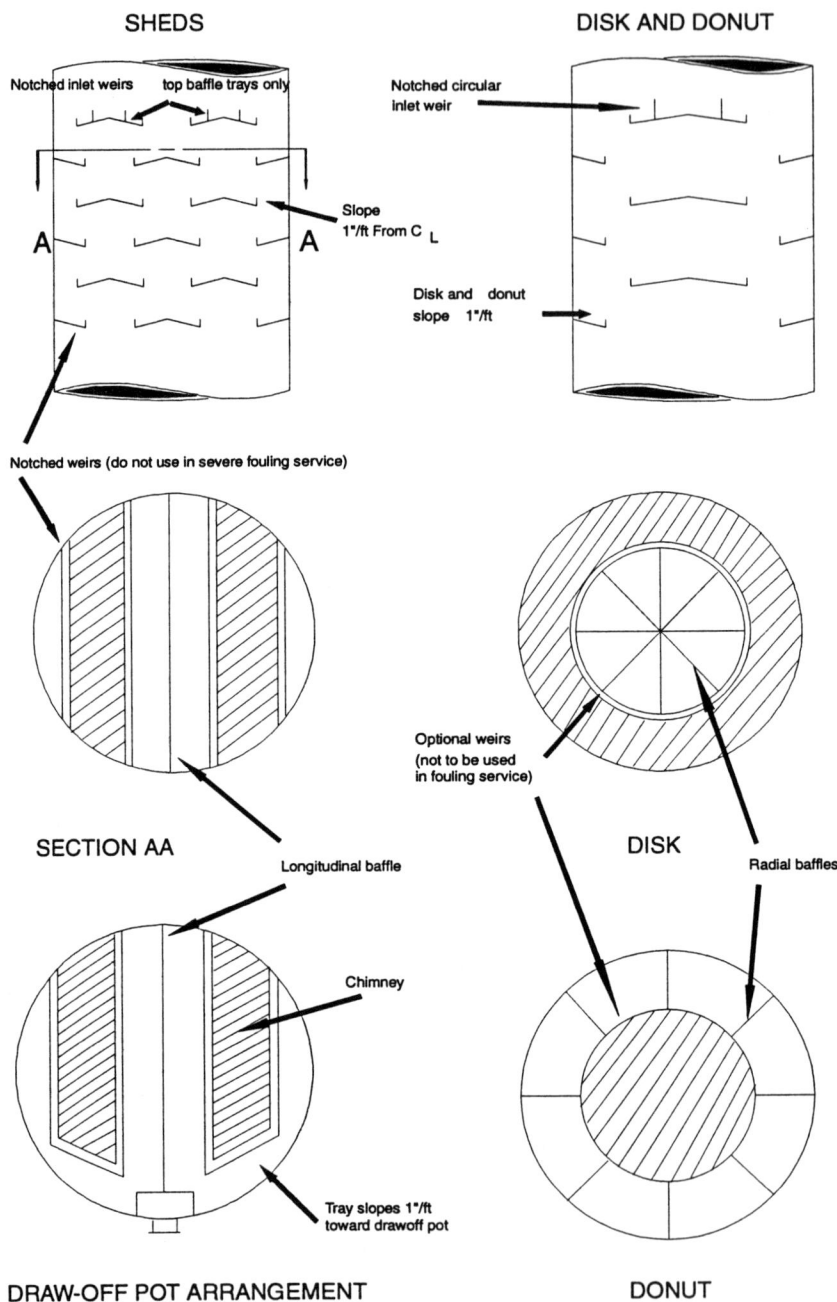

Figure 11.22. Disk and donut shed-type surfaces

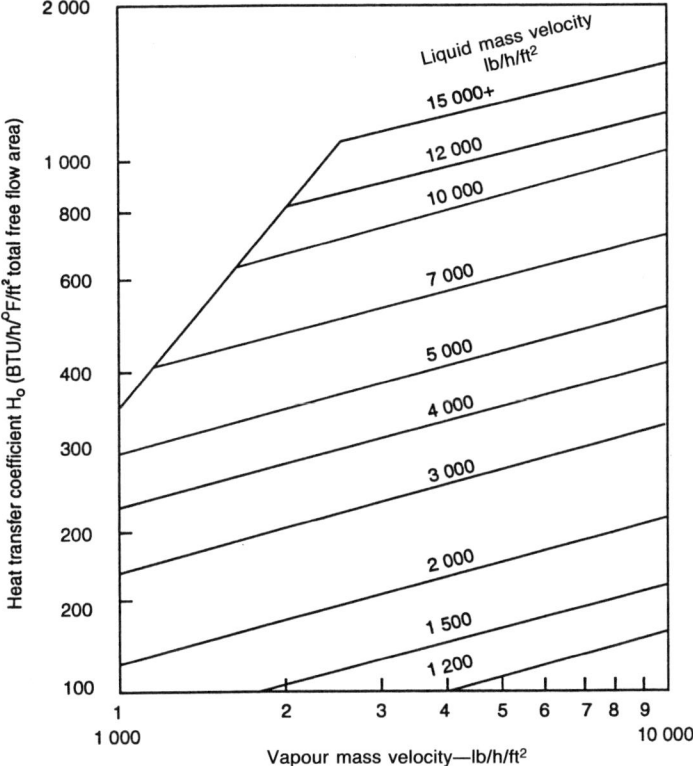

Figure 11.23. Disk and donut and shed baffle heat transfer

● *Step 8*   From the temperatures in and out of the quench section calculate the log mean temperature difference. Use the expression

$$\text{Total free area } A \;=\; \frac{Q}{H_0 \, \Delta t_{\mathrm{m}}}$$

where

Q = the total duty of the quench section (Steps 1 and 2)

H_{o} = overall heat transfer coefficient (Step 7)

Δt_{m} = log mean temperature difference

A = the total horizontal free area in the section

● *Step 9*   Calculate total horizontal free area divided by that calculated for overflow (Step 6) to give number of rows required for heat transfer. Compare with equipment data.

- *Step 10*   From liquid/vapour traffic over the bottom row of baffles calculate actual density of the vapour and liquid in lb/ft³.
- *Step 11*   Calculate the critical vapour velocity for the trays by the expression

$$V_c = 0.157 \sqrt{\frac{\rho_L - \rho_v}{\rho_v}} \text{ ft/s}$$

- *Step 12*   Using the vapour velocity in cubic feet per second calculate the actual linear velocity based on tower cross-section. This should be less than 170% of V_c to ensure no entrainment.

An example calculation now follows.

Example Calculation

1. Calculate heat balance over quench section

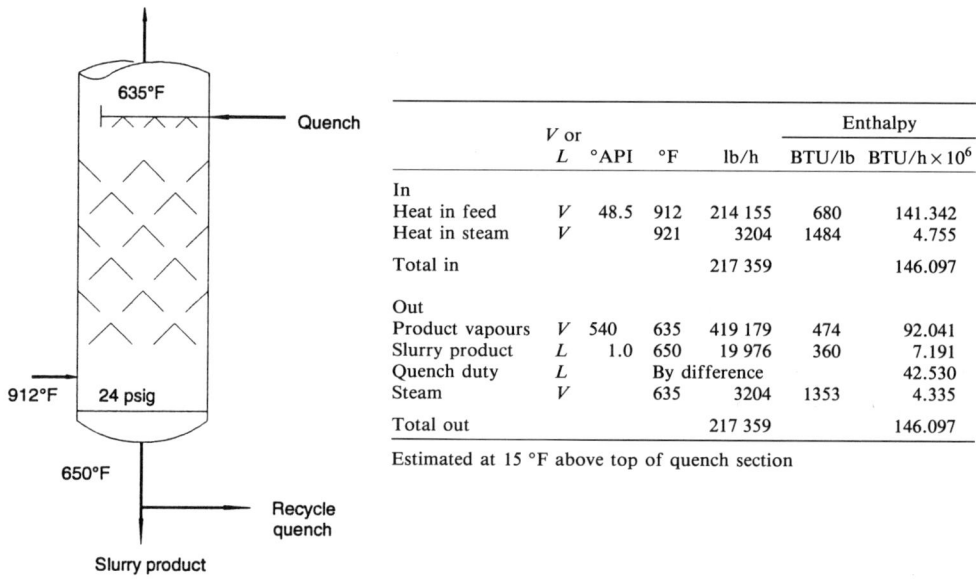

	V or L	°API	°F	lb/h	Enthalpy BTU/lb	Enthalpy BTU/h × 10⁶
In						
Heat in feed	V	48.5	912	214 155	680	141.342
Heat in steam	V		921	3204	1484	4.755
Total in				217 359		146.097
Out						
Product vapours	V	540	635	419 179	474	92.041
Slurry product	L	1.0	650	19 976	360	7.191
Quench duty	L			By difference		42.530
Steam	V		635	3204	1353	4.335
Total out				217 359		146.097

Estimated at 15 °F above top of quench section

2. Calculate rate of quench flow

	V or L	°API	°F	lb/h	Enthalpy BTU/lb	Enthalpy BTU/h × 10⁶
In						
Quench leaving tower	L	1.0	650	X	360	360 X
Total in				X		360 X
Out						
Quench returning	L	1.0	480	X	214	214 X
Quench duty						42.530
Total out				X		42.530 + 214 X

Return temp fixed at 480 °F

$$X = \frac{42.530 \times 10^6}{360 - 214} = 291\,301 \text{ lb/h}$$

Type of trays in quench section – sheds.
No weirs.

Equipment data sheet shows diameter of quench section is 10 ft. It contains eight shed baffle rows with the following configuration:

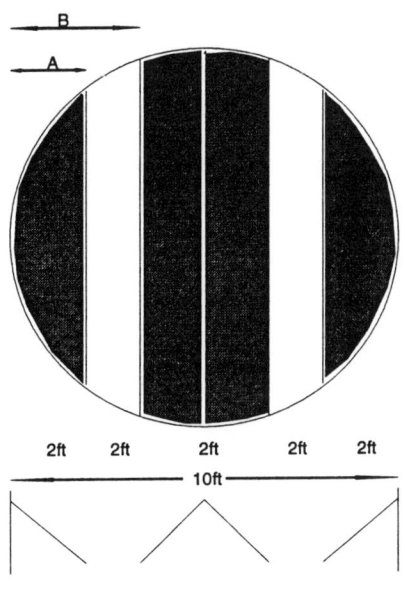

To calculate horizontal free area

From the weir length and downcomer (Table A1.1 in Appendix 1) calculate τ for segment A.

$$= \frac{2}{10} = 0.2$$

Corresponding $A_D = 0.142$

$$A_D = \frac{\text{Segment area}}{\text{Tower area}}$$

Segment area $= 0.142 \times 78.5$

$ = 11.15 \text{ ft}^2$

Repeat for segment B

Area $= 29.36 \text{ ft}^2$

Total horizontal free area

$ = 2\,(29.36 + 11.15)$

$ = 36.42 \text{ ft}^2$

Liquid mass velocity lb/h/ft² (free area)

$$= \frac{311\,277}{36.42} = 8547 \text{ lb/h/ft}^2$$

Vapour mass velocity lb/h/ft² (free area)

$$= \frac{214\,155}{36.42} = 5880 \text{ lb/h/ft}^2$$

Using disk and donut and shed baffle heat transfer

Correlation curves in the appendix read off H_0 in BTU/h/ft² $= 760$ BTU/h/°F. ft²

Calculate log mean temp difference (ΔT_M)

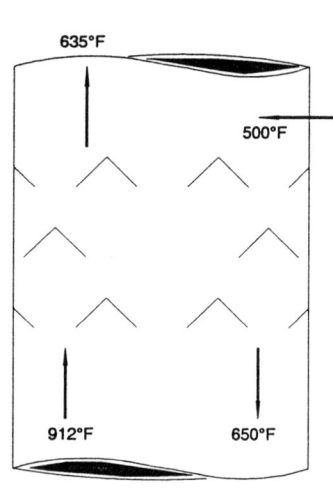

$$
\begin{array}{lll}
912 & \rightarrow & 635 \\
\underline{650} & \leftarrow & \underline{500} \\
272 & & 135
\end{array}
$$

$$\Delta T_M = \frac{272 - 135}{\mathrm{Log}_e(272/135)}$$

$$= \quad 195.6\,°\mathrm{F}$$

Total free area required

$$= \quad \frac{Q}{H_o \Delta T_M}$$

$$= \quad \frac{42.52 \times 10^8}{760 \times 195.6}$$

$$= \quad 287\ \mathrm{ft}^2$$

Total area available

$$
\begin{array}{ll}
= & 36.42 \times 8 \\
= & 291.4
\end{array}
$$

Heat transfer area satisfactory

Check tray loadings (Bottom baffle row)

Liquid flow	=	$301\,301 + 14\,476 = 311\,277$ lb/h
Sg at 60 °F	=	1.071
Sg at 650 °F	=	$0.773 = 6.44$ lb/gal
gal/h at 650 °F	=	$48\,335$
CFS at 650 °F	=	1.79
ρ_L at 650 °F	=	48.3 lb/ft^3

Vapour flow at 912 °F $\quad = \quad$ 2442.5 moles/h 214 155 lb/hr

$$\mathrm{ACFS} \quad = \quad \frac{2442.5 \times 378 \times 1372 \times 14.7}{3600 \times 38.7 \times 520}$$

$$= \quad 257\ \mathrm{CFS}$$

$$\rho_v \qquad\qquad = \quad \underline{0.231\ \mathrm{lb/ft}^3}$$

Critical velocity (tower diameter)

Shed baffle trays can operate at a maximum of 170% of critical velocity.

$$\text{Critical velocity } V_c \quad = \quad 0.157 \sqrt{\frac{\rho_L - \rho_V}{\rho_V}}$$

$$V_c \quad = \quad 0.157 \sqrt{\frac{48.3 - 0.231}{0.231}}$$

$$= \quad 2.265 \text{ ft/s}$$

$$\text{Actual vapour velocity } = \frac{257}{78.5} = 3.27 \text{ ft/s}$$

which is 145% of V_L and is acceptable. Quench section should perform well under these conditions.

12 DISTILLATE HYDROCRACKING

12.1 Process Description of a Distillate Hydrocracker

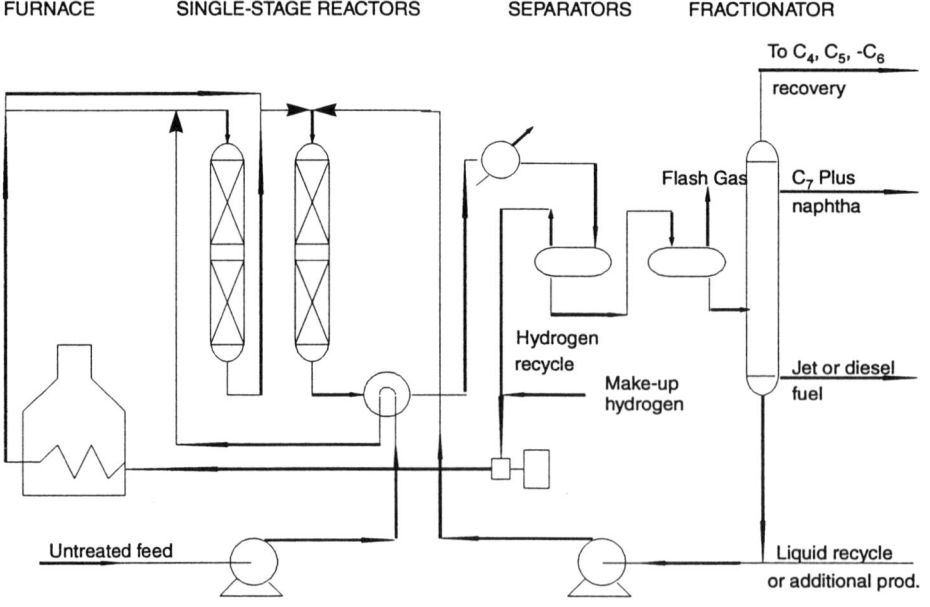

Figure 12.1. Distillate hydrocracking

Figure 12.1 shows a typical hydrocracking unit in which the fresh feed is pumped through a series of feed/effluent exchangers to the inlet of the first of two reactors in series. The preheated feed is mixed with hot recycle gas at the inlet to the first reactor. In this case only the recycle gas passes through the fired heater, although in some processes the combined feed and recycle gas are heated. The first reactor is usually filled with a hydrotreating catalyst for the partial desulphurization and denitrogenation of the fresh feed. The catalyst employed for the hydrotreating is an alumina-based type containing cobalt and molybdenum. To protect the catalyst from fouling by iron compounds and any salt in the feedstock, the first few feet of the reactor is usually used as a guard bed. Some processes utilize an external guard reactor.

The hydrogenated feedstock is then mixed with additional hydrogen and passes to the hydrocracking reactor, where cracking to the desired products takes place. The number of hydrocracking reactor stages will either be one or two, depending upon the products required. The hydrocracker reactor effluent is cooled by exchange with fresh feed and recycle hydrogen and then flashed in the high-pressure separator. The liquid from the high-pressure separator passes to a further low-pressure separator while the gas stream, which is rich in hydrogen, is recycled to the reactors. Make-up hydrogen is added as required.

The liquid from the low-pressure separator is then pumped to the fractionation train where products are separated. The unconverted portion of the fresh feed may then be either recycled for further cracking or used as product. Flash gas from the low-pressure separator is usually treated to remove the acidic components before being sent to the refinery fuel gas system.

Typical feeds include atmospheric and vacuum virgin gas oils, catalytic cycle oils, deasphalted oil, coker gas oil, thermal gas oil and paraffin raffinates from an SO_2 kero plant. Heavy feedstocks such as vacuum gas oils are usually limited to an end point of about 1050 °F.

Products range from LPG, naphthas, kero, and middle distillates through to lube oils, the more usual being naphthas, kero and diesel fuel. It is possible to vary the products from a hydrocracking unit by merely changing the reactor operating conditions and the fractionation cut points. This flexibility of operation is one of the major factors in favour of hydrocracking.

The overall hydrocracking reaction is exothermic. Temperature control of the reactor is accomplished by the introduction of cold hydrogen quench streams at the appropriate locations in the reactor.

12.2 Typical Distillate Hydrocracker Reactor Side Conditions and the Mechanism of Hydrocracking

The reactor side of a hydrocracker is always a licensed unit and the data concerning the reactor side are therefore proprietary to the licensor. As in the case of all licensed units, however, data are provided to clients under a licensing agreement to use in the operation and maintenance of the unit. Such data are limited to the client's personnel only and disclosure of these data to third parties is prohibited by secrecy agreements. The kind of data made available to licencees are in the form of graphs or curves giving operating variables and the like. Sufficient data exist, however, in the public domain to give a reasonable picture as to the range of operating conditions under which distillate hydrocrackers normally function. These are discussed below.

REACTOR TEMPERATURE

Unlike thermal cracking and catalytic cracking, the temperature levels for hydro-cracking are relatively low. Normally, reactor inlet temperatures are between 700 °F and 750 °F, at the beginning of a run rising to a maximum of about 950 °F towards the end. These temperatures depend on the age of the catalyst and the type of operation (i.e. naphtha production, diesel production or jet fuel, etc.). Strict temperature control is required in hydocracking, however, due to the exothermal nature of the reactions. As with all catalytic reactions, the rate of reaction increases with increase in temperature. Without this temperature control therefore the unit would soon experience a temperature-runaway situation. The control is effected by injection of cold hydrogen streams into predetermined sections of the reactor. A series of strategically placed thermocouples in the reactor beds monitor the temperature profile through the reactor(s).

REACTOR PRESSURE

The reactor pressure is set by the requirement of a hydrogen partial pressure that the licensor and catalyst manufacturer has determined for the proper function of the catalyst. Normally, this partial pressure at the outlet of the reactor is required to be in excess of 1000 psig for most operations. This figure in turn is dependent on the reactor pressure, the purity of the hydrogen at the outlet and the degree of hydro-carbon vapours present at this point. The reactor pressure therefore becomes a variable in establishing and maintaining the hydrogen partial pressure and which can be set on the plant and be controlled. For most operations the reactor pressure is set between 2000 and 2500 psig at the reactor inlet.

SPACE VELOCITY

This is really a measure of the reactant's residence time in contact with the catalyst and at the reaction conditions. The space velocity is set by the catalyst manufacturer and will vary for the type of operation, the feed quality, and the reactor temperature. Generally, the space velocity is lowest in processing to meet the lighter product make. In naphtha production space velocities of about 1.0 v/v/h would be used while for diesel production a space velocity of about 1.5–2.0 is common.

CATALYSTS

Catalyst properties are being continually modified and a number of higher-activity catalysts have been developed in recent years. This enables existing plants to obtain longer run lengths between regenerations. It also enables new plants to be designed with smaller reactors—that is, to be able to operate at higher space velocities. Present-day catalysts are unaffected by nitrogen and this has enabled the series operation in two-stage hydrocracking. In this method the effluent from the first reactor can be sent directly to the second without having to remove the products of nitrogen and H_2S.

HYDROGEN RECYCLE

The high hydrogen content gas from the high-pressure separator is recycled to the reactors and together with make-up hydrogen it satisfies the partial pressure requirements of the process. Normally, this recycle stream will have a hydrogen content of 75–80% mol. The make-up stream is usually in excess of 95% mol. The recycle rate is set by the licensor to satisfy the partial pressure requirements and to ensure a satisfactory flow of hydrogen to 'wash' the catalyst. The recycle rate depends on the type of operation and the catalyst type. Rates range from about 700 Scf/Bbl for middle distillates to 1000+ Scf/Bbl for light distillate production.

THE EFFECT OF IMPURITIES IN THE FEED

Most hydrocracker feeds contain undesirable impurities. The major ones normally encountered are:

- Sulphur
- Organic nitrogen compounds
- Metals
- Free carbon (Conradson carbon)

High concentrations of nitrogen and the metals poison the catalyst and significantly reduce its performance. High carbon is not so disastrous. It will, however, shorten cycle runs requiring more frequent catalyst regeneration. The effect of sulphur on the catalyst is not significant. Desulphurization of the sulphur compounds occurs under the severe conditions of hydrocracking and even the heavy thiophenes are converted to H_2S. Of course, the H_2S formed in the process has to be removed and this adds to the cost of the plant. The most undesirable effect of sulphur and nitrogen is that the ensuing H_2S and ammonia combine to form ammonium bisulphide, which precipitates out of the oil in the reactor effluent at a temperature of about 400 °F. These salt crystals are particularly abrasive and, unless properly handled, lead to equipment errosion and plugging. The poisoning of the catalyst by metals is usually countered by the installation of a guard reactor containing CoMo molybdenum catalyst.

Undesirably high concentrations of impurities in the feed can be considered as being in the range of:

- Sulphur   2–3% wt
- Nickel and vanadium   40+ppm by wt
- Conradson carbon   0.4%+ by wt
- Nitrogen   1000+ppm by wt

THE MECHANISM OF HYDROCRACKING

The mechanism of hydrocracking is complex as several reactions occur in the process almost simultaneously. The major ones are as follows:

(1) Long-chain paraffins are isomerized and then cracked into light isoparaffins.

(2) Monocyclic naphthenes are dealkylated and then isomerized into light isoparaffins and low molecular weight naphthenes.

(3) Bicyclic naphthenes are reacted to open one ring to alkyl-substituted monocyclic naphthenes which then follow reaction 2.

(4) Alkylbenzenes are dealkylated to form aromatics. Isoparaffins are produced by combining the alkyl groups removed from the alkylbenzenes. Alkylbenzenes are also hydrogenated to form monocyclic naphthenes which also follow reaction 2.

(5) Benzenaphthenes react primarily by opening the naphthene ring to produce alkylbenzenes which react as described in reaction 4. Benzonaphthenes also undergo hydrogenation of the benzene ring to form bicyclic naphthenes which follow reaction 2.

(6) Polyaromatics (two or more benzene rings) first undergo hydrogenation of one ring to form benzonaphthenes. These in turn react as described in reaction 5.

As can be seen, most of the reactions move away from the aromatic compounds to favour the production of the paraffin isomers and naphthenes. This provides products which are excellent cat reformer stock for gasolines and, in petrochemicals, the BTX (benzene, toluene, xylene) precursors. The breakdown of the polyaromatics and the heavier aromatics to essentially paraffins and naphthenes provides the basis for good jet fuel (low smoke point) and diesel stock (high diesel index).

12.3   Building Up the Recovery Side's Main Fractionator Feed from Lab Data

The feed to the distillate hydrocracker recovery side is the liquid phase from the low-pressure separator. This stream is still high in hydrogen and the light ends from the hydrocracking reactions. These light ends must be removed before the main fractionator can be used to separate the product distillates. A debutanizer column is used to effect this light ends removal. This fractionator also recovers a butane and lighter as overhead gas and distillate streams from which propane and butane LPG products can be obtained.

The bottom product from the debutanizer is the feed to the main fractionator. In this unit the main fractionator performs as a crude distillation unit. It operates with a small positive pressure in the overhead drum (say 5–10 psig). The overhead product is a full-range naphtha similar to the crude unit. The side-streams, also similar to the crude unit, are usually kero and diesel cuts. Normally, there will not be a heavier side-stream distillate than diesel. There will be a bottoms product which will be the unconverted portion of the reactor feed. This is sometimes recycled back to the second reactor for further hydrocracking or, in some cases, taken off and sent to storage for blending into fuel oil or heavy gas oil. A third option for the disposal of this stream is to blend it into FCCU feed.

This describes the development of the main fractionator feed TBP curve. The development of these data follows the section steps given below:

- *Step 1* Calculate the TBP curves for light naphtha, heavy naphtha, kero, diesel and unconverted oil (UCO) from their respective ASTM curves (lab data).
- *Step 2* Plot each of the TBP curves as shown in Figure 12.2

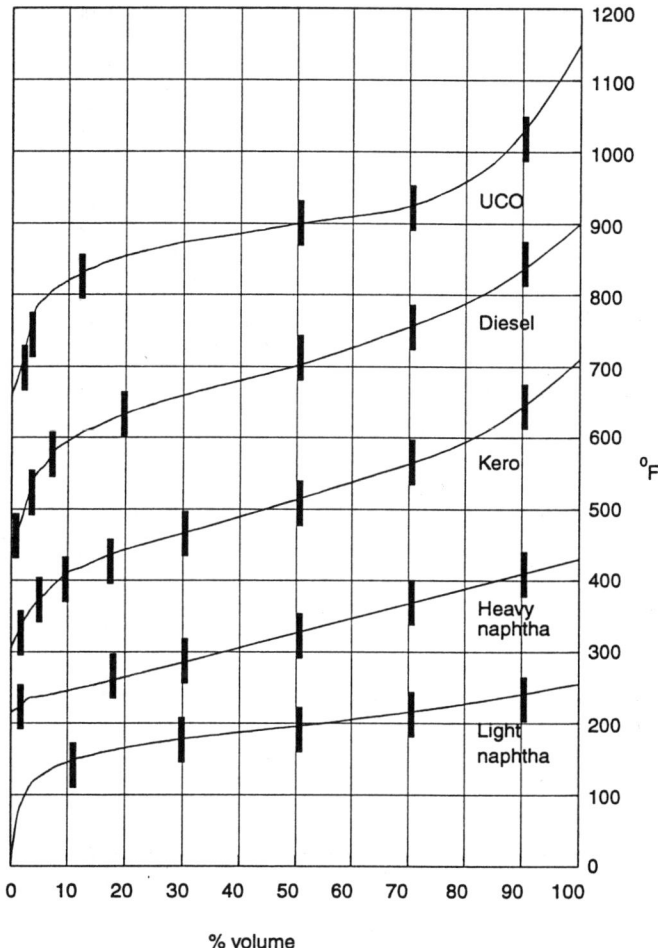

Figure 12.2. Main fractionator TBP curves

- *Step 3* Starting with the light naphtha, divide each of the TBP curves into about six boiling-point fractions.
- *Step 4* In making these fractions a boiling range chosen in one product TBP must also be represented in adjacent products if that product includes the boiling range (see also Section 6.3).
- *Step 5* From plant flow data calculate the yield of each product as a volume percentage on total feed.
- *Step 6* List the cut points for each fraction developed in Step 3. Against each fraction calculate its percentage on total feed. Do this for each of the product steams that contain the respective fractions.
- *Step 7* Add the individual yields of the cut point fractions horizontally for each product stream. The total gives the yield on total feed of all the chosen cut point ranges.
- *Step 8* Plot the cumulative volume per cent from Step 7 against the respective cut point. This is the main fractionator feed TBP curve and is drawn as Figure 12.3.

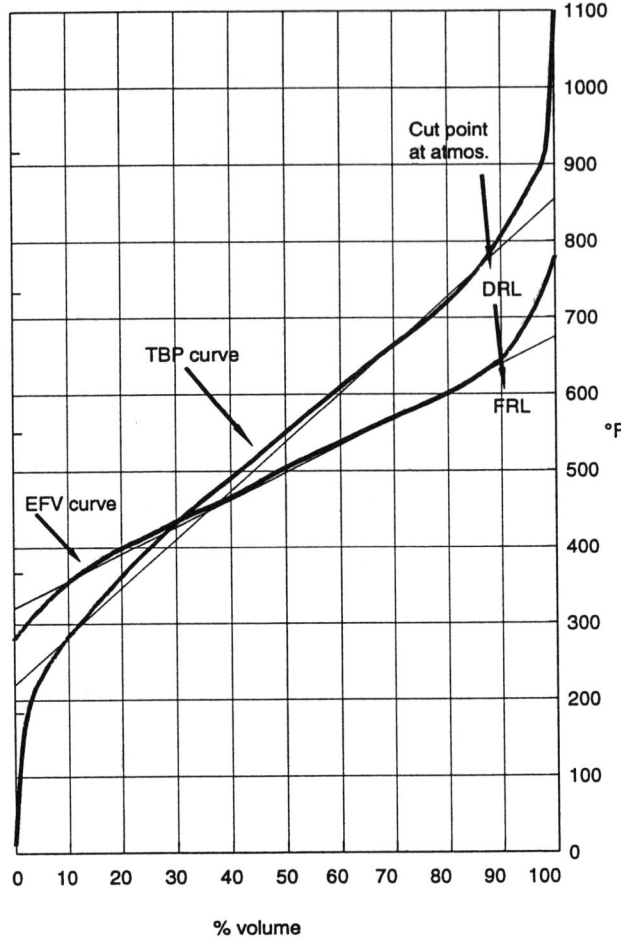

Figure 12.3. Debutanized liquid effluent ex hot flash-TBP curve

An example calculation now follows:

Example Calculation

The following are lab plant data from a test run on the recovery of the DHC.

Lab results

	ASTM distillation (°F)			All converted to D86 (°F)		
	Light nap	Heavy nap.	Kero	Diesel	UCO	Total
IBP	27.9	131.5	193.3	261	365.5	
10	79.7	139.2	231.3	324	433	
30	88.8	152.8	247.0	346	447.5	
50	94.1	164.9	264.0	361	455.1	
70	100.7	176.9	283.3	381	463.0	
90	109.2	194.8	317.7	424	510.0	
FBP	114.8	210.3	354.8	455	N/A	
°API	72.6	60.7	46.3	28.3	16.5	
Plant data flows m³/h	3.38	13.06	24.74	22.78	2.90	66.86

Calculate Liquid TBP Curve

% LV yield

Light naphtha	5.06
Heavy naphtha	19.53
Kero	37.00
Diesel	34.07
UCO	4.34
Total	100.00

Light naphtha

	ASTM		TBP	
	°F	ΔT	ΔT	°F
IBP	82			12
10	176	94	134	146
30	192	16	32	178
50	201	9	17	195
70	214	13	22	217
90	228	14	20	237
FBP	239	11	14	251

Heavy naphtha

	ASTM		TBP	
	°F	ΔT	ΔT	°F
IBP	270			216
10	282	12	27	243
30	307	25	46	289
50	329	22	37	326
70	351	22	34	360
90	383	32	42	402
FBP	410	27	31	433

Kero

	ASTM		TBP	
	°F	ΔT	ΔT	°F
IBP	379			310
10	448	69	104	414
30	477	29	51	465
50	507	30	47	512
70	541	34	48	560
90	604	63	73	633
FBP	671	67	74	707

Diesel

	ASTM		TBP	
	°F	ΔT	ΔT	°F
IBP	502			439
10	615	113	154	593
30	655	40	65	658
50	682	27	43	701
70	718	36	51	752
90	795	77	85	837
FBP	851	56	61	898

UCO

	ASTM		TBP	
	°F	ΔT	ΔT	°F
IBP	690			662
10	811	121	162	824
30	838	27	49	873
50	851	13	24	897
70	865	14	22	919
90	950	85	93	1012
FBP	00			00

Cut points (°F)	% Volume on feed						Cumulative total
	LT nap.	HY nap.	Kero	Diesel	UCO	Total	
12 to 149	0.51					0.51	0.51
to 180	1.01					1.01	1.52
to 190	1.01					1.01	2.53
to 220	1.01					1.01	3.54
to 240	1.01	0.59				1.60	5.14
to 255	0.51	2.72				3.23	8.37
to 285		2.54				2.54	10.91
to 320		3.91	0.37			4.28	15.19
to 365		3.91	1.11			5.02	20.21
to 405		3.91	1.85			5.76	25.97
to 432		1.95	2.59			4.54	30.51
to 465			5.18	0.34		5.52	36.03
to 510			7.40	0.34		7.74	43.77
to 560			7.40	1.36		8.76	52.53
to 635			7.40	4.43		11.83	64.36
to 710			3.70	11.25	0.04	14.99	79.35
to 755				6.13	0.09	6.22	85.57
to 835				6.81	0.39	7.20	92.77
to 900				3.41	1.65	5.06	97.83
to 920					0.87	0.87	98.70
to 1010					0.87	0.87	99.57
to 1160					0.43	0.43	100.00
Totals	5.06	19.53	37.00	34.07	4.34	100.00	

This is plotted as Figure 12.3.

12.4 A Procedure for Determining Liquid/vapour Separation from High-pressure and Low-pressure Flash

The reactor effluent from the distillate hydrocracker (DHC) reactor is successively cooled and condensed in a series of heat exchangers before entering a high-pressure flash drum. This drum operates at slightly lower than the reactor pressure at 2518 psia. The effluent entering this drum has been cooled to 120 °F and much of the heavy oil and LPG fractions condensed. The remaining uncondensed gas is separated from the liquid portion of the effluent in this drum. The uncondensed vapour will be high in hydrogen content, and after treating for H_2S removal, will be recycled to the reactor.

The liquid portion from the high-pressure separation will still contain an appreciable amount of hydrogen and light hydrocarbons. This stream is routed to a second drum at considerably lower pressure. Here more of the light ends come out of solution under lower pressure. This vapour stream is routed to fuel while the liquid stream is fed to a debutanizer in the recovery side of the unit.

Knowing the composition of the reactor effluent, the quantity and composition of each stream leaving the two flash drums can be predicted by flash vaporization calculations. This is described in the following steps:

- *Step 1* From the lab data on the composition of the reactor effluent and plant data for its flow rate calculate moles/h for each component. These will be hydrogen through C_6^+, the C_6^+ portion is the product oil which is assumed liquid throughout all the separation stages.
- *Step 2* Check plant data for the temperature and pressure conditions of the high-pressure flash drum. This should be 2000 psia plus and at about 100–120 °F temperature.
- *Step 3* Because of the high pressure use equilibrium constants K at 5000 psia convergence pressure if using K's based on convergence pressures. List the K in the table against each component and their quantity.
- *Step 4* The flash vaporization calculation is 'trial and error' using guesses for the ratio V/L (moles vapour to moles liquid). This ratio is used in the expression:

$$L = \frac{X_F}{1 + (V/L)K}$$

for each component liquid portion. The sum of all liquid portion is the total liquid moles. The vapour moles is the feed minus L. The calculated V/L is solved when 'assumed V/L' equals 'calculated V/L'. A straight-line plot on a LogLog graph helps to obtain a final guess. (See Sections 8.7, 9.5, 10.5).
- *Step 5* Take the liquid moles/hour calculated in Step 4 and use this as the feed in a flash vaporization calculation at the low-pressure drum conditions. Repeat Step 4 for this but use K values at 1500 psia convergence, if using K's based on convergence pressure.

An example calculation now follows. The effluent composition and quantity are fictional. They were, however, predicted by successive bubble point calculations starting with the debutanizer feed.

Example calculation

High-pressure flash vaporization

Flash drum operates at 120 °F and 2503 psig

Comp.	Reactor effluent moles/h	K $\dfrac{2518\ psia}{120\ °F}$	1st trial $V/L = 10$		2nd trial $V/L = 5.0$		3rd trial $V/L = 22.0$		Moles vapour	% mol.
			$V/L\ K$	$L = \dfrac{X_F}{1+(V/L)K}$	$V/L\ K$	$L = \dfrac{X_F}{1+(V/L)K}$	$V/L\ K$	$L = \dfrac{X_F}{1+(V/L)K}$		
H_2	4574.9	8.2	82	55.1	41	108.9	180	25.3	4549.6	88%
C_1	273.4	1.9	19	13.7	9.5	26.0	42	6.4	267.0	
C_2	75.5	0.9	9	7.6	4.5	13.7	19.8	3.8	71.9	
H_2S	272.0	0.82	8.2	29.6	4.1	53.3	18.0	14.3	257.7	
C_3	69.0	0.54	5.4	10.8	2.7	18.6	11.9	5.3	63.7	
iC_4	47.0	0.40	4.0	9.4	2.0	15.7	8.8	4.8	42.2	
nC_4	43.3	0.34	3.4	9.8	1.7	16.0	7.5	5.1	38.2	
iC_5	3.0	0.24	2.4	0.9	1.2	1.4	5.3	0.5	2.5	
nC_5	3.7	0.20	2.0	1.2	1.0	1.9	4.4	0.7	3.0	
C_6^+	209.1	—	—	209.1	—	209.1	—	209.1	Nil	
Total	5570.9			347.2		464.6		275.1	5295.8	
Calculated V/L				15.0		11.0		19.3		

Convergence pressure 5000 psia
Vapour phase is recycle hydrogen stream
After H_2S scrubbing H_2 mol. % recycle $= 90\%$
Design H_2 content $= 88\%$

Low-pressure flash drum

This flash drum operates at 120 °F and 225 psig $= 240$ psia

Use 1500 psia convergence pressure

Comp.	Liquid Ey HF moles/h	K $\dfrac{240\ psia}{120\ °F}$	1st trial $V/L = 0.5$		2nd trial $V/L = 0.1$		3rd trial $V/L = 0.125$	
			$V/L\ K$	$L = \dfrac{X_F}{1+(V/L)K}$	$V/L\ K$	$L = \dfrac{X_F}{1+(V/L)K}$	V/LK	$L = \dfrac{X_F}{1+(V/L)K}$
H_2	25.3	74	37	0.67	7.4	3.0	9.25	2.47
C_1	6.4	11.5	5.8	0.94	1.15	2.98	1.44	2.62
C_2	3.6	2.9	1.5	1.44	0.29	2.79	0.36	2.65
H_2S	14.3	2.3	1.15	6.65	0.23	11.63	0.29	11.08
C_3	5.3	1.0	0.5	3.53	0.10	4.82	0.125	4.71
iC_4	4.8	0.5	0.25	3.84	0.05	4.57	0.063	4.52
nC_4	5.1	0.36	0.18	4.32	0.036	4.92	0.045	4.88
iC_5	0.5	0.18	0.09	0.46	0.018	0.49	0.023	0.49
nC_5	0.7	0.14	0.07	0.65	0.014	0.69	0.018	0.69
C_6^+	209.1		209.1	209.10		209.1		209.1
Total	275.1			231.6		244.99		243.21
Calculated V/L				0.19		0.12		0.13

Liquid portion is feed to recovery side.

12.5 Calculating Tray Liquid and Vapour Flows in the Debutanizer

The recovery side debutanizer takes feed from the DHC's LP separator, which is preheated by exchange with hot reactor effluent stream before entering the debutanizer column. In this column butanes and lighter are taken off as a vapour and liquid overhead stream while the C_5^+ oil stream leaves as bottom product.

This section demonstrates the method used to predict tower conditions and to calculate tray loading for top and bottom trays. These calculations are described by the following steps:

- *Step 1* From the composition and flow of the low-pressure separator liquid calculate its mol wt and gravity.
- *Step 2* Break down the C_6^+ portion of the feed to pseudo-component. This would be accomplished from the respective TBP curves (see Section 12.3).
- *Step 3* Build up a unit component and material balance.
- *Step 4* Because the feed contains a significant amount of hydrogen and methane the overhead condenser will be a partial condenser. Carry out a flash vaporization calculation of the overhead total product at the reflux drum temperature and pressure. (See Section 11.4 for this method.)
- *Step 5* Calculate mol. weight and gravity for the vapour leaving the drum and for the liquid distillate.
- *Step 6* Calculate the tower-top conditions for the partial condenser. Set the reflux ratio (use about 3.1 or higher in this case) and the tower-top pressure. This pressure will be reflux drum pressure plus, say, 5 psi for condenser.
- *Step 7* Add mol/h of each component in the total overhead product to the corresponding component in the reflux stream. Calculate the mol. fraction components in the composite stream.
- *Step 8* Carry out a dew point calculation at tower-top pressure on the composite stream. The solution will be the tower-top temperature.
- *Step 9* Allocating mol. wts, gravity to the X column from the dew point calculation sum to establish the mol. wt and gravity of the overflow liquid from top try.
- *Step 10* Using the fixed external reflux rate and top product rates calculate a heat balance over the condenser. This gives the condenser duty.
- *Step 11* Using the condenser duty, calculate heat balance over the tower-top tray. The unknown X will be the weight per hour of the overflow from the top tray. Estimate the vapour from tray 2 in this calculation to be 5 °F above tower top-temperature.
- *Step 12* Calculate tower-bottom pressure by adding 0.2 psi per tray to the tower-top pressure.
- *Step 13* Establish tower-bottom temperature by carrying out a bubble point calculation on the bottom product at the bottoms pressure.
- *Step 14* The Y factor of the bubble point calculation in Step 13 is composition of the vapour going to the bottom tray. Calculate its mol. wt and gravity.
- *Step 15* Carry out an overall heat balance over the tower. The unknown in this case is the reboiler duty. Establish this by difference.

• *Step 16* Calculate a heat balance over the reboiler. The feed will be bottom product plus the vapour to bottom tray as a liquid from the bottom tray. The unknown is the pounds/hour of the vapour to the bottom tray. Solve by equating heat in equals heat out.

• *Step 17* Summarize the vapour and liquid flows on the top and bottom tray.

An example calculation now follows.

Example Calculation

Debutanizer feed includes DHC low-pressure flash liquid, kero and absorber liquid recycle. Debutanizer feed composition is as follows:

	Moles					
	Moles/h	Mol. frac.	MW	lb/hr	lbs/gal	gal/h
H_2	15.81	0.015	2	32	—	—
C_1	19.86	0.019	16	318	—	—
C_2	24.10	0.023	30	723	2.96	244
H_2S	111.07	0.105	34	3776	6.55	576
C_3	51.12	0.048	44	2249	4.22	533
iC_4	52.93	0.050	58	3070	4.69	655
nC_4	56.59	0.053	58	3282	4.87	674
iC_5	4.92	0.0005	72	354	5.29	68
nC_5	6.50	0.006	72	468	5.26	89
C_6^+	717.31	0.676	220.3	158 023	6.72	23 531
Total	1060.21	1.000	162.5	172 295	6.53	26 370

C_5^+ portion of feed is broken down as follows:

	Moles/h	Mol. wt	lb/gal	lb/h	gal/h
NBP 194 °F	74.9	96	5.832	7191	1233
302 °F	106.0	130	6.122	12 812	2256
412 °F	158.7	170	6.409	27 054	4221
516 °F	151.2	220	6.666	33 330	5000
651 °F	161.7	297	6.943	48 000	6913
815 °F	54.5	426	7.268	23 227	3187
921 °F	10.3	525	7.50	5409	721
	717.31	220.3	6.72	158 023	25 531

Debutanizer material balance

	Feed			Total overheads			Bottom product			MW	lb/gal
	Moles/h	lb/h	gal/h	Moles/h	lb/h	gal/h	Moles/h	lb/h	gal/h		
H_2	15.81	32	—	15.81	32	—				2	—
C_1	19.86	318	—	19.86	318	—				16	—
C_2	24.10	723	244	24.10	723	244				30	2.96
H_2S	11.07	3776	576	111.00	3774	575	0.07	2	1	34	6.55
C_3	51.12	2249	533	51.04	2245	532	0.08	3	1	44	4.22
iC_4	52.93	3070	655	52.63	3053	651	0.30	17	4	58	4.89
nC_4	56.59	3282	674	56.18	3258	689	0.41	24	5	58	4.57
iC_5	4.92	354	68	4.76	344	55	0.14	10	2	72	5.21
nC_5	6.50	458	89	6.25	451	86	0.24	17	3	72	5.25
NBP 194 °F	74.90	7191	1233	1.89	181	31	73.01	7010	1202	96	5.83
NBP 302 °F	106.00	13 812	2256				106.00	13 812	2256	130	6.12
NBP 412 °F	158.70	27 054	4221				158.70	27 054	4221	170	6.41
NBP 516 °F	151.20	33 330	5000				151.20	33 330	5000	110	6.67
NBP 651 °F	161.70	48 000	6913				161.70	48 000	8913	297	6.94
NBP 615 °F	54.50	23 227	3187				54.50	23 227	3187	426	7.29
NBP 921 °F	10.30	5409	721				10.30	5409	721	525	7.50
Total	1050.20	172 295	26 370	343.55	14 380	2854	716.65	157 915	23 516	162.5	8.53

Carry out flash calculation on overheads

Set drum at 90 °F 180 psig

	Moles/h	$K\frac{90\,°F}{195\,psia}$	1st trial $V/L=1.0$		2nd trial $V/L=0.5$		3rd trial $V/L=1.4$		Liquid	
			V/LK	$L=\frac{X_F}{1+(V/L)K}$	V/LK	$L=\frac{X_F}{1+(V/L)K}$	V/LK	$L=\frac{X_F}{1+(V/L)K}$	lbs/h	gal/h
H_2	15.81	98	98	0.2	49					
C_1	19.86	12.5	12.5	1.5	6.25	0.32	137.2			
C_2	24.10	2.8	2.8	6.3	1.4	2.74				
H_2S	111.00	2.1	2.1	35.8	1.05	10.04	17.5	0.11	—	—
C_3	51.04	0.9	0.9	26.9	0.45	54.15	3.92	1.07	17	—
iC_4	52.63	0.41	0.41	37.3	0.2	35.20	2.94	4.90	147	50
nC_4	56.18	0.30	0.30	43.2	05	43.68	1.26	28.17	958	146
iC_5	4.78	0.12	0.12	4.2	0.15	48.85	0.57	22.58	994	236
nC_5	6.25	0.10	0.10	5.7	0.06	4.51	0.42	33.52	1944	414
NP194	1.89	0.01	0.01	1.87	0.05	5.96	0.17	39.56	2294	471
						1.89	0.14	4.09	294	58
							0.0	5.49	395	75
							14	1.86	179	31
Total	343.55			162.97		207.34		141.35	7222	1479
Calculated V/L				1.11		0.66		1.43		

Overhead distillate liquid = 51.1 mol. wt
= 4.8 lb./hr = 110.1 °API

Overhead vapour = 35.4 mol. wt
= 5.22 lb./gal = 94.1 °API

Tower-top conditions

Set reflux ration to 3.2:1.0

Moles/h reflux = 343.55 × 3.2 = 1099

Composition and overhead dew point (pressure = 185 psig) 2000 psia

Comp.	Prod.	Reflux	Total	Mol. frac. γ	1st trial at 170 °F		2nd trial at 182 °F		Weight factor	Volume factor
					K	$X = Y/K$	K	$X = Y/K$		
H_2	15.81	0.85	16.66	0.012	76	NEG		NEG	—	—
C_1	19.86	8.32	23.18	0.320	13	0.002	13.5	0.001	0.016	—
C_2	24.10	38.09	62.19	0.043	4.5	0.010	5.0	0.009	0.270	0.09
H_2S	111.00	219.02	330.02	0.229	0.5	0.065	3.8	0.060	2.040	0.31
C_3	51.04	175.56	226.60	0.157	1.6	0.037	2.0	0.079	3.476	0.82
iC_4	52.63	260.82	313.25	0.217	0.98	0.221	1.1	0.197	11.426	2.44
nC_4	56.18	307.56	363.75	0.252	0.74	0.341	0.83	0.304	7.632	3.52
iC_5	4.78	31.80	36.53	0.025	0.36	0.066	0.46	0.054	3.888	0.75
nC_5	6.25	42.68	49.94	0.034	0.31	0.110	0.36	0.089	6.408	1.22
NB194	1.89	14.46	16.35	0.011	0.05	0.220	0.062	0.177	16.992	2.81
Total	343.55	1083.98	1442.43	1.000		1.122		0.970	62.143	12.10

Actual temp.   = 181 °F
Mol. wt of liquid overflow   = 64
lb/gal of liquid overflow   = 5.1
°API of liquid overflow   = 100
K 12.3

Calculate condenser duty and top tray loading

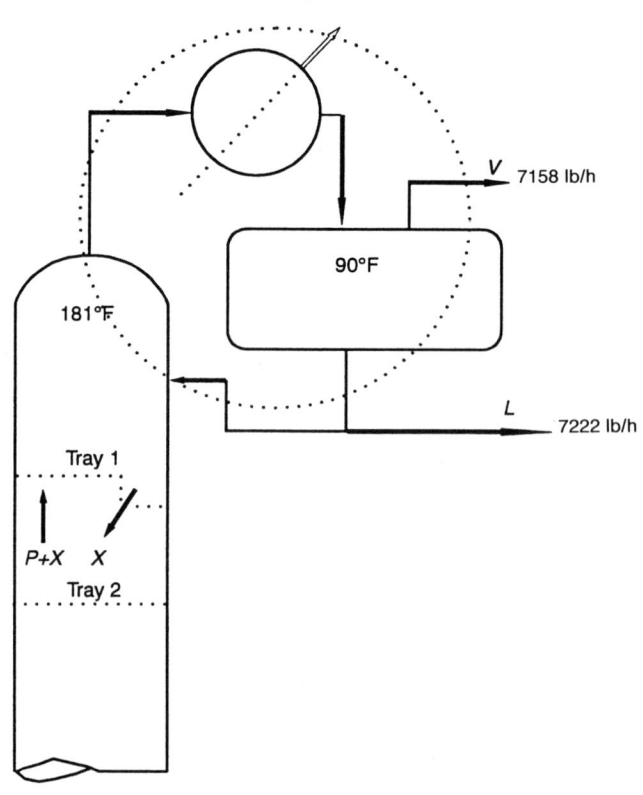

	V or L	°API	°F	lb/h	Enthalpy BTU/lb	Enthalpy BTU/h × 10⁶
In						
Product vapour	V	102	181	14 380	254	3653
Reflux vapour	V	110	181	58 151	260	14 599
Total in				70 531		18 252
Out						
Product vapour	V	94	90	7158	206	1.475
Product liquid	L	110	90	7222	52	0.376
Reflux	L	110	90	56 151	52	2.920
Condenser				By difference		13.481
Total out				70 531		18.252

Calculate vapour/liquid load on tray 1

	V or L	°API	°F	lb/h	Enthalpy BTU/lb	Enthalpy BTU/h × 10⁶
In						
Product vapour	V	102	186	14 380	257	3.696
Overflow	V	100	186	X	255	$255\,X$
Total in				$14\,380 + X$		$3.596 + 255\,X$
Out						
Product vapour	V	94	90	7158	206	1.475
Product liquid	V	110	90	7222	52	0.378
Overflow	L	100	181	X	102	$102\,X$
Condenser						13.481
Total out				$14\,380 + X$		$15.332 + 102\,X$

$$X = \frac{11\,636\,000}{153} = 76\,052 \text{ lb/h}$$

Vapour to tray $= 90\,432$ lb/h
$= 1\,531.9$ mole/h

Calculate bottom temperature

Tower top pressure $\quad = 185$ psig
Number of trays $\quad\quad = 30$
Tower bottom pressure $= (30 \times 0.2) + 185$
$\quad\quad\quad\quad\quad\quad$ 191 psig
$\quad\quad\quad\quad\quad\quad$ 206 psia

| Comp. | Mol. frac. X_F | 1st trial at 650 °F | | 2nd trial at 630 °F | | Vapour | |
		K	$Y = K_X$	K	$Y = K_X$	Weight factor	Volume factor
H_2S	NEG						
C_3	NEG						
iC_4	0.0004	8.8	0.0035	8.0	0.0032	0.136	0.0397
nC_4	0.0006	7.8	0.0047	6.5	0.0039	0.226	0.0464
iC_5	0.0002	5.8	0.0012	4.50	0.0009	0.065	0.0125
nC_5	0.0003	5.0	0.0015	4.00	0.0012	0.086	0.0163
NB194	0.1019	6.31	0.6430	4.59	0.4677	44.899	7.7014
NB302	0.1479	2.62	0.3875	1.80	0.2662	34.605	5.5546
NB412	0.2215	1.02	0.2259	0.68	0.1506	25.602	3.9941
NB516	0.2110	0.48	0.0802	0.24	0.0505	11.132	1.6690
NB651	0.2258	0.107	0.0242	0.056	0.0126	3.742	0.5392
NB815	0.0760	0.024	0.0018	0.006	0.0005	0.213	0.0292
NB921	0.0144	0.002	NEG	—	NEG		
Total	1.0000		1.3735		0.9574	120.757	19.7024

Actual temp.   = 635 °F
Mol. wt of vapour = 126.1
lb/gal. of vapour = 6.13
°API of vapour   = 60.5

Calculate reboiler duty

Enthalpy in this feed $= 27.862 \times 10^b$ BTU/h (previously calculated)

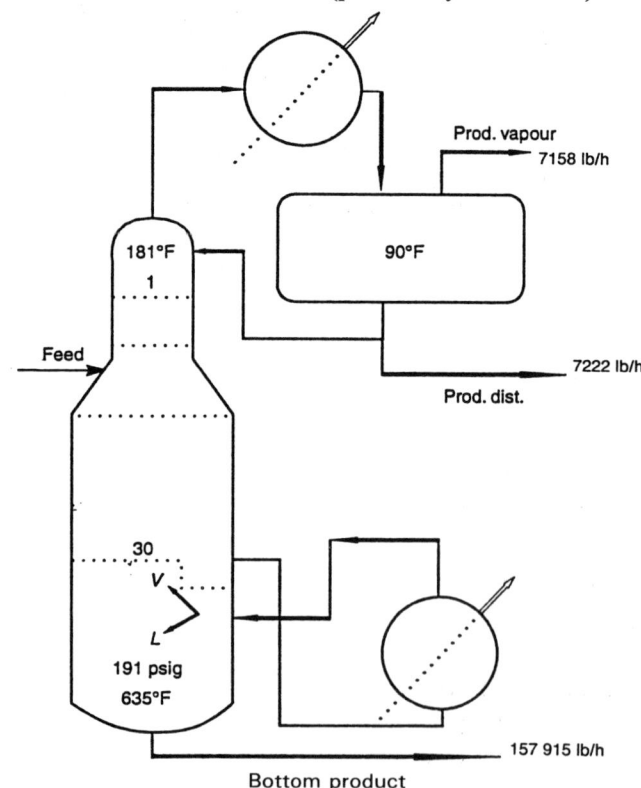

1. Overall heat balance

	V or L	°API	°F	lb/hr	Enthalpy BTU/lb	Enthalpy BTU/h $\times 10^6$
In						
Feed	$V+L$		186	172 295		27.862
Reboiler				By difference		48.109
Total in				172 295		75.971
Out						
Product vapour	V	94	90	7158	208	1.475
Product liquid	L	110	90	7222	52	0.376
Bottom product	L	44	635	157 915	384	60.639
Condenser						13.481
Total out				172 295		75.971

2. Calculate vapour/liquid flows to bottom tray

	V or L	°API	°F	lb/h	Enthalpy BTU/lb	Enthalpy BTU/h $\times 10^6$
In						
Vapour feed*	L	605	620[†]	V	399	399 V
Liquid feed*	L	44	620	157 915	374	59.060
Reboiler						48.109
Total in				$V+157\,915$		$107.169 + 395\,V$
Out						
Bottom product	L	44	635	157 915	384	60.638
Vapour to tray 30	V	60.5	635	V	481	481 V
Total out				$157\,915 + V$		$60.638 + 481\,V$

*Feed to the reboiler
[†]Estimated

$$V = \frac{46\,530\,000}{82} = 567\,439 \text{ lb/h}$$

$$= 4499.9 \text{ Moles/h}$$

and 92 568 gal/h at 60 °F

Vapour and liquid traffic in the debutanizer

Vapour flow	Top tray (1)	Bottom tray (30)
Temp. (°F)	186	635
Pressure (psia)	200	206
Moles/h	1531.9	4499.9
ACFS	14.68	71.0
lb/h	90 432	567 439
lb/ft³ρ_V (at cond.)	1.71	2.22
Liquid flow		
Temp. of liquid (°F)	181	620
gal/h at 60 °F	14 192	116 034
gal/h at temp.	17 836	209 026
CFS at temp.	76 052	725 354
lb/ft³ ρ_L (at cond.)	32.0	25.96

12.6 Main Fractionator Calculations

This section deals with establishing the flash zone conditions and the condenser duty of the main DHC fractionator. The flash zone calculation in this case is complicated by the presence of a bottom product reboiler. The flash zone acts as a liquid and vapour separator together with the washing of the unconverted oil stream (bottom product). The calculation is set out in the following steps:

- *Step 1* Set column reflux drum conditions. In this case this is taken from plant readings.
- *Step 2* Calculate the mean average boiling points and mol. weight for light naphtha, heavy naphtha, kero, diesel and UCO. The ASTM distillation curves from Section 12.3 are used in the method described in the introduction to this book.
- *Step 3* Develop the material balance for the unit using yield data given in Section 12.3.
- *Step 4* Calculate the whole feed EFV from the TBP curve (Figure 12.3). Establish the atmospheric flash temperature.
- *Step 5* Calculate the pressure for the flash zone. This is reflux drum pressure plus condenser pressure drop plus 0.25 psi pressure drop per tray.
- *Step 6* Calculate the provisional partial pressure of hydrocarbons in the flash zone. This is moles vapour plus overflash divided by moles hydrocarbon plus steam multiplied by total pressure.
- *Step 7* Refer to the vapour pressure curves in Figure A1.6 in Appendix 1 and read off the corresponding temperature to atmospheric flash temperature (Step 4) at the partial pressure.
- *Step 8* The reboiler adds heat to the tower. Calculate this, assuming that the total feed to it is vaporized at the stream temperature. Using this assumption, calculate moles/h of additional hydrocarbon vapour entering the flash zone.

● *Step 9*   This hydrocarbon vapour will be quenched by reflux entering flash zone. Therefore partial pressure now becomes

$$\frac{\text{Product moles}}{\text{Product moles} + \text{reboiler moles} + \text{steam}} \times \frac{\text{total}}{\text{pressure}}$$

● *Step 10*   Read off new temperature at the partial pressure calculated in Step 9 from vapour pressure curves. This is the estimated flash zone temperature.
● *Step 11*   From plant data establish side-stream (bottom of strippers) temperatures. Calculate an overall heat balance over the fractionator to establish the condenser duty.

An example calculation now follows.

Example Calculation

Column will operate at a reflux drum pressure of 9 psig (24 psia) and 122 °F

Total liquid naphtha (overhead) $\begin{cases} \text{LT} = & 5.06 \\ \text{HY} = & 19.53 \end{cases}$
$\qquad\qquad\qquad\qquad\qquad\qquad$ 24.59% vol. on feed

Kero = 37.00
Diesel = 34.07
UCO = 4.34

The material balance is given on page 296
From ASTM curves (see Section 12.3) calculate the ME AVBP thus:

Light naphtha

VABP $\qquad = \dfrac{27.9 + (2 \times 94.1) + 101.2}{4}$

$\qquad\qquad = 81.33\,°\text{C} \quad = \quad 178\,°\text{F}$

Slope $\qquad = \dfrac{229 - 175.5}{80} = 0.67\,°\text{F}\%$

$Me_{\text{AVBP}} \quad = 178 - 2 \quad = \quad 176\,°\text{F}$

Mol. wt $\quad = 96$

Naphtha

VABP $\qquad = 331\,°\text{F}$
Slope $\qquad = 2.05\,°\text{F}$
$Me_{\text{AVBP}} \quad = \quad 32\,°\text{F} \qquad MW = 142$

Kero

VABP   = 516.5 °F
Slope   = 1.9 °F/%
Me_{AVBP} = 509 °F   MW = 211

Diesel

VABP   = 694
Slope   = 2.25 °F/%
Me_{AVBP} = 684 °F   MW = 288

UCO

VABP   = 866
Slope   = 1.7 °F/%
Me_{AVBP} = 861 °F   MW = 420

Calculate whole feed EFV at atmos.

From Figure 12.3

Slope of DRL   = 6.25 °F/%
Slope of FRL   = 3.45 °F/% (from Figure A1.1)
ΔT_{50} (DRL − FRL) = 38 °F
DRL 50%   = 535 °F (from Figure A1.1)
FRL 50%   = 497 °F

% off	TBP (°F)			Δ (Flash)	EFV (°F)		
	TBP	DRL	ΔT (TBP)	ΔT (TBP)	ΔT FRL	FRL	EFV
0	12	225	− 213	0.2	43	325	282
10	285	285	0	0.4	0	355	355
20	350	365	+ 15	0.36	+ 5.4	390	395
30	435	410	+ 25	0.34	+ 8.5	425	434
40	490	475	+ 15	0.34	+ 5.1	460	465
50	550	540	+ 10	0.34	+ 3.4	497	500
60	610	600	+ 10	0.34	+ 3.4	532	535
70	665	665	0	0.34	0	567	567
80	720	730	− 10	0.34	− 34	605	602
90	810	795	+ 15	0.34	+ 5.1	640	645
100	1100	805	+ 295	0.34	+ 100	675	775

From the atmospheric EFV curve flash temp. = 695 °F

First pass

Total moles HC vapour	$= 564.9$
Total moles steam	$= 55.1$

Tower pressure $= 24$ psia $+ 3$ psi (condenser) $+ (28 \times 0.25)$
$\qquad\qquad = 34.0$ psi

Hydrocarbon partial pressure $\qquad\qquad = \dfrac{564.9}{620} \times 34.0$ psia

$\qquad\qquad\qquad\qquad\qquad\qquad = 31.0$ psia

Provisional flash zone temp. $\qquad\qquad = 765\,°F$

Pseudo-outlet temp. for UCO $\qquad\qquad = 745\,°F$

But reboiler adds 4.929×10^6 BTU/h

UCO temperature to reboiler $\qquad\qquad = 923\,°F$ (from model)

Assume all material circulation is vaporized at $923\,°F$

The moles HC to be quenched in flash zone:

$$\frac{4.929 \times 10^6}{(642 - 588)\ \text{BTU/lb}} = 91\,278\ \text{lb/h}$$

Additional HC moles in flash zone $= 217.3$

Partial pressure $= \dfrac{564.9}{837.3} \times 34.0$ new $°F$ zone temp. $= 23$ psia

New flash zone temp. $= 735\,°F$ (plant data gave $745\,°F$)

Draw-off temperature taken at stripper bottoms

Kero $\quad = 283\,°C = 541\,°F$
Diesel $= 339\,°C = 642\,°F$

Carry out overall heat balance to find condenser duty

	V or L	°API	°F	lb/h	Enthalpy BTU/lb	Enthalpy BTU/h × 10⁶
In						
Vapour feed	V	42	735	115 425	532	61.310
Liquid feed	L	16	735	4712	430	2.026
Main reboil						4.929
Kero reboil						1.667
Steam	V			1543	1265	1.952
Total in				121 500		71.784
Out						
Naphtha	L	69.5	122	26 298	61	1.604
Kero	L	48.3	541	43 262	315	13.628
Diesel	L	28.3	642	44 292	372	16.477
UCO	L	16.5	823	6105	588	3.590
Water	L			1543	122	0.188
Condenser				By difference		36.297
Total out				121 500		71.784

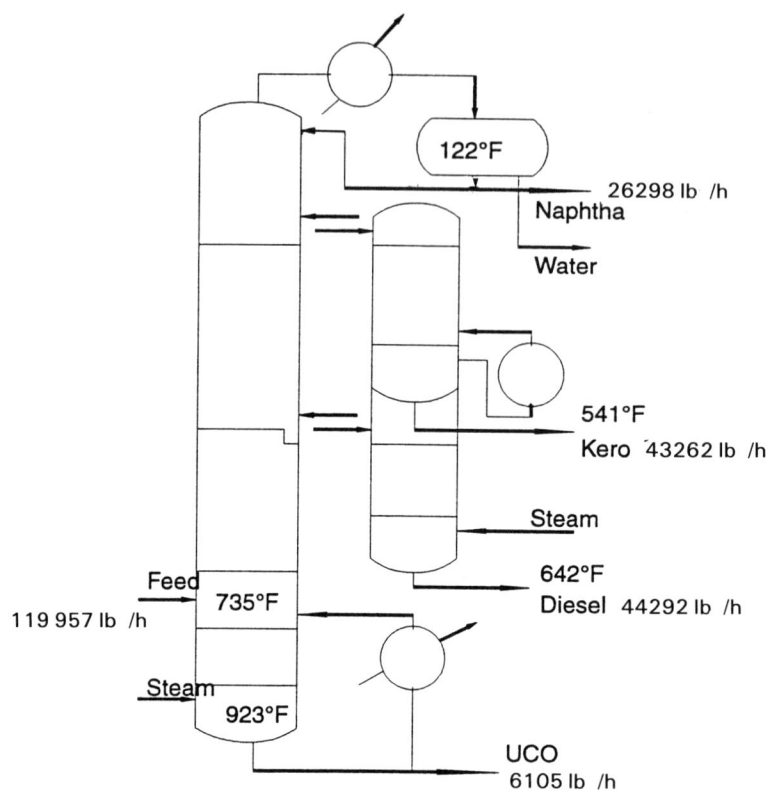

Main fractionator mat. balance

	Cut point (°F)	Volume			Gravity			Moles			
		% vol.	gal/h	BPSD	°API	lb/gal	lb/h	MW	Moles/h	% mole	
LT naptha	IBP −240	5.06	894	511	72.6	5.76	5149	96	53.6	9.31	} Total
HY naphtha	−400	19.53	3450	1971	60.7	6.13	21 149	142	148.9	25.86	} o/head
Kero	−618	37.00	6535	3734	46.3	6.62	43 262	211	205.0	35.60	distillate
Diesel	−872	34.07	6018	3438	28.3	7.36	44 292	288	153.8	25.71	
UCO	+872	4.34	767	438	16.5	7.96	6105	420	14.5	2.52	
Total		100.00	17 664	10 092	42.0	6.79	119 957	208	575.8	100.00	

Flash zone material balance

	Cut point (°F)	% vol.	gal/h	BPSD	°API	lb/gal	lb/h	MW	Moles/h	% mole	
Total product vapours	−872	95.66	16 897	9854	43.5	6.73	113 852	202.8	561.3	97.48	
O/flash	−888	1.00	177	101	18	7.87	1393	385	3.8	0.63	
UCO	+886	3.34	590	437	15.8	7.99	4712	432	10.9	1.89	
Total		100.00	17 664	10 092	42.0	6.79	19 957	200	575.8	100.00	

13 RESIDUE CONVERSION PROCESSES

13.1   Process Description—a Residue Conversion Unit

Figure 13.1 shows the general configuration of a vacuum residue conversion unit. This particular configuration consists of a fixed bed residue hydrocracker and a thermal cracker (or visbreaker) in series.

Bitumen feed from the crude vacuum distillation unit enters the hydrocracking section of the plant to be preheated by hot flash vapours in shell and tube exchangers. A recycle and make-up hydrogen stream is similarly heated by exchange with hot flash vapours. The hydrogen stream is mixed with the hot bitumen stream before entering the hydrocracker heater. The feed streams are raised to the reactor temperature in the heater and leave to enter the top of the reactor vessel. They flow downwards through the catalyst beds contained in the reactor. Additional cold hydrogen is injected at various sections of the reactor to provide temperature control as the hydrocracking process is exothermal.

The reactor effluent leaves the reactor to enter a hot flash drum. Here the heavy bituminous portion of the effluent leaves from the bottom of the drum while the lighter oil and gas phase leaves as a vapour from the top of the drum. This vapour is subsequently cooled by heat exchange with the feed and further cooled and partially condensed by air or water cooling. This cooled stream then enters a cold separator operating at a pressure only slightly lower than that of the reactor. A hydrogen-rich gas stream is removed from this drum to be amine treated and returned as recycle gas to the process. The liquid phase leaves from the bottom of the separator to join a

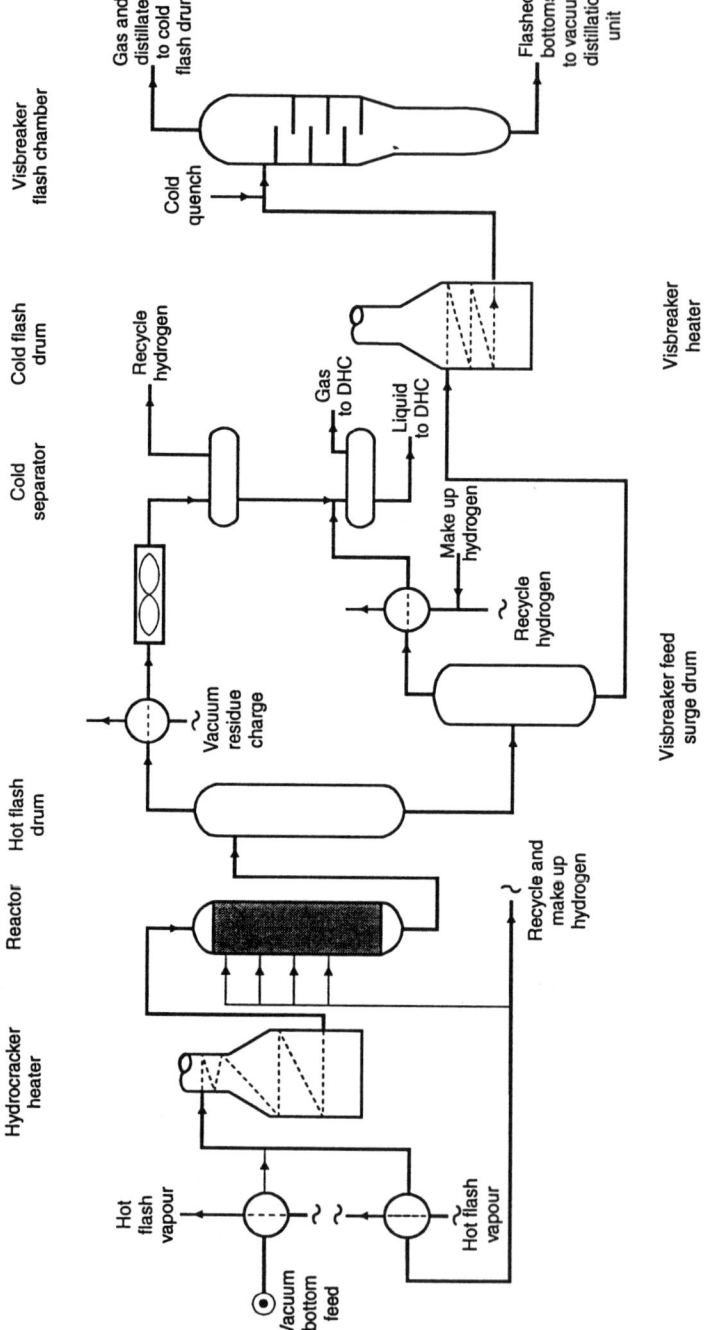

Figure 13.1. Residue cracking and visbreaker units

vapour stream from the hot flash surge drum (visbreaker feed surge drum). Both these streams enter the cold flash drum which operates at a much lower pressure than the upstream equipment. A gas stream is removed from the drum and this may be routed to an absorber or other suitable unit which can recover light naphtha and LPG. The liquid distillate from the drum is routed to a debutanizer and thence to a fractionator for the recovery of the normal distillate products such as naphtha, kero and diesel.

The visbreaker section of the unit takes as feed the heavy bituminous liquid from the hot flash drum. This enters the visbreaker furnace via a surge drum. The visbreaker heater in this case has two parallel coils, and the oil feed enters these coils to be thermally cracked to form some lighter products. The stream leaving the heater is quenched before entering a flash chamber. This vessel contains some baffled trays and a light gas and oil vapour stream leaves overhead. This stream is subsequently cooled and the distillate formed routed to the cold flash drum. The bottoms from the flash chamber is fed to the visbreaker vacuum distillation unit where vacuum gas oil is removed as possible feed to the FCCU or cut back in fuel oil blending. The bottom product from this tower may be sent to bitumen processing (air blowing) or to fuel.

13.2   Calculating the Composite Hot Flash Liquid TBP Curve

The hot flash liquid is the feed to the visbreaker and the properties of this stream will need to be known in order to examine the visbreaking process. This section provides a method of developing the TBP curve for this feed stream. The calculation is similar to that given in Section 6.3 and it is not proposed to repeat these calculation steps here. However, if the process configuration described in Section 13.1 exists and it is required to evaluate the process performance then the following data are needed:

● The ASTM curves for LVGO (light vacuum gas oil)
● The ASTM curves for MVGO (medium vacuum gas oil)
● The ASTM curves for HVGO (heavy vacuum gas oil)
● The yields of these streams contained in the flash liquid (this could be obtained from the visbreaker vacuum unit plant data when this column is taking feed directly from the flash drum and the visbreaker heater is closed down)

Where this calculation is to be developed as a design basis the above information would be provided from pilot plant data or from licensors' files. The following calculation is an example of this procedure, and the result is plotted as Figures 13.2 and 13.3.

Example Calculation

Predicting the effluent TBP

LVGO	Lab data	ASTM listed below	Yield 16.3% vol.
MVGO	Lab data	ASTM listed below	Yield 27.5
HVGO	Lab data	ASTM listed below	Yield 7.2
		Residue	Yield 49.0

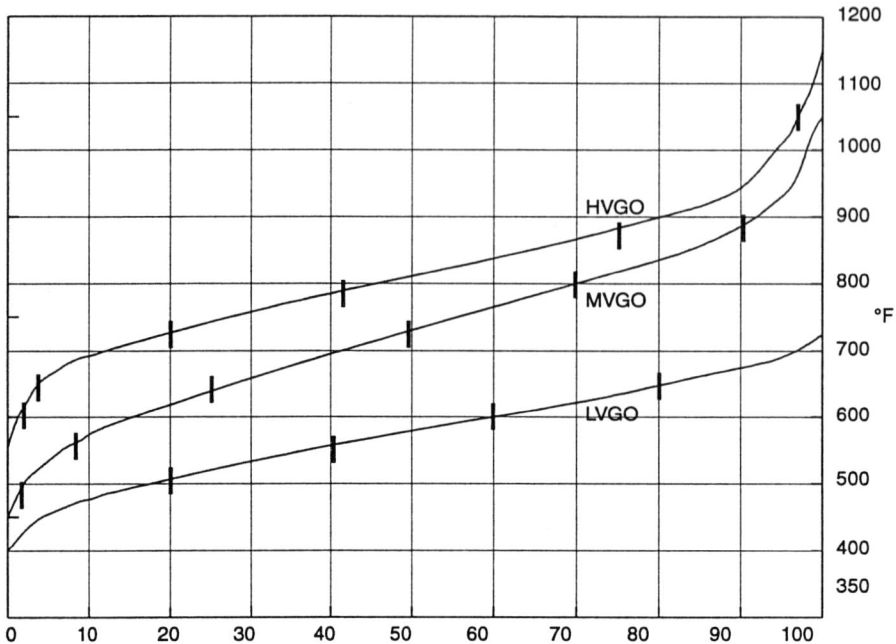

Figure 13.2 Residue cracker vacuum unit Product TBP curves

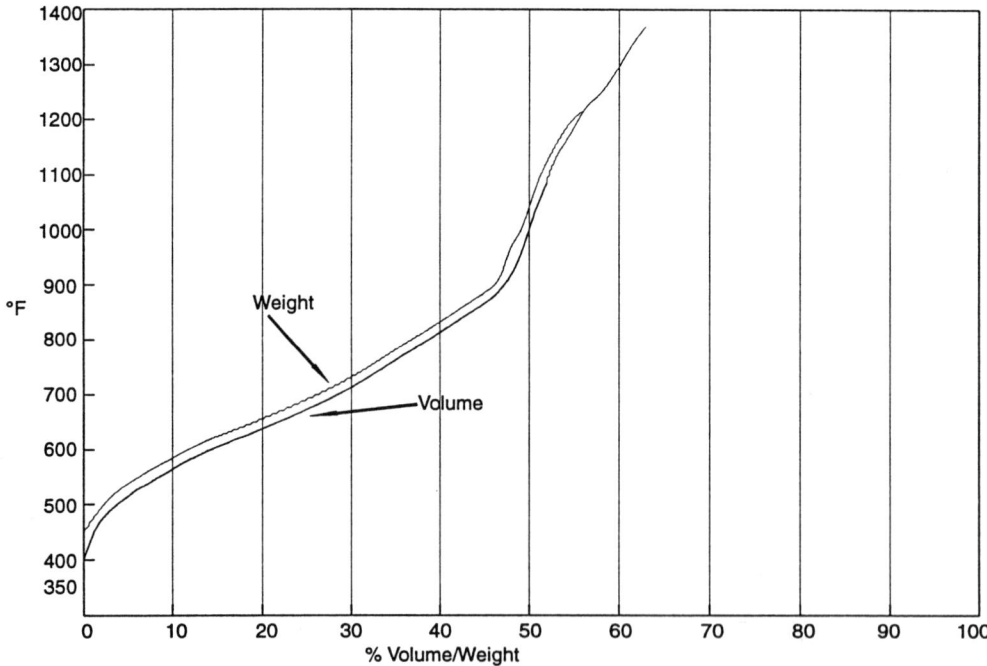

Figure 13.3. Liquid effluent ex res HCK hot flash TBP curves

LVGO

% vol.	ASTM		TBP	
	°F	Δ°F	Δ°F	°F
0	460	50	81	398
10	510	35	59	479
30	545	25	41	538
50	570	30	44	579
70	600	36	48	623
90	638	51	46	671
100	689			727

MVGO

0	510	90	130	447
10	600	60	85	577
30	660	50	70	662
50	710	50	66	732
70	760	80	88	798
90	840	111	162	886
100	951			1048

HVGO

0	690	93	133	559
10	783	42	67	692
30	825	35	53	759
50	860	40	55	812
70	900	50	61	867
90	950	—	—	928
100	∞			∞

Cut points	LVGO		MVGO		HVGO		Residue		Total on feed	Cumulative total
	% vol.	% vol. on feed	% vol.	% vol. on feed	% vol.	% vol. on feed	% vol.	% vol. on feed		
IBP to 505	20.0	3.26	2.0	0.55					3.81	3.81
555	20.0	3.26	5.0	1.36					4.64	8.45
600	20.0	3.26	8.0	2.20	1.0	0.07			5.53	13.98
645	20.0	3.26	12.0	3.30	2.0	0.14			6.70	20.68
725	20.0	3.26	22.0	6.05	17.0	1.22			10.53	31.21
800			21.0	5.77	23.0	1.66			7.43	38.64
885			20.0	5.50	32.0	2.31			7.81	40.45
1050			10.0	2.75	22.0	1.58			4.33	50.78
+1050					3.0	0.22	100	49.0	49.22	100.0
Total	100.0	16.3	100	27.5	100.0	7.2	100	49.0	100.0	—

Mid-BPT   610 °F °F = 30.9 °API
           825 °F °F = 22.1 °API
          1025 °F °F = 18.1 °API

13.3 A Procedure to Predict the Cold Flash Separator Operation

In the residue hydrocracker the reactor effluent is first flashed at the reactor outlet conditions. The heavy residue portion of the effluent is separated in this hot flash from the lighter products of the hydrocracking and the hydrogen stream. This lighter product leaves the hot separator as vapour to be cooled and partially condensed before entering the cold separator. A hydrogen stream is separated from the condensed light hydrocarbon stream at approximately the reactor pressure but at a temperature of about 120 °F. The liquid from this separator is routed to the cold flash drum operating at a much lower pressure than the cold separator. A fuel gas stream is separated from the hydrocarbon stream in this vessel.

Knowing the quantity and composition of this hydrocarbon stream, the other gas cold flash separator can be calculated. This calculation is described by the following steps:

- *Step 1* From laboratory analysis obtain the composition of the liquid stream from the cold flash drum. Obtain its flow rate from plant data.
- *Step 2* Calculate the composition of this liquid stream in moles per hour. This liquid is in equilibrium with a gas stream leaving the cold flash drum.
- *Step 3* Calculate the gas stream composition and quantity by a bubble point calculation using the liquid composition from Step 2. This bubble point calculation will be at the flash drum conditions of temperature and pressure.
- *Step 4* The Y column calculated in Step 3 is, by definition, the composition and quantity of the vapour leaving the drum. Adding this to the liquid portion (Step 2) gives the quantity and composition of the feed to the drum.
- *Step 5* Using the composition and quantity of the feed to the cold flash drum, calculate its bubble point at the cold separator conditions.
- *Step 6* The Y column of this bubble point calculation is the quantity and composition of the vapour leaving the cold separator. Some of this stream is used as the hydrogen recycle in the process.
- *Step 7* Adding the vapour stream calculated in Step 6 to the feed to the cold flash drum calculated in Step 4 gives the composition and quantity of the vapour leaving the hot flash drum.
- *Step 8* A complete material balance over the separator and flash drum can now be developed. Note that the recycle gas is about 91% mole hydrogen. This compares well with plant data.

An example calculation now follows.

Example Calculation

This drum operates at 225 psig and 120 °F. The cold flash liquid (CFL) has the following composition.

	kg moles/h	lb moles/h
H_2	1.14	2.51
C_1	1.33	2.93
C_2	1.61	3.55
H_2S	7.15	15.76
C_2	3.57	7.87
iC_4	3.92	8.64
nC_4	4.17	9.19
iC_5	0.38	0.84
nC_5	0.50	1.10
C_6^+	39.87	87.90
Total	63.64	140.29

As this liquid is in equilibrium with a gas phase at 225 psig and 120 °F a bubble point calculation will define the gas phase quantity and composition thus:

	X_F	$K_{120\,°F}^{240\,psia}$	$Y=XK$	Feed to CF drum moles/h	$K_{120\,°F}^{3018\,psia}$	$Y=KX$	Feed to hot flash drum moles/h	lbs/h
H_2	2.51	75	188.25	190.76	7.0	1335.32	1526.08	3052.15
C_1	2.93	11	32.23	35.16	1.6	56.26	91.42	1462.72
C_2	3.55	2.8	9.94	13.43	0.86	11.87	25.36	760.80
H_2S	16.78	2.7	42.55	58.31	0.84	48.98	107.29	3647.85
C_3	7.87	1.0	7.87	15.74	0.52	8.18	23.92	1052.48
iC_4	8.64	0.5	4.32	12.96	0.35	4.92	17.65	1037.04
nC_4	9.19	0.37	3.40	12.59	0.34	4.28	16.87	978.46
iC_5	0.84	0.15	0.15	0.99	0.24	0.24	1.23	88.56
nC_5	1.10	0.14	0.15	1.25	0.21	0.26	1.51	108.72
C_6^+	87.90	—	Nil	87.90	—	Nil	87.90	20 856.50
Total	140.29		288.86	429.15		1470.31	1899.46	32 845.30

Cold separator operates at 3000 psig and 120 °F

Cold separator (based on liquid feed of 10 360 BPSD)

	Feed ex hot separator					Flashed gas (recycle)				Liquid to CF drum		
Comp.	Moles/h	Mol. wt	lb/h	lb/gal	gal/h	Moles/h	lb/h	gal/h	scf × 10^3	Moles/h	lb/h	gal/h
H_2	1526.08	2	3052	—	—	1335.32	2671	—	504.7	190.76	381	—
C_1	91.42	16	1483	—	—	56.26	900	—	21.3	35.16	563	—
C_2	25.36	30	751	2.96	257	11.67	356	120	4.5	13.49	405	137
H_2S	107.29	34	3648	6.55	557	48.98	1665	254	18.5	58.31	1983	303
C_3	23.92	44	1052	4.22	249	8.18	360	85		15.74	693	164
iC_4	17.88	58	1037	4.69	221	4.92	285	61		12.96	752	190
nC_4	16.87	58	976	4.87	201	4.28	248	51	6.8	12.59	730	150
iC_5	1.23	72	89	5.21	17	0.24	17	3		0.99	71	14
nC_5	1.51	72	109	5.26	21	0.26	19	4		1.25	90	17
C_6^+	87.90	235	20 657	6.82	4816	Nil	Nil	Nil		87.90	20 657	4816
Total	1899.46	17.2	32 846	5.18	6339	1470.31	6521	578	555.9	429.15	26 325	3761

Cold flash drum

	Feed ex cold separator			Gas to fuel				Liquid		
	Mole/h	lb/h	gal/h	Mole/h	lb/h	gal/h	scf $\times 10^3$	Mole/h	lb/h	gal/h
H_2	190.76	381		188.25	377	—		2.51	5	—
C_1	35.16	563		32.23	516	—		2.93	47	—
C_2	13.49	405	137	9.94	298	101		3.55	107	36
H_2S	58.31	1983	303	42.55	1447	221		15.76	536	82
C_3	15.74	893	164	7.87	345	82		7.87	346	82
iC_4	12.96	752	160	4.32	251	54		8.64	501	107
nC_4	12.59	730	150	3.40	197	40		9.19	533	109
iC_5	0.99	71	14	0.15	11	2		0.84	60	12
nC_5	1.25	90	17	0.15	11	2		1.10	79	15
C_6^+	87.90	20 657	4816	Nil	Nil	Nil	Nil	87.90	20 657	4816
Total	429.15	26 325	5761	288.86	3454	502	108.189	140.29	22 671	5258

13.4 Visbreaking and Thermal Cracking—Soaking Volume Factor Concept

The design of a visbreaker or a thermal cracker is keyed to the configuration and temperature profile across the heater and soaking drum or soaking coil. The degree of cracking is dependent on this temperature profile and the residence time of the oil under these conditions.

The thermal cracking (or visbreaking) reaction is accepted as being of the first order. Thus it complies with the equation:

$$\ln (\text{conv}) = (A)e^{-E/RT} \times t$$

where

$$
\begin{aligned}
A, E, R &= \text{constants} \\
t &= \text{reaction time} \\
T &= \text{reaction temperature}
\end{aligned}
$$

The thermal cracking reaction occurs in the heater along a curve of increasing temperature. In this concept of design the cracking progression is expressed by a soaking volume factor which is defined by the following equation:

$$F = 1/D \int_0^t \frac{K_t}{K_0} .dv$$

where

V = coil volume
D = feed rate
K_t = reaction rate constant at any given temperature
K_0 = standard reference value for K_t

The standard reference temperature for visbreaking and thermal cracking is taken as 800 °F. A curve giving values of K_t/K_0 at a typical visbreaking heater is given in Figure 13.4 (Section 13.5). The curve was produced from experimental results using a normal accepted pressure drop profile across the heater. For simplicity, the curve is related to temperature versus the factor K_t/K 800 °F.

THE SOAKING VOLUME FACTOR

The calculation of the soaking volume factor is given in Section 13.5. This factor is related to product yields and the degree of conversion:

- *The degree of conversion* The relationship of the soaking volume factor to the degree of conversion is given by Figure 13.7 in Section 13.6. These curves were the result of experimental data from the laboratory cracking of many feedstocks. The family of curves given in Figure 13.7 demonstrate the comparative ease of cracking the large molecular structure of short residue to that of increasing wax distillate content. Conversion is measured by the result of gas and gasoline (to a cut point of 390 °F) produced.
- *Product yields* A family of curves given in Section 13.6 show the relationship of the soaking volume factor (SVF) to the yield of products of thermal cracking or visbreaking. Figure 13.8 (Section 13.6) shows the yield of gas, gasoline, and gas oil when cracking the wax distillate (662–1022 °F cut points) portion of the feed. Figure 13.9 shows the yields when cracking the bitumen portion of the feed (+ 1022 °F cut point).
- *The zone of critical decomposition* Experimental data shows that cracking and stability of the cracked product varies with the characteristics of the crude source material and boiling range. Figure 7.2 in Chapter 7 of this book demonstrates this criterion as a relationship with the Watson characteristic factor K of the feedstock. The shaded area of the figure is the range in which the major portion of cracking occurs. Above this area the cracked or visbroken residue becomes unstable and precipitates sediment when stored. It is undesirable to operate above this zone in most cases.

DISCUSSION OF THE CONCEPT

This is not a new concept since it has certainly been accepted as a basis for design since the late 1950s. To date, there are at least six visbreakers or thermal crackers to the author's knowledge that were designed using these parameters and are still in commercial use. In evaluating the calculation method given in the following Sections 13.5 and 13.6 it is interesting to note the following test run data compared with those calculated by SVF.

	Test run actual data	Calculated
Feedstock	Kuwait residue + 350 °C	
Conversion % wt	11.07	10.5
Feed rate	11 500	12 000
Healer inlet press (psig)	512	Not calculated
Outlet press (psig)	252	Not calculated
Conversion inlet (°F)	548	590
Conversion outlet (°F)	651	655
Heater outlet (°F)	880	860
Soaking drum outlet press (psig)	242	250
Inlet temperature (°F)	870	850
Outlet temperature (°F)	849	830
Yields		
Gas % wt	4.4	4.58
Total distillate % wt*	33.59	32.43
Residue % wt	62.01	62.99

*This is total distillate C_5 to 350 °C EP.

It can be seen that the data calculated compares well with those actually observed under test conditions. One interesting point in the test run data is the high coil outlet temperature that was used. As this is slightly above the critical decomposition zone for this type of crude it would lend one to suspect that the visbroken residue would be unstable.

13.5 Calculating the Soaking Volume Factor

This section provides a stepwise procedure for calculating the soaking volume factor. In this case average plant inlet and outlet temperature conditions have been selected. The heater configuration and the temperature profile are, of course, fictitious:

- *Step 1* Obtain from equipment data the configuration of the visbreaker heater. Obtain also an average temperature profile across the heater.
- *Step 2* Calculate the volume per tube. List the temperature at the end of each section and the number of tubes per section.
- *Step 3* From the number of tubes per section and the volume per tube, calculate the volume of coils per section. Plot volume of coil versus temperature (Figure 13.4).
- *Step 4* List the relative KT/K800 for each section of the coil. Use the curve in Figure 13.5 for this.
- *Step 5* Plot coil temperature versus KT/K800 (Figure 13.6). The area under this curve shown shaded in Figure 13.6 is the soaking volume. Note that the transfer line before quenching must also be included in this soaking volume.
- *Step 6* Most visbreakers and thermal crackers inject high-pressure steam or boiler feedwater into the heater coil. This provides turbulence to prevent the laydown of coke and to enhance heat transfer. If this is so, the soaking volume calculated must reflect the volume taken up by steam. As a rule of thumb, assume this to be 20%.

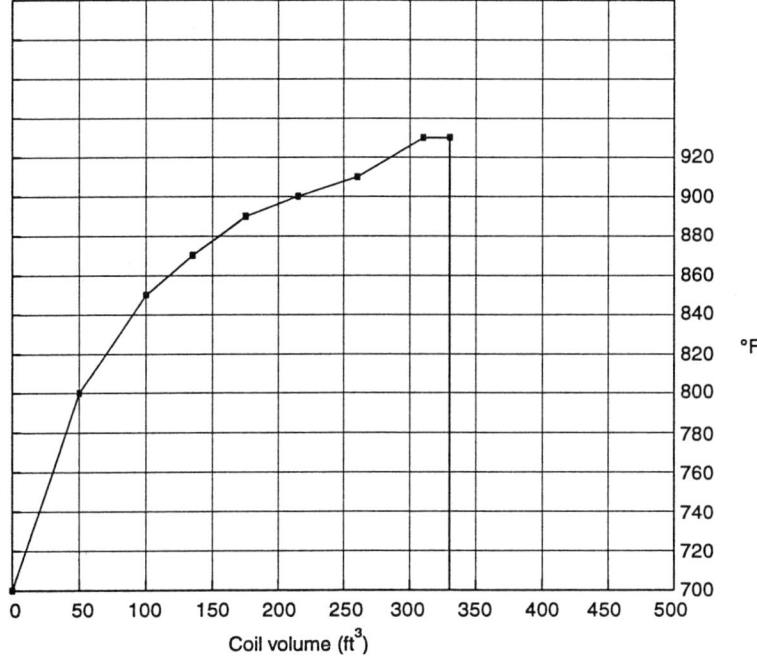

Figure 13.4. Coil volume versus temperature

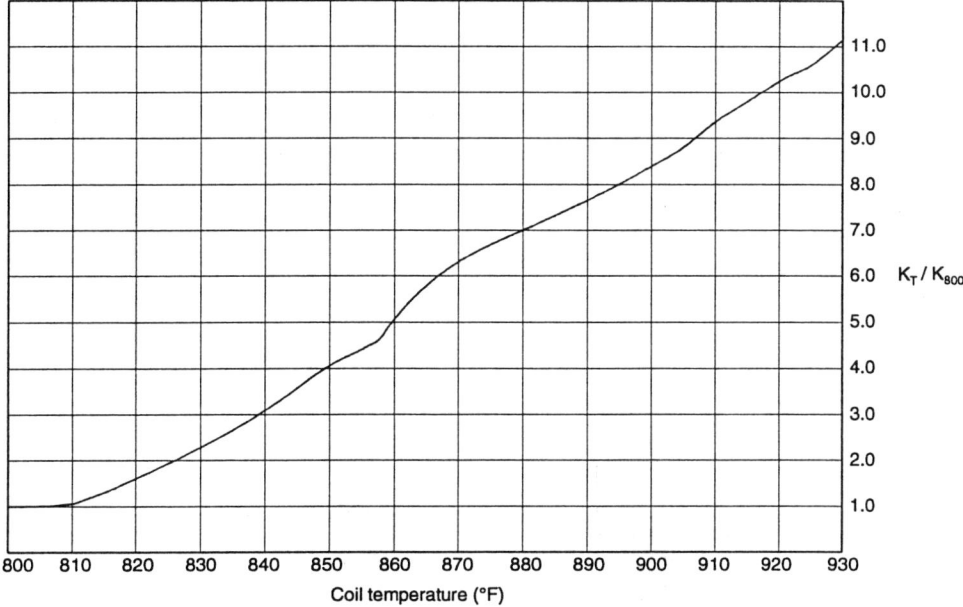

Figure 13.5. Visbreaking curve of coil temperature versus KT/K800

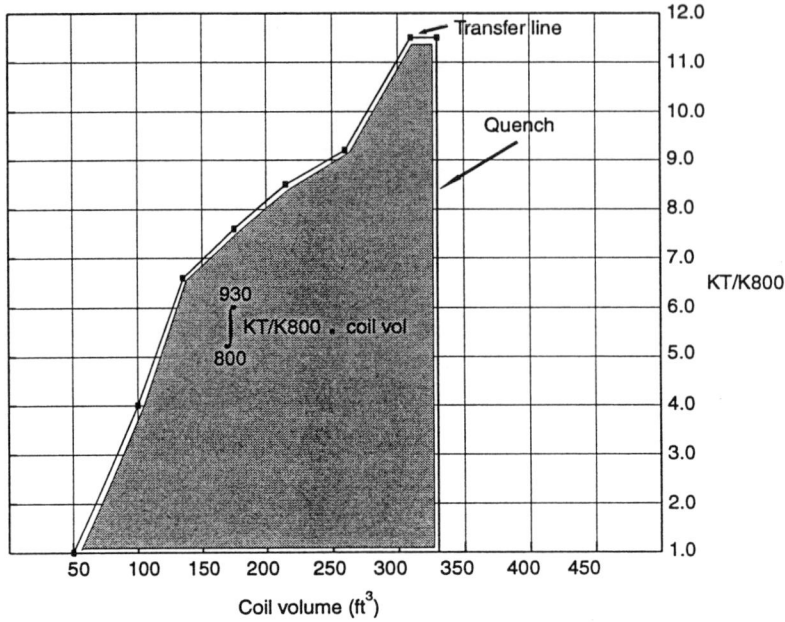

Figure 13.6. Coil volume versus KT/K800

● *Step 7*   Measure the area under the curve and divide this by the feed rate in BPSD. The result is the soaking volume factor (SVF).

An example calculation now follows.

Example Calculation

Throughput to be 8000 BPSD of hot flash liquid. The heater configuration and proposed temperature profile is shown in the diagram below.

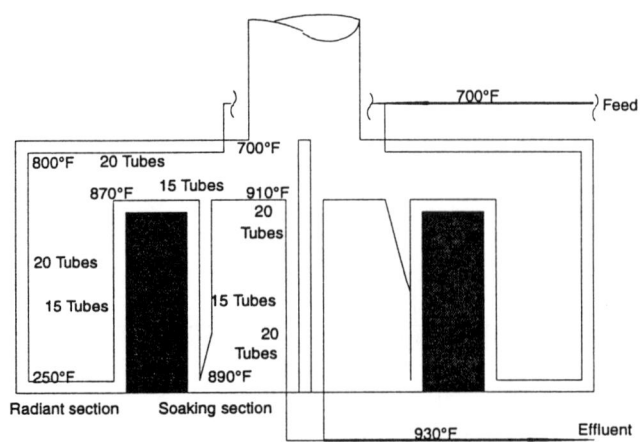

Radiant section has 55 tubes. Soaking section has 70 tubes

Temperature versus tube volume (volume/tube $= 2.5$ ft^3)

Temp. at end of section	No. of tubes	Volume (ft^3)		Relative KT/K800
		Per sect.	Cumulative	
800	20	50	50	1.0
850	20	50	100	4.1
870	15	37.5	137.5	6.6
890	15	37.5	175	7.65
900	15	37.5	212.5	8.40
910	20	50	262.5	9.20
930	20	50	312.5	11.45

Temperature versus coil volume per cell (there are two symmetrical cells) is plotted in Figure 13.5.

The transfer line is 14/h (Sched. 80) with a length of 60 ft.
Volume equivalent $= 16$ ft^3
Area under curve in Figure 13.4 is 1947
Now steam is injected after the first radiant pass and again at the first soaker pass. This takes up soaking volume. As a rule of thumb this amounts to 20%.

$$\text{Soaking volume} \quad = \int_{800}^{930} KT/K800. \text{ Coil vol. in ft}^3$$

$$= 1947$$

$$\text{Corrected for steam} \quad = 1947 \times 0.8$$

$$= 1557.6$$

$$\text{Soaking volume factor} = \frac{1557.6}{\text{BPSD}}$$

$$\text{BPSD per cell} \quad = 4000$$
$$\text{SVF} \quad = 0.389$$

13.6 Predicting Visbreaker Yields Using the Soaking Volume Factor

The soaking volume factor is used to evaluate the performance of the visbreaker by measuring conversion and predicting product yields. This factor is changed if temperature profiles or feed rate changes. This section provides a calculation procedure to predict product yields from the soaking volume factor. It is described by the following steps:

● *Step 1* Using the data calculated in Section 13.2 (TBP curve Figure 13.3) develop a weight percentage TBP curve. This is necessary because yields for cracking process are usually in percentage weights. Feedstocks in cracking processes undergo change in composition in the process and either increase or decrease in volume.

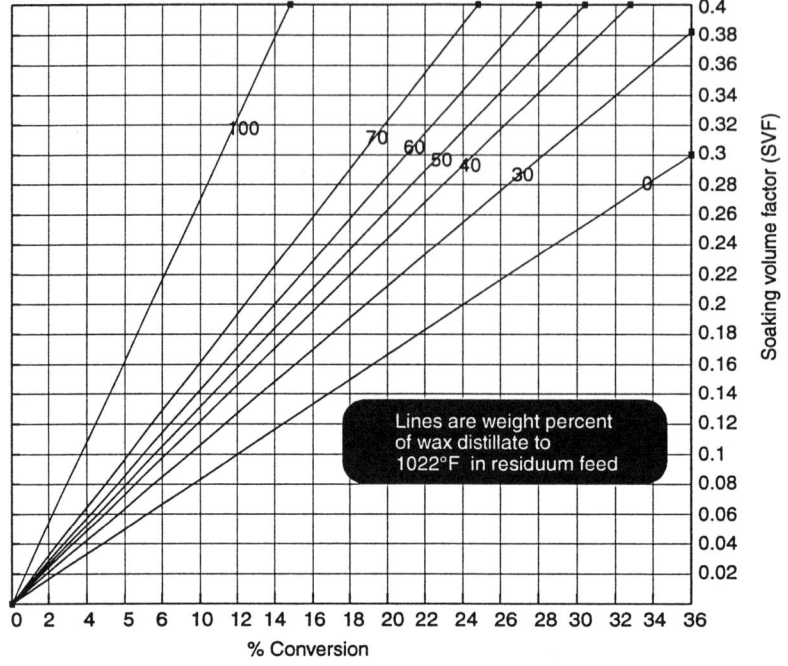

Figure 13.7. SVF versus conversion (conversion is % vol. gas + gasoline to 257 °F cut point)

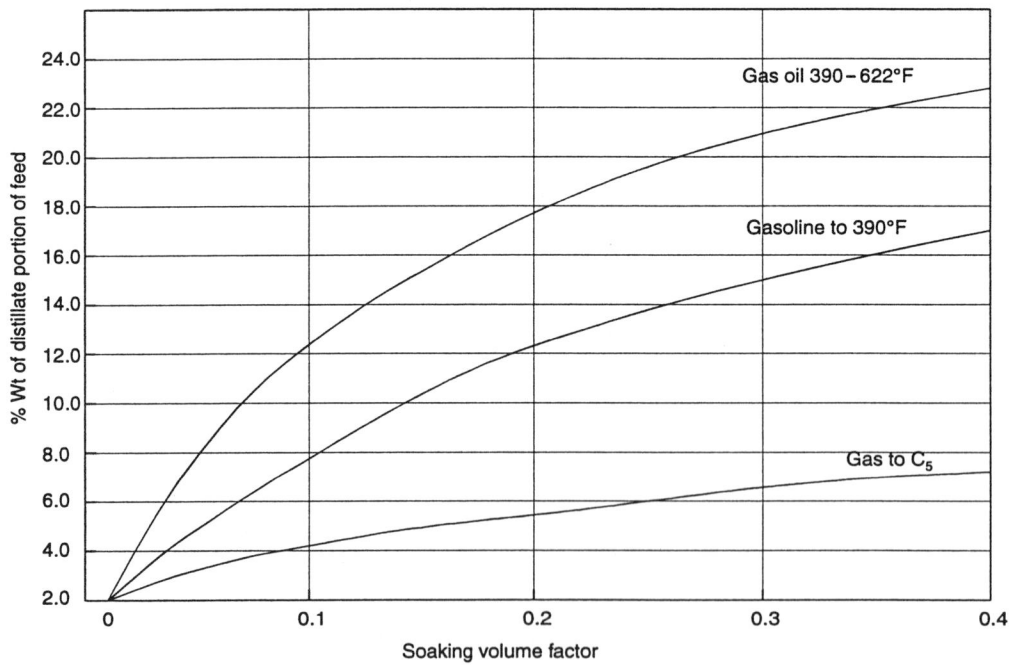

Figure 13.8. Visbreaking and thermal cracking. Yields % wt versus soaking volume factor. Waxy distillate portion of feed (662–1022 °F)

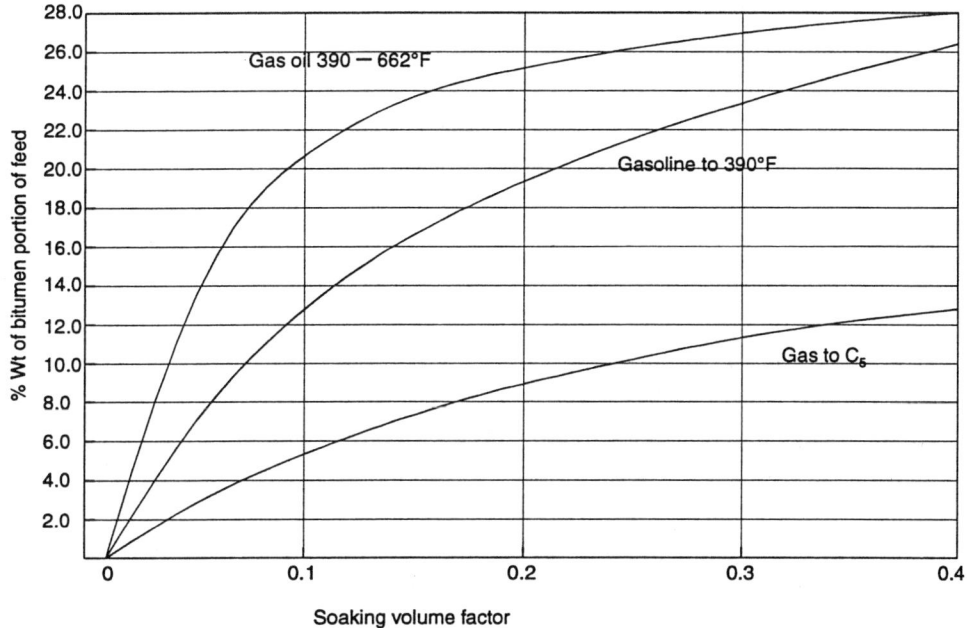

Figure 13.9. Visbreaking and thermal cracking. Yields % wt versus soaking volume factor. Bitumen portion of feed (+1022 °F)

- *Step 2* Establish the percentage weight of the waxy distillate (to 1022 °F) in the feed and the bitumen portion of the feed (+1022 °F). Using Figure 13.7 establish the conversion in terms of gas and gasoline to 257 °F cut point.
- *Step 3* Using Figure 13.8, read off the percentage weight of gas, gasoline and gas oil relating to the SVF calculated in Section 13.5.
- *Step 4* Figure 13.8 refers to the wax distillate portion of the feed. Then multiply the percentage weights read in Section 13.3 by the total weight of wax distillate from Step 2. This answer will give the yield attributed to the wax distillate portion of the feed.
- *Step 5* Repeat Step 3 using Figure 13.9 and Step 4 for the bitumen portion of the feed.
- *Step 6* Add the composition calculated in Step 4 to that in Step 5 to give the overall yield of products. Check that the per cent gas + gasoline is close to conversion given the Figure 13.7.

An example calculation now follows.

Example Calculation

The feed to the visbreaker is 8000 BPSD whose TBP curve is given in Figure 13.3. Calculate the weight % boiling curve.

SG of hot flash liquid = 0.943 (lab data)

Then weight curve

	% vol.	Volume		Weight			
		BPSD	gal/h	lb/h	% wt	lb/gal	°API
LVGO	16.3	1304	2285	16 578	15.1	7.255	30.9
MVGO	27.5	2200	3854	29 564	26.8	7.671	22.1
HVGO	7.2	576	1009	7948	7.2	7.877	18.1
Residue	49.0	3920	6868	56 020	50.9	8.157	13.0
Total	100.0	8000	14 016	110 110	100.0	7.856	18.5

Percentage weight of wax distillate to 1022 °F cut point in visbreaker feed = 48.6%

Then lb/h of wax distillate portion = 53513
lb/h of bitumen portion   = 56597

Yield of products are as follows (from Figures 13.8 and 13.9) based on the calculated SVF of 0.389 (Section 13.5) and a conversion to 257 °F of 29% wt

	Distillate		Bitumen		Total			Volume		
	% wt	lb/h	% wt	lb/h	lb/h	% wt	lb/gal	gal/h	BPSD	% vol.
Gas to C_5	5.5	2943	12.6	7131	10 074	9.1	4.5	2239	1279	13.4
$C_5 \cdot 390$ °F	14.9	7973	25.9	14 659	22 632	20.6	5.34	4236	2422	25.4
Gas/oil to 662	20.6	11 024	28.0	15 847	26 871	24.4	6.89	3900	2229	23.3
Wax dist. to 1022 °F	59.0	31 573	—	—	31 573	28.7	7.89	4002	2287	24.0
Residuum + 1022 °F	—	—	33.5	18 960	18 960	17.2	8.16	2324	1328	13.9
Total	100.0	53 513	100.0	56 597	110 110	100.0		16 703	9545	100.0

Lab analysis: (actual predicted product analysis from *Advances in Petroleum Chemistry and Refining* by T. A. Cooper and W. P. Ballard)[11]

Gas Analysis

	% mol.	Weight factor	lb/gal	Vol. factor
H_2	26.5	53.0		
C_1				
C_2	20.0	465.6	2.96	157.3
C_3	15.5	682.0	4.22	161.6
iC_4	10.0	580.0	4.69	123.7
nC_4	12.5	730.8	4.87	150.1
iC_5	7.8	561.6	5.21	107.8
nC_5	7.6	547.2	5.26	104.0
	100.0	3620.2	4.50	804.5

Naphtha ASTM

IBP	70	Me AvBP = 215.5
10	120	
30		°API = 89 (5.341 lb/gal)
50	240	
70		Mol. wt = 114
90	334	
FBP	410	
Gas oil	390–662	

ASTM

IBP	330	°API = 39.6 (6.886 lb/gal)
10	400	
50	487	Me AvBP = 475 °F
90	590	
FBP	720	Mol. wt = 188

Wax distribution to 1022 °F =

662–1022   °API = 18.0 (7.89)
(mid-BPT = 842 °F)
              Mol. wt = 405

13.7 Establishing the Flash Chamber Conditions and Developing the Vacuum Unit Material Balance

This section presents a procedure for setting the visbreaker flash chamber conditions and therefore producing the vacuum unit material balance. The effluent from the visbreaker heater contains light gas, gasoline, gas oil, waxy distillates and residue. A flash chamber is used to separate the gas and light distillate material. The separation is controlled by the temperature in the chamber. This in turn is controlled by introducing a cool quench stream. The procedure for developing this material balance is given by the following steps:

- *Step 1* Develop the effluent TBP curve from the data given by Section 13.6. Plot this as Figure 13.10
- *Step 2* Establish or obtain from plant data the flash chamber pressure.
- *Step 3* Obtain the steam rate from plant data. Some stripping of the heavy oil is done in this chamber.
- *Step 4* Develop the FRL from the TBP curve. It is not necessary to be so precise as to develop the EFV curve in this case. We will use the FRL.
- *Step 5* From Section 13.6 calculate moles of distillate and gas flashed. Using this, calculate the hydrocarbon partial pressure using the distillate to 622 °F cut point.
- *Step 6* From the cut point percentage volume on the FRL read off the flash temperature at atmospheric pressure.
- *Step 7* Using the vapour pressure curves read off the flash temperature at the hydrocarbon partial pressure. This will be controlled using quench flow and temperature.
- *Step 8* Take the liquid portion of the flash and develop its volume per cent cut for LVGO, HVGO, HVGO and residue using data in Section 13.2

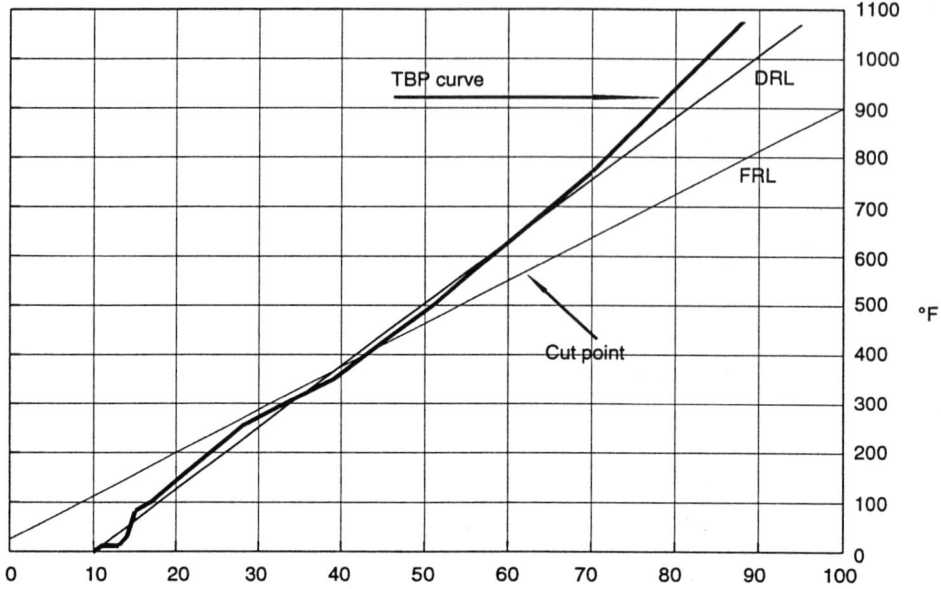

Figure 13.10. Visbreaker effluent TBP and flash curves

● *Step 9*   Build up the material balance using data from Step 8.

An example calculation now follows.

Example Calculation

Calculate visbreaker flash chamber conditions

The TBP curve of the visbreaker effluent stream is given in Figure 13.10. This was developed from calculations in Section 13.6.

Pressure at flash chamber will be:

Pressure flash vapour receiver   = 50 psig
There are two coolers in series (say, 7 psi)
Line ΔP (say, 3 psi)
Total pressure at chamber   = 60 psig
                                      = 75 psia

Let steam to chamber be 1.140 lb/h
                                      = 63 moles/h
Mol. wt Gas   = 36
Naphtha   = 114
Gas oil   = 188

Moles distillate flashed   = 621.2

Slope of DRL	$= 13\ °F/\%$
Slope of FRL	$= 9.2\ °F/\%$
50% point of FRL	$= 460\ °F$
Cut temperature at atmos.	$= 555\ °F$

Partial pressure $\dfrac{621.2}{621.2 + 63} \times 75$ $= 68$ psia

Flash zone temperature $= 700\ °F$
Effluent will be quenched to 700 °F

Calculate material balance over the visbreaker vacuum unit

	Cut °F	Volume					Moles	
		% vol.	BPSD	gal/h	lb/gal	lb/h	MW	Moles/h
Feed	622 +	100.0	3938	6892	7.76	53 457	444	120.3
LVGO	to 730	17.9	705	1234	7.30	9008	260	34.6
MVGO	1000	43.6	1717	3005	7.67	23 048	405	56.9
HVGO	1049	12.8	504	882	7.88	6950	620	11.2
Residue	+ 1049	25.7	1012	1771	8.16	14 451	820	17.6
Total		100.0	3938	6892	7.76	53 457	444	120.3

13.8 Evaluating the Performance of the Visbreaker Heater

This section demonstrates a procedure to evaluate the performance of the visbreaker heater in terms of its heat flux. The procedure is similar to that described for the crude unit in Section 6.16 with one notable difference. In the case of visbreaking the process requires heat of reaction to be included. This is quite significant in calculating the heat absorbed by the oil. Thus the thermal items to be considered in this case are:

- Heat to raise the feed temperature to the cracking temperature
- Heat to vaporize the effluent
- Heat of reaction to convert the feed

The procedure is described by the following steps:

- *Step 1*   Calculate the heat of reaction as the difference between the heat content of the feed and the sum of the heat content of all the products. (Nelson, *Petroleum Refining*, Third Edition,[12] p. 639, was used for this.)
- *Step 2*   Multiply the weight in lb/h of the feed by its heat content in BTU/lb to give heat content at 60 °F of the feed in BTU/h.
- *Step 3*   Do the same for all the product streams as with the feed in Step 2.
- *Step 4*   Calculate the heater duty by a heat balance over the heater inserting the heat of reaction as 'heat out'. The heater duty is found by difference.

- *Step 5*   For the heat balance all the distillate products from the flash drum are vapour. Assume also that 20% (mol.) of the residue from the flash drum (feed to vacuum unit) is also vapour at the coil outlet conditions.
- *Step 6*   From Section 13.5 (or from heater drawings) calculate the internal surface area of the heater coils. The coil is assumed in this case to be 4 in. sched. 120 11/13 chrome. (Remember to include both cells of the heater.)
- *Step 7*   Divide the heater duty by the surface area calculated in Step 6. This gives the heat flux in BTU/h/ft². Check with normal heat fluxes given in Section 6.16. If it is below the acceptable limit then there probably is fouling inside the coil.

An example calculation now follows.

Example Calculation

To calculate the duty of the visbreaker heater

The heater will provide:

- Heat to raise temperature of feed
- Vaporization of the effluent
- Heat of reaction to convert the feed to product

Heat of reaction (per cell)

This is based on the difference of heat energy contained in the feed to that contained in the total products (calorific value). (Data taken from Nelson, *Petroleum Refining*, p. 639.)

Total heat contained in feed $= 18\,990$ BTU/lb
From Section 13.6 $= 18\,990 \times 110\,110$ BTU/h
                          $= 2091 \times 10^6$ BTU/h

Total heat contained in products

Gas	$= 22\,250 \times 10\,074 =$	224.147×10^6 BTU/h
C_5—390	$= 21\,000 \times 22\,632 =$	475.272×10^6 BTU/h
Gas oil	$= 19\,520 \times 26\,871 =$	524.522×10^6 BTU/h
Wax dist.	$= 18\,900 \times 31\,573 =$	596.730×10^6 BTU/h
Residue	$= 18\,900 \times 18\,960 =$	358.344×10^5 BTU/h
	Total	$= 2179.015 \times 10^6$ BTU/h

Heat of reaction $= (2179.015 - 2091) \times 10^6$
                 $= 88.015 \times 10^6$ BTU/h

Heat balance over the heater

	V or L	°API	°F	lb/h	Enthalpy	
					BTU/lb	BTU/h $\times 10^6$
In						
Heat in feed	L	20.3	120	110 110	50	5.506
Heater duty			By difference			150.157
Total in				110 110		155.157
Out						
Product vapour	V	17.1	930	892 731*	628	56.063
Residue	L	11.5	930	20 837	556	11.585
Heat of reaction						88.015
Total out				110 110		155.663

*Pressure at coil outlet (say, 100 psig)

Assume 20 mol. % of residue is vaporized at this point.

Total moles product vapour   = 102.7
Total moles residue in vapour =   3.5

Total   = 106.2

$$PP = \frac{106.2}{169.2} \times 115 = 72.1 \text{ psia}$$

Draw curve at 72.1 psia parallel to FRL in Figure 13.10. Where it cuts 930 °F note cut yield. This is 86% vol. vapour or 14% vol. residue.

The assumption of 20% mol. residue vaporized is about right.
% weight will be about 30% vaporized

Heater duty $= 150.157 \times 10^6$ BTU/h

Heater tubes are 4 in. sched. 120 11/13 chrome

Internal cross-section area $= 10.315$ in^2
                                      $= 0.0715$ ft^2

Tube length $= \dfrac{2.5}{0.0716}$ cuft (see Section 13.5)

                                      $= 35$ ft

Length of tube per ft^2 of internal surface
                                      $= 1.05$ ft (tables—Cameron hydraulic data)

Ft^2 internal surface $= \dfrac{35}{1.05}$ ft^2

	$= 33.3 \text{ ft}^2$
Total number of tubes	$= 125 \times 2$
	(See Section 13.5)
	$= 250$
Total surface area	$= 250 \times 33.3$
	$= 8325 \text{ ft}^2$

$$\text{Heat flux} = \frac{150.157 \times 10^6}{8325}$$

$$= 18037 \text{ BTU/h/ft}^2$$

This is within acceptable limits (see Section 6.16).

14   AN ALKYLATION PROCESS

14.1   Process description of an alkylation plant

Figure 14.1 illustrates the flow of a typical hydrofluoric acid (HF) alkylation unit in simplified form. Olefin feed from the FCC gas concentration unit is treated in an extractive process unit (e.g. a merox unit) for removal of sulphur compounds. After drying, the olefinic feed is mixed with fresh and recycle isobutane and enters the reactor which is maintained at roughly 90 °F. After passing through a time tank, the reactor effluent is settled into two phases: HF, which is pumped back to the reactor, and hydrocarbon, which goes to the isostripper. Here alkylate and excess butane are separated from isobutane. Reboiler duty of this tower represents the major utility cost of the process.

Propane is rejected from the unit via the overhead of the depropanizer. It is freed of acid in the HF stripper and is treated for removal of combined fluorides. The acid regenerator processes the system acid to maintain a purity in the 75–90% area by fractionating acid from tar and a constant boiling mixture of acid and water. Acid-containing vent streams are routed to a vent gas scrubber for contact with lime water or KOH.

Olefin feed should be dry and very low in non-condensables. Normally, the latter is taken care of in the gas-concentration unit and water removal is accomplished by drying with alumina or molecular sieves. While a small quantity of water (say 0.5 wt% in the acid) contributes to catalyst performance, its presence is minimized

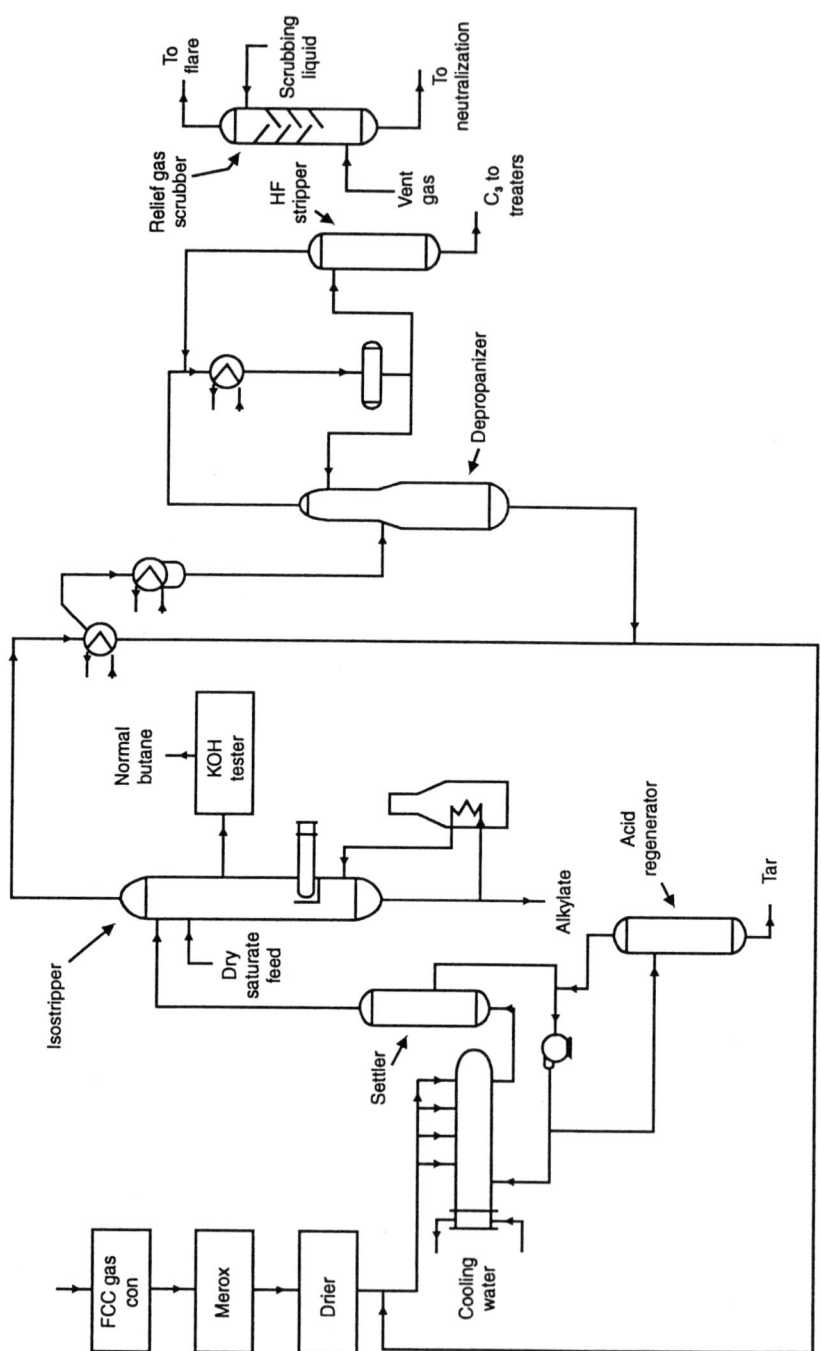

Figure 14.1. Process flow diagram of a typical HF alkylation unit

to control corrosion and acid consumption. When excess ethane ($>0.1\%$) is present it will probably be most satisfactory to remove it and water by fractionation. Ethane will concentrate in the recycle streams, causing condensation problems in the depropanizer overhead system. Venting, of course, gets rid of the ethane, but carries HF with it, thus contributing to acid losses and neutralization costs.

The reactor accomplishes intimate contacting of acid and hydrocarbon which gives the desired reaction and, at the same time, removes the substantial heat of reaction. Over the years the reactor design has evolved through several configurations (internally mixed, gravity circulation, perforated deck, emulsion circulation) to what is now called 'pumped, settled acid'.

Hydrocarbon feed enters through a multiplicity of inlets designed to disperse the hydrocarbons uniformly into the acid phase. This provides good utilization of the acid's heat sink properties and of the heat transfer surface, and substantial inferfacial area between acid and hydrocarbon across which the important mass transfer race occurs between isobutane and olefin. Spacing of the inlets several baffles apart permits recooling of the acid between each exposure to olefin.

The settled acid pump returns acid from the settler at a controlled rate, which ensures that the desired acid-to-hydrocarbon ratio is maintained. In addition, this pump overcomes the pressure drop in the reactor (permitting utilization in the design of high heat transfer coefficients) and the pressure drop which ensures good interfacial contact in the time tank or reaction soaker.

The final component of the reactor section is the time tank which, by allowing additional soaking time, permits use of lower-strength acid (desirable for increasing octane), reduces combined fluorides in the products and provides an acid sink which essentially eliminates the chance of an acid runaway.

On the recovery side of the process the isostripper is a very important part of the plant in that it generates the recycle isobutane previously shown to be important for product quality. It also controls the vapour pressure of the alkylate, rejects excess N-butane, and can fractionate isobutane for the reaction from low-purity butanes. The cost of heat associated with these tasks represents roughly 75% of the utility costs of the unit and, hence, the operation of this tower warrants a great deal of attention.

Reactor effluent is preheated and fed to the top tray from which a stream rich in isobutane is taken overhead. Further down the tower it is common to feed a low-purity iC4/nC4 stream to supply a portion of the required isobutane make-up. Vapour rates in the tower are based on generating the desired isobutane-to-olefin ratio in the reactor and not primarily on stripping the alkylate. It is possible therefore (by slightly increasing tower diameter) to handle significant quantities of side-feed without penalizing the separation efficiency. In many cases complete integration is achieved by adding a butane isomerate stream from an isomeration unit as a feed to the isostripper.

The next item of interest is the upper reboiler, which can supply nearly 50% of the isostripper heat duty from exhaust steam or other low-level heat sources. This feature is of growing importance as the cost of heat increases. Finally, the remainder of the heat is supplied by high-pressure steam in the lower reboiler.

The next tower of importance is the depropanizer, the function of which is to maintain a low concentration of propane in the recycle while at the same time rejecting propane at high (99%) purity. The bottoms product is high-purity, acid-free isobutane, which is a premium recycle stream as well as a good material for pump flush. In addition, this stream is useful during start-up and shut-down as a means of transporting isobutane to and from storage free of acid. Before propane can be yielded from the unit, it is freed of acid in the HF stripper and treated with alumina and KOH for fluoride removal. It is then suitable for sale as LPG. The alkylate leaving the bottom of the isostripper tower is debutanized in a debutanizer fractionation tower (not shown in Figure 14.1) before being routed to storage.

The vent gas scrubber provides treatment for acid-containing, gaseous streams leaving the unit. Using either lime water scrubbing or potassium hydroxide scrubbing followed by lime regeneration, HF is kept out of the flare system and leaves the unit as calcium fluoride which can be used for landfill.

The process is completed by the inclusion of an acid regenerator. It operates as needed to reject tar (polymer) and water (as a constant boiling mixture—CBM) by fractionation and thus maintains acid strength at roughly 85% with less than 1% water. Neutralized polymer can be burned in a special burner, blended off with asphalt or residuum, or can be charged to the crude or the FCC unit. CBM is neutralized with lime in the pit associated with the vent gas scrubber.

14.2 The Chemistry of Alkylation

Knowing the basic chemistry of alkylation is useful both in understanding fundamental mechanisms (which can lead to process improvement) and in defining upper limits or targets for commercial performance. This section is aimed at providing a general understanding of how feedstock composition and operating variables affect product quality.

Propylene follows the general rule that what should be simple usually isn't. One might expect something like 2,2-dimethylpentane (2,2-DMP) as the primary product:

$$
\begin{array}{ccc}
\text{C} & & \text{C} \\
| & & | \\
\text{C}-\text{C}+\text{C}=\text{C--1} \dashrightarrow & \text{C}-\text{C}-\text{C}-\text{C}-\text{C} & \text{(93 RON)} \\
| & & | \\
\text{C} & & \text{C}
\end{array}
$$

Instead, a susbstantial yield of 2,3-DMP (91 RON) and 2,4-DMP (83 RON) results, which is explained by carbonium ion instability and methyl shifts. To confuse the issue further, a hydrogen transfer reaction occurs to some extent which yields a C_8 product:

$$ \text{C}=\text{C--C} + \quad 2\text{IC}_4 \dashrightarrow \quad \text{C--C--C} + C_8 $$

The product is that produced from isobutylene which will be shown later to be largely 2,2,4-trimethylpentane (2,2,4-TMP). This reaction can be enhanced where a plentiful supply of isobutane and a ready market for the by-product propane exist.

In addition, polymerization, cracking, and similar side-reactions lead to the production of small quantities of C_5, C_6 and C_9^+ materials. Depending on reaction conditions, octanes will range from the high 80s to the mid-90s.

Butylene chemistry is complicated by the existence of four isomers isobutylene, 1-butene, and *cis* and *trans* 2-butene, both of which alkylate in the same way. Equilibrium composition at alkylation conditions favours substantial conversion of 1-butene to 2-butene. The competition between isomerization and alkylation is:

$$C=C\text{–}C\text{–}C + IC_4 \dashrightarrow \text{ dimethylhexane (DMH)}$$

$$C\text{–}C=C\text{–}C + IC_4 \dashrightarrow \text{ TMP} \quad \begin{array}{l} \text{heat of reaction} \\ \text{600 BTU/lb Olefin} \end{array}$$

Temperature and catalyst composition are selected such that 2-butene, and hence TMP, is maximized. Isobutylene does not enter these isomerization reactions, but rather goes to 2,2,4-TMP in a straightforward fashion.

A typical reaction of 2-butanes is:

thus explaining other isomers of TMP.

All the olefins show some tendency to polymerize. Alkylation and polymerization proceed along essentially the following routes:

Initiation	$C_4 = + HF \dashrightarrow$	$C_4^+ + F-$
Alkylation	$C_4^+ + C_4 = \dashrightarrow$	C_8^+
Saturation and continuation	$C_8^+ + iC_4 \dashrightarrow$	$C_8 + iC4^+$
Polymerization	$C_8^+ + C_4 = \dashrightarrow C_{12}^+$	
Cracking	$C_{12} \dashrightarrow iC_5 + C_7 =$	

The last two reactions explain the variety of side-products typically occurring at either end of the boiling range of primary products. Amylenes react by a combination of straightforward alkylation to a C_9 and hydrogen transfer yielding C_8 alkylate and a saturated C_5.

The above reaction mechanisms are confirmed by bench-scale work on pure components. The significance of this work is that maximum possible octane ratings from propylene is about 95 and from FCC butylenes about 98. Therefore, a maximum octane of 96.5 on 50/50 FCC C_3/C_4 with hydrogen transfer enhancement (or about 95 without) is achieved.

Since the by-products are generally lower in octane rating than the primary products, the process is designed to enhance the primary reactions. This is achieved by maintaining low temperatures, high isobutane concentration, and good dispersion of hydrocarbon into acid. Reduction of the polymerization reaction is the major function of excess isobutane and is the reason also for the importance of keeping the molar ratio of isobutane to olefin as high as can be economically justified. Simple kinetics indicates that the relative rates of alkylation and polymerization can be shown by the following relationship:

$$\frac{K_1 (C_8^+) (iC4)}{K_2 (C_8^+) (C4=)}$$

which reduces to:

$$K_3 \frac{(iso)}{(olefin)}$$

thus explaining the significance of the familiar iso-to-olefin ratio.

14.3   Calculating the Material Balance Over the Isostripper

This section is one of the most important in the alkylation process. The performance of this unit with its associated depropanizer provides the proper isobutane circulation required in the process (see Section 14.2). The feeds to this unit in this case are:

- The reaction effluent
- Isomerate from a butane isomerization unit
- A saturated C_4 stream

A procedure for obtaining a firm material balance over this unit is given by the following steps:

- *Step 1*   Obtain stream analysis for the isobutane recycle, the alkylate, the isomerate and the standard butanes from lab data.
- *Step 2*   From plant data obtain stream flows and other data corresponding in time to the lab data.
- *Step 3*   From the lab data (in component per cent mols) calculate the weight factor for each component using its molecular weight.
- *Step 4*   Calculate volume factor from weight factor and specific gravity or lb/gal. (Weight factor multiplied by specific gravity gives volume factor.)
- *Step 5*   Calculate mol. wt and specific gravity from the sum of component mol. per cent, weight factors and volume factors for each stream.

- *Step 6* From plant flow data (usually m³/h) fraction SG and mol. wt calculate weight of fraction in lb/h and its number of moles in mol/h.
- *Step 7* Develop the material balance over the stripper in terms of total feed, overheads and bottom product.

An example calculation follows.

Example Calculation

Feeds to the isostripper are:

(1) Alkylate plus iso C_4 recycle
(2) Isomerate
(3) Saturated C_4

The feed rates are:

Alkylate $= 19.43 \text{ m}^3/\text{h}$
iC_4 recycle $= 260.00 \text{ m}^3/\text{h}$
Isomerate $= 17.04 \text{ m}^3/\text{h}$
Sat. C_4 $= 15.10 \text{ m}^3/\text{h}$

TOTAL $= 311.57 \text{ m}^3/\text{h}$

The component content of the feed and the component balance over the isostripper is as follows:

			Alkylate			iC$_4$ recycle			Isomerate			Sat. C$_4$'s		
	MW	Sg at 60	Mol. %	Wt fact.	Vol. fact.	Mol. %	Wt fact.	Vol. fact.	Mol. %	Wt fact.	Vol. fact.	Mol. %	Wt fact.	Vol. fact.
C$_3$	44	0.5075				10.65	469	238	0.88	39	20	0.82	40	20
iC$_4$	58	0.583				82.62	4792	2698	48.40	2807	1580	26.64	1545	870
nC$_4$	58	0.584				6.73	390	228	48.55	2816	1645	70.02	4061	2372
iC$_5$	72	0.624	3.20	230	144				1.69	122	76	2.13	158	95
nC$_5$	72	0.001	4.02	289	182				0.45	35	22	0.29	21	13
C$_6$	86	0.664	2.79	240	159									
C$_7$	100	0.586	80.55	6055	4166									
C$_8$	114	0.707	23.08	2631	1860									
C$_9$	128	0.722	2.69	344	246									
C$_{10}$	142	0.734	2.18	310	228									
C$_{11}$	156	0.746	1.49	232	173									
Total			100.0	10 331	7160	100.00	5651	3164	100.00	5819	3343	100.00	5820	3370

	Alkylate	iC$_4$ recycle	Isomerate	Sat. C$_4$'s	Total
m³/h	19.43	260.00	17.04	15.10	311.57
Sg	0.6931	0.5500	0.5745	0.5790	0.570
kg/h	13 467	145 600	9789	8743	177 599
lb/h	29 680	320 990	21 581	19 275	39 1535
MW	103.3	5.65	58.2	58.2	58.7
lb moles/h	287.4	5681.2	370.8	331.2	6670.6

Overall column material balance

| | Feed | | | | iC$_4$ recycle (top) | | | Alkylate feed to DEB | | |
	Moles/h	lb/hr	lb/gal	gal/h	Moles/h	lb/h	gal/h	Moles/h	lb/h	gal/h
C$_3$	611.38	26 900	4.24	6344	605.05	26 622	6279	6.31	278	85
iC$_4$	4961.51	287 716	4.70	61 216	4693.31	272 192	57 913	267.70	15 524	3303
nC$_4$	794.27	46 068	4.88	9440	382.34	22 176	4544	411.93	23 892	4896
iC$_5$	22.52	1621	5.18	313				22.52	1621	313
nC$_5$	14.29	1029	5.25	196				14.29	1029	196
C$_6$	8.02	690	5.53	125				8.02	690	125
C$_7$	174.02	17 402	5.73	3037				174.02	17 402	3037
C$_8$	66.33	7562	5.89	1284				66.33	7562	1284
C$_9$	7.73	989	6.02	164				7.73	989	164
C$_{10}$	6.27	890	6.10	146				6.27	890	146
C$_{11}$	4.28	668	6.21	108				4.28	668	108
Total	6670.60	391 535	4.75	82 373	5681.2	320 990	68 736	989.40	70 545	13 637
Mol. wt			58.7			56.5			71.3	
lb/gal			4.75			4.67			5.17	
API			116.7			121.1			96.3	

14.4 Determining Critical Tray Loadings on the Isostripper

The isostripper is an unique fractionator as it has two reboilers internally located in separate locations in the tower. This provides a distributed heat load and consequently a distributed tray loading in that area of the tower. This section describes a procedure for calculating the reboiler duties and the critical tower vapour and liquid loadings. The following steps gives this procedure:

- *Step 1*   From plant data obtain the stream temperature to the isostripper and from the unit. These should correspond in time to the data used for the material balance (see Section 14.3).
- *Step 2*   From the material balance and temperatures calculate the overall heat balance over the isostripper. The enthalpies for the reactor effluent feed stream and the saturated C$_4$ stream are obtained by calculating their vapour flash equilibrium at plant conditions (see Section 3.7).
- *Step 3*   The unknown in the heat balance in Step 2 will be the total reboiler duties. Calculate this by difference of 'heat in' and 'heat out'.
- *Step 4*   From plant data using the steam rate to the bottom reboiler calculate its duty (heat in steam—heat in condensate).
- *Step 5*   By difference from total reboiler duty calculated in Step 3 calculate the top reboiler duty. (Check with steam rate to this reboiler.)
- *Step 6*   Plot the tower temperature profile (again from plant data). (Figure 14.2 on page 338.)
- *Step 7*   Carry out heat balances over the bottom reboiler, top reboiler below tray 28 (isomerate feed tray) and top trays. The molecular weights of vapour and liquid in each case can be estimated here as there are so few components in mixed phase. The enthalpy data from Maxwell's *Hydrocarbon Data Book* are used here.

- *Step 8* The unknowns in each of these heat balances will be the overflow liquid to each section. The heat balances equate to provide these figures in lb/h.
- *Step 9* Determine the vapour and liquid loads to and from the tray using the calculated overflow and vapour/liquid product. Summarize these. These data are then used to evaluate tray performance in terms of flood/entrainment conditions (see Sections 6.14 and 8.8).

An example calculation now follows.

Example calculation

To calculate reboiler duties and critical loading of feed trays of isostripper

From plant data

Temperature of alkylate $+ iC_4$ feed (F1)

	$= 65\,°C = 149\,°F$
Isomerate to tray 28	$= 109\,°C = 228\,°F$ (F2)
Saturated butanes	$= 82\,°C = 180\,°F$ (F3)
Top temperature	$= 71\,°C = 160\,°F$
Bottom temperature	$= 127\,°C = 261\,°F$

Overall heat balance

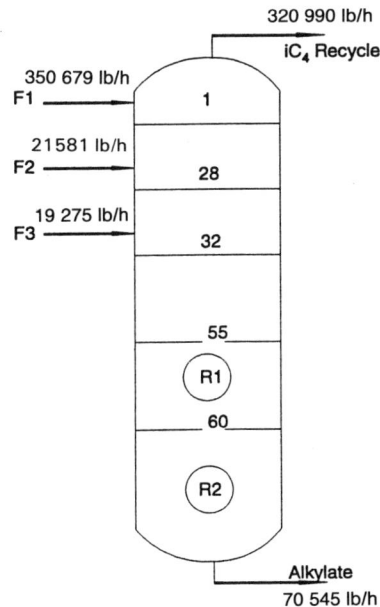

	V or L	MW	°F	lb/h	Enthalpy BTU/lb	BTU/h × 10⁶
In						
Heat in F1	$V+L$	58.8	149	350 679	(1)	69.701
Heat in F2	L	58.2	228	21 581	237	5.115
Heat in F3	$V+L$	58.2	180	19 275	(2)	5.003
Reboilers		By difference				36.072
Total in						115.891
Out						
Heat in o/head	V	58.5	160	320 990	305	97.902
Heat in bottoms	L	71.3	261	70 545	255	17.988
Total out						115.891

Note from the figure:

320 990 lb/h — iC₄ Recycle

350 679 lb/h — F1

21 581 lb/h — F2

19 275 lb/h — F3

Trays: 1, 28, 32, 55, R1, 60, R2

Alkylate 70 545 lb/h

(1) Calulate enthalpy of F1 inlet pressure = 164 psia

	Moles/h	K 162 psia 149 °F	1st trial $V/L=1$ V/LK	$L=\dfrac{F}{1+(V/L)K}$	-2nd trial $V/L=0.5$ V/LK	$L=\dfrac{F}{1+(V/L)K}$	3rd trial $V/L=0.15$ V/LK	$L=\dfrac{F}{1+(V/L)K}$	lb/h liquid	lb/h vapour
C_3	605.05	1.8	1.8	216.09	0.9	318.45	0.27	476.42	20 962	
iC_4	4693.81	0.92	0.92	2444.69	0.465	3203.97	0.138	4124.61	239 227	
nC_4	382.34	0.68	0.68	227.58	0.34	285.33	0.102	346.95	20 123	
iC_5	9.20	0.34	0.34	6.87	0.17	7.86	0.051	8.75	630	
nC_5	11.55	0.25	0.28	9.02	0.14	10.13	0.042	11.08	798	
C_6	8.02	0.11	0.11	7.23	0.055	7.60	0.0165	7.89	679	
C_7	174.02	0.046	0.046	166.37	0.023	170.10	0.0089	172.83	17 283	
C_8	68.33	0.019	0.019	65.09	0.0095	65.71	0.0029	66.14	7540	
C_9	7.73	0.009	0.009	7.66	0.0045	7.70	0.0014	7.72	988	
C_{10}	6.27	0.004	0.004	6.25	0.0020	6.26	0.0006	6.27	890	
C_{11}	4.28	—	—	4.28		4.28	—	4.28	668	
Total	5988.6			3161.13		4087.39		5232.94	309 788	40 891
Calculated V/L				0.89		0.46		0.141		

Mol. wt liquid $= 59.2$
Mol. wt vapour $= 55.6$

Enthalpy at 149 °F $= 309\,788 \times 185 = 57.311$ BTU/h
$40\,891 \times 303 = 12.390$
$= 69.701 \times 10^6$ BTU/h

Calculate enthalpy in sat C_4 stream

	Moles/h X_F	K 164 psia 180 °F	1st trial $V/L=1.0$ V/LK	$L=\dfrac{F}{1+(V/L)K}$	Liquid lb/h	Vapour lb/h
C_3	3.05	2.30	2.30	0.92	40	
iC_4	88.23	1.25	1.25	39.21	2274	
nC_4	231.91	0.92	0.92	120.79	7006	
iC_5	7.05	0.48	0.48	4.76	343	
nC_5	0.96	0.40	0.40	0.69	50	
Total	331.20			166.37	9713	9562
Calculated V/L				0.991		

Mol. wt liquid $= 58.4$
Mol. wt vapour $= 58.0$

Enthalpy liquid $= 9713 \times 205 = 1.991 \times 10^6$
vapour $= 9562 \times 315 = \underline{3.012 \times 10^6}$

Total $= 5.003 \times 10^6$

From plant data

600 # steam to bottom reboiler = 12 320 kg/h
                                = 27 161 lb/h

Assume condensing

Heat/lb of steam	= 1351 BTU/lb
(800 °F superheated)	
Heat/lb condensate	= 486 BTU/lb
Heat duty	= 27 161 × 865 BTU/h
	= 23.494 × 10⁶ BTU/h
Top reboiler duty	= 12.578 × 10⁶ BTU/h

Heat balances

1.   Bottom reboiler

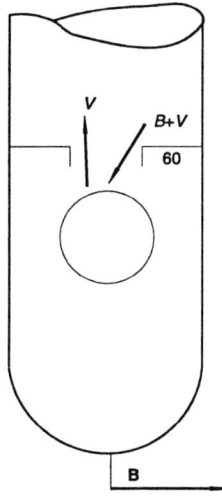

	V or L	MW	°F	lb/h	Enthalpy BTU/lb	Enthalpy BTU/h × 10⁶
In						
Reboil liquid	L	72[(1)]	258	V	247	247 V
Bottom prod.	L	72	258	70545	236	16.649
Reboiler						247 V + 40.143
Total in				70545 + V		247 V + 40.143
Out						
Reboil vapour	V	72	261	V	370	370 V
Bottom prod.	L	72	261	70545	255	17.989
Total out				70545 + V		370 V + 17.989

(1) Estimated

$$V = \frac{22\,154\,000}{123} = 180\,114 \text{ lb/h}$$

2. Top reboiler section

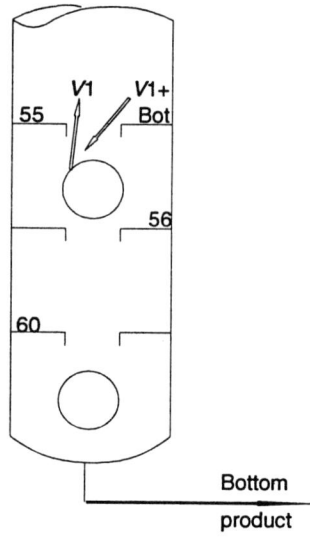

Bottom product

	V or L	MW	°F	lb/h	BTU/lb	Enthalpy BTU/h × 10⁶
In						
Reboil liquid	L	72	200	$V1$	215	$215\,V$
Bottom prod.	L	72	200	70545	210	14.814
Reboiler duty			Both top and bottom			36.072
Total in				$135\,845 + V1$		$251\,V + 50.886$
Out						
Reboil vapour	V	72	223	$V1$	353	$353\,V1$
Bottom prod.	L	72	261	70545	255	17.989
Total in				$135\,845 + V1$		$353\,V1 + 17.989$

$$V1 = \frac{32\,897\,000}{138} = 238\,384 \text{ lb/h}$$

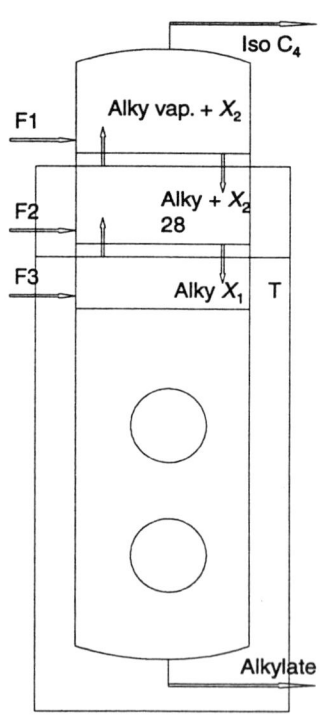

3. Overflow from tray 28

	V or L	MW	°F	lb/h	BTU/lb	Enthalpy BTU/h $\times 10^6$
In						
Heat with F3d	$V+L$	—	180	19 275	—	5.003
Overflow 28	L	72	177	$X1$	203	203 $X1$
Alkylate liquid	L	72	177	70 545	203	14.321
Total reboilers	L					36.072
Total in				89 320 + $X1$		53.396 + 203 X
Out						
Prod. vapour (F3)	V	72	181	19 275	336	6.515
Overflow vapour	V	72	181	$X1$	336	338 $X1$
Alkylate product	L	72	261	70545	255	17.989
Total out				89 820 + $X1$		24.504 + 338 X

$$X1 = \frac{30\,892\,000}{135} = 228\,830\ \text{lb/h}$$

2. Overflow from tray 1

	V or L	MW	°F	lb/h	BTU/lb	Enthalpy BTU/h $\times 10^6$
In						
Heat with F2	L	—	228	21 581	237	5.115
Heat with F3	V/L	—		19 275		5.003
Overflow tray 1	L	72	163	$X2$	190	190 $X2$
Alkylate prod.	L	72	163	70 545	190	13.404
Total reboilers						36.072
Total in				111 401 + $X2$		59.594 + 190 $X2$
Out						
Prod. vapour (F2)	V	58	168	21 581	330	7.122
Prod. vapour (F3)	V	58	156	19 275	330	6.361
Overflow vapour	V	72	168	$X2$	335	335 $X2$
Bottom product	L	72	261	70 545	255	17.989
Total out				111 401 + $X2$		31.472 + 335 $X2$

$$X2 = \frac{28\,122\,000}{145} = 193\,945\ \text{lb/h}$$

Tray no.	lb/h	Temp (°F)	Vapour to tray Press. (psia)	Moles/h	ACFS	lb/ft^3 ρ_V	lb/h	Sg at 80 °F	Temp (°F)	Liquid on tray Sg at T1	lb/gal T1	GPH (hot)	CPS (hot)	lb/ft^3 ρ_L
1 (top)	234 801	168	154	3398	38.4	1.70	391 535	0.570	163	0.495	4.14	24 574		
							193 945	0.621		0.558	4.66	41 619		
							585 480					136 193	5.06	32.14
28	248 105													
55 (top reboiler)	238 384	223	175	3311	38.4	1.72	308 929	0.621	200	0.533	4.44	69 579	2.58	33.26
60 (bottom reboiler)	180 114	261	176	2502	30.4	1.65	250 659	0.621	258	0.500	4.16	60 255	2.24	31.02

The above data can now be applied to the tray geometry to determine percentage of flood, etc.

14.5 Determining the Duty of the Isostripper Overhead Condenser

This condenser is different from a conventional fractionator condenser in so far as it allows a propane vapour stream to leave from the shell. This section describes a method to determine the duty of the condenser. This is provided by the procedure described by the follow steps:

- *Step 1* Obtain from plant data the flow of iso recycle from the overhead drum and the feed to the depropanizer. These two streams constitute the feed to the overhead condenser from the isostripper column.
- *Step 2* Check the dew point and bubble point of this overhead condenser feed stream. (This calculation procedure is found in Section 6.6.)
- *Step 3* Calculate the flash vaporization of the feed stream to the condenser at the condenser temperature and pressure (see Section 8.7 for this procedure).
- *Step 4* Using the temperature and quality of the isostripper overhead stream, the condenser temperature and the liquid vapour separation from Step 2, calculate the heat balance over the condenser. The unknown in this case is the condenser duty.
- *Step 5* Check the duty by calculating heat in and out with the cooling water from plant data.

An example calculation now follows.

Example Calculation

To calculate isostripper overhead condenser

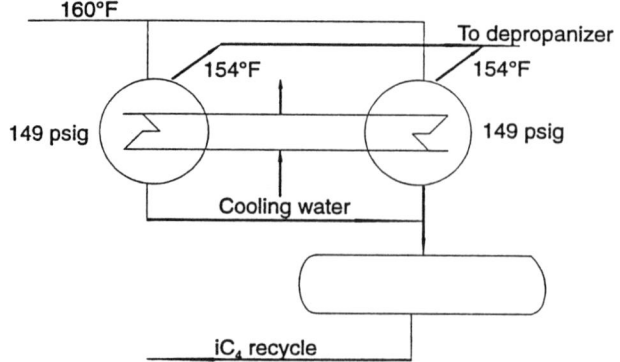

Total iC_4 overhead from isostripper = 320 990 lb/h
Plus 4.9 m³/h of C_3 (quantity was small and neglected in loading calculations)

Additional C_3 = 5488 lb/h
$\qquad\qquad\quad$ = 12 417 moles

Total overhead to condenser =

		Mol. Fract.
C_3	= 729.75	0.1257
iC_4	=4693.81	0.8084
nC_4	= 382.34	0.0659
Total	5805.90	1.000

Carry out dew point calculation at 164 psia

			1st trial at 160 °F	
	Y_F		K	$X = Y/K$
C_3	0.1257		1.80	0.0698
iC_4	0.8084		1.00	0.8084
nC_4	0.0659		0.78	0.0845
Total				0.9627

This is close enough. Actual dewpoint = 158 °F.

Carry out bubble point calculation at 164 psia

	X	1st trial at 129 °F		2nd trial at 147 °F	
		K	$Y = XK$	K	$Y = XK$
C_3	0.1257	1.50	0.1886	1.70	0.2137
iC_4	0.8084	0.75	0.6063	0.903	0.7300
nC_4	0.0654	0.54	0.0356	0.65	0.0428
Total	1.000		0.8305		0.9865

Actual bubble point = 149 °F.

Carry out flash vaporization at 154 °F

				1st trial $V/L = 0.1$		
	X_F	K 154 °F 164 psia	V/LK	$L = \dfrac{X_F}{1+(V/L)K}$	Liquid (lb/h)
C_3	729.75	1.85	0.185	615.8	27 095
iC_4	4693.81	0.98	0.098	4274.9	247 944
nC_4	382.34	0.72	0.072	356.7	20 689
Total	5805.90			5247.4	295 728
Calculated V/L				0.106	

lb vapour $\quad\;\;$ = 30 750
Moles vapour = 55 815
MW $\qquad\quad\;$ = 55.1

This vapour is fed to the depropanizer.

Duty of the condenser

	V or L	°API	°F	lb/h	BTU/lb	Enthalpy BTU/h × 10₆
In						
Isostripper overhead	V	56.2	160	326 478	330	107.738
Total in				326 478		107.738
Out						
Propane stream	V	5511	154	30 750	325	9.994
iC₄ stream	L	5614	154	295 728	193	57.076
Condenser			By difference			40.663
Total out				326 478		107.738

14.6 Evaluating the Depropanizer Performance

The depropanizer is an integral part of the iso C_4 recycle system. It in turn plays a major role in the proper operation of the alkylation process. This section describes the procedure for evaluating the depropanizer performance in terms of its tray efficiencies. This procedure is given by the following steps (see also Section 8.4):

- *Step 1* From lab data and plant flow data establish the purity of the propane stream and the quantity of propane in the iso C_4 recycle stream leaving the depropanizer.
- *Step 2* Calculate the moles/h of iC_4 in the depropanizer overhead stream and the C_3 in the depropanizer bottom stream.
- *Step 3* Develop the material balance over the tower in moles/h, lb/h and gallons per hour.
- *Step 4* Using the Fenske equation calculate the minimum number of theoretical stages required to give the split. The Fenske equation is:

$$N_{M+1} = \text{Log}\left[\left(\frac{X \text{ LT Ley}}{X \text{ HY Key}} \right)_D \cdot \left(\frac{X \text{ HY Key}}{X \text{ LT Key}} \right)_W \right] \div \text{Log } \phi_{\text{ave}}$$

where

N_{M+1} = minimal number of theoretical trays at total reflux plus 1 for reboiler
X LT Key = C_3
X HY Key = iC_4
D = distillate
W = bottoms product

ϕ_{ave} = relative volatility = $\dfrac{K \text{ LT Key}}{K \text{ HY Key}}$ at average tower temperature and pressure

- *Step 5* Calculate the minimum reflux to achieve the split using the Underwood equation (see Section 8.5).

- *Step 6* From the Underwood equation calculate ratio. *R* is *RM* times a factor 1.2 to 1.5. In this case it will be 1.2
- *Step 7* Use the ratio from Step 6 in the Gilliland curve (Figure 8.2 in Section 8.4) to give the ratio:

$$\frac{N-N_{\mathrm{m}}}{N+1}$$

where *N* is the theoretical number of stages.
- *Step 8* Solve Step 7 for *N*. The overall tray efficiency will be

$$\frac{N}{\text{Actual number of trays}} \times 100$$

An example calculation now follows.

Example Calculation

Setting the split

Lab data showed propane purity to be 97.6% C_3 mole.
C_3 in bottoms $= 2.7\%$.

Feed is as follows:

	Moles/h	lb/h	GPH	m³/hr
C_3	113.95	5014	1183	4.48
iC_4	418.91	24 297	5170	19.57
nC_4	25.64	1487	305	1.15
Total	558.50	30 798	6658	25.20

Plant data gives propane stream to be 4.1 m³/h *per 100 moles overhead*

$C_3 = 97.6 \text{ moles} = 4294 \text{ lb} = 1012 \text{ gal} = 3.83 \text{ m}^3$
$iC_4 = 2.4 \text{ moles} = 139.2 \text{ lb} = 29.6 \text{ gal} = 0.11 \text{ m}^3$

$$\text{Moles/m}^3 = \frac{100}{3.94} = 25.38$$

Total moles propane stream $= 104.06$ moles/h

$C_3 = 101.6$
$iC_4 = 2.46$

Bottom product $= C_3 = 12.35$ moles/ h $\qquad 2.7\%$
$\phantom{\text{Bottom product} = } iC_4 = 416.45$ moles/h
$\phantom{\text{Bottom product} = } nC_4 = 25.64$ moles/h

The material balance is as follows:
Per hour

	Feed			Overheads			Bottoms		
	Moles	lb	gal	Moles	lb	gal	Moles	lb	gal
C_3	113.95	5014	1183	101.60	4470	1054	12.35	544	129
iC_4	418.91	24 297	5170	2.46	143	30	416.45	24 154	5140
nC_4	25.64	1487	305	<—	Nil	—>	25.64	1487	305
Total	558.50	30 798	6658	104.06	4613	1084	454.44	26 185	5574

Calculate minimum trays at total reflux

The Fenske equation gives

$$N_{M+1} = \text{Log}\left[\left(\frac{X\ LT\ Ley}{X\ HY\ Key}\right)_D \cdot \left(\frac{X\ HY\ Key}{X\ LT\ Key}\right)_W\right] \div \text{Log}\ \phi_{ave}$$

Ave temp $= 186\ °F$
Ave press. $= 297\ psig = 312\ psia$
LT Key $= C_3$
HY Key $= iC_4$

$\phi_{ave} \quad = \dfrac{1.4}{0.8} = 1.75$

$$N_{M+1} = \text{Log}\left[\left(\frac{0.98}{0.02}\right)_D \cdot \left(\frac{0.916}{0.027}\right)_W\right] \div \text{Log}\ 1.75$$

$$\text{Log}\ [1662.4] \div \text{Log}\ 1.75$$

$$= \frac{3.22}{0.243} = 13.25$$

$N_M \qquad = \underline{12.25}$

Calculate minimum reflux at infinite trays

Using the Underwood equation.

Calculating for function 'B'

				$\sum \dfrac{X_F - \phi_{ave}}{\phi_{ave} - B} = 0$		
Comp	Fract. feed X_F	ϕ_{ave}	$(X_F)(\phi_{ave})$	$B = 1.5$	$B = 1.55$	$B = 1.513$
C_3	0.204	1.75	0.357	1.428	1.785	1.506
iC_4	0.750	1.00	0.750	−1.500	1.364	−1.462
nC_4	0.046	0.80	0.037	−0.053	−0.049	−0.059
Total				−0.125	0.372	−0.015

B function $= 1.513$

Take K's at $186\,°F$ and $312\,psia$

Calculating for R_{M+1}

Comp.	Fract. in distillate X_D	ϕ_{ave}	$\dfrac{X_D}{\phi_{ave}}$	$R_M+1 \sum \dfrac{X_D \cdot \phi_{ave}}{\phi_{ave}-B}$
C_3	0.976	1.75	1.708	7.207
iC_4	0.024	1.00	0.024	0.047
Total	1.000			7.160

$R_{MIN} = 7.160 - 1.000 = 6.16$
$R \quad = 6.16 \times 1.2 \quad = 7.39$

$$\frac{R-R_M}{R+1} = \frac{1.23}{8.39} = 0.147$$

$$\frac{N-M_M}{N+1} = 0.52 \text{ (from curve Section 8.4)}$$

$$\frac{N-12.25}{N+1} = 0.52$$

$$N-12.25 = 0.52\,N + 0.52$$

$$N \quad = \frac{12.87}{0.48} = 26.8$$

Actual number of trays $= 42$

$$\text{Efficiency} = \frac{26.8}{4.2} = 64\%$$

14.7 Developing Pseudo-components for the Alkylate

The alkylate leaving the debutanizer is routed to the gasoline pool. It is often required to detail its properties for blending and other requirements. This section provides a method to obtain these data through the development of psuedo-components making up the stream. The method is described by the following steps:

- *Step 1* Obtain the ASTM curve of the debutanized alkylate.
- *Step 2* Develop the TBP curve using the method by Edmister and described in Section 1.3.
- *Step 3* Using the ASTM curve calculate the mean average boiling point using the method described in the introductory chapter of this book.

- *Step 4*   Read off the curve in Appendix 1 (Figure A1.5), the corresponding molecular weight and characterization factor K for the alkylate. Lab data for its SG are used in this case.
- *Step 5*   Split the TBP curve into about six to eight mid-boiling point cuts using the method described in Section 1.4.
- *Step 6*   Using the mid-boiling point for each component and the characterization factor K for the whole cut (Step 4). Read off API the mole weight from the curve in Appendix 1 (Figure A1.5) for the pseudo-components.
- *Step 7*   From the components percentage volume on cut and the component SG calculate a weight factor for each of the component. Weight factor in this case is SG times 1000 times percentage volume.
- *Step 8*   Add the weight factors to give the SG or the alkylate when the sum total is divided by 100 000.
- *Step 9*   Using the mole weight for each component calculate a mole factor for the component. The mole factor is weight factor divided by the mole weight.
- *Step 10*   Add the mole factors and divide the weight factor by the total mole factor to give the alkylate mole weight.

An example calculation now follows.

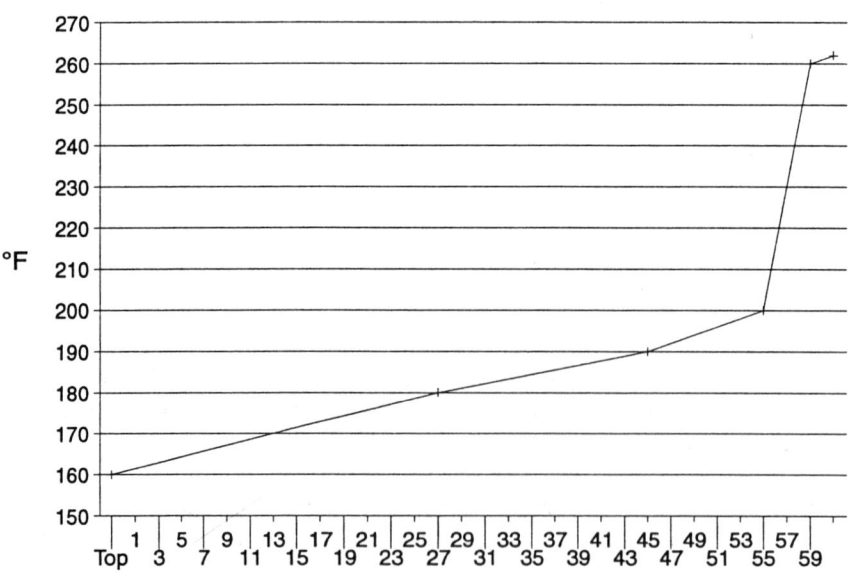

Figure 14.2. Estimated tower profile of the isostripper

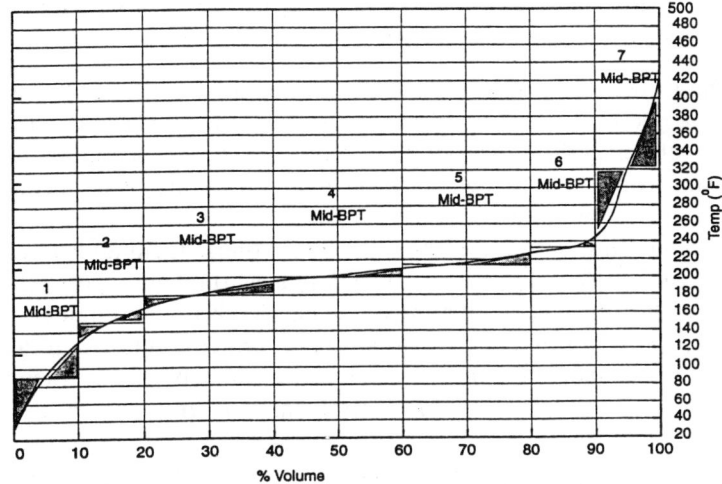

Figure 14.3. Typical alkylate TBp curve and pseudo components

Example Calculation

From lab results for ASTM distillation the TBP is calculated as follows:

% vol.	°C	ASTM °F	TBP ΔF	ΔF	°F	
IBP	38	100			29	
			69	104		Plotted on Figure 14.3
10	76	169			133	
			29	51		VABP $= \dfrac{169 + (2 \times 207) + 235}{4}$
30	92	198			184	
			9	17		$= 204.5$ °F
50	97	207			201	
			7	12		Slope $= 0.825$
70	101	214			213	
			21	30		MEABP $= 200.5$ °F
90	113	235			243	
			116	186		SG $= 0.6922$ (lab) $= 72.9$
FBP	177	351			429	MW $= 104\ K = 12.6$

Components

	% vol.	Mid-BPT	°API	SG	Weight factor	MW	Mole factor
1	10	91	79.3	0.672	6720	70	95.4
2	10	150	78	0.676	6760	83.1	81.3
3	20	182	75.3	0.684	13 690	102.4	133.6
4	20	202	73.0	0.691	13 820	110	125.8
5	20	216	71.5	0.696	13 920	113	123.1
6	10	231	70.0	0.702	7020	119	58.8
7	10	322	62.0	0.730	7300	154	47.5
Total	100		72.9	0.6922	69 220	104	665.6

14.8   Developing the Tray Traffic (Vapour and Liquid) for an Isomerate Stabilizer

Although the isomeration unit is not always integrated with an alkylation plant it does provide a good source of iC_4's which enhances the alkylation process. Because of this many refineries do integrate the two processes. This section is included here to provide data from a simple tower operation for further tray loading analysis which will be given in Section 14.9.

The feed to the isomerate stabilizer is the liquid phase from an isomerizer reactor separator. It contains the isomerate with light hydrocarbon and hydrogen in solution. The purpose of the stabilizer is to remove the light ends (C_2 and lighter) and to leave the isomerate stable and rich in C_4s. Because the feed contains a significant amount of hydrogen and methane the stabilizer can operate with a partial overhead condenser. The overhead product will be a vapour stream and only the reflux stream will be condensed.

This section describes a method of calculating the tower liquid and vapour traffic over the top tray and the bottom tray. This procedure is as follows:

- *Step 1*   From lab data obtain the moles/h of overhead vapour. Check that this composition is the dew point at the plant conditions for the reflux drum.
- *Step 2*   If Step 1 shows this composition is not dew point then calculate the amount of C_4s that will be added to make it dew point.
- *Step 3*   Let iC_4 be x moles/h to be added. Then nC_4 will be the proportion of C_4s in term of x.
- *Step 4*   Calculate the dew point at reflux drum conditions using x moles/h for C_4s.
- *Step 5*   By definition, the moles liquid in equilibrium at dew point with the overhead vapour will equal moles vapour. Equate and solve for x in moles/h.
- *Step 6*   Construct the material balance using the value calculated in Step 5 and plant/lab data for the isomerate.
- *Step 7*   Using plant data for the external reflux flow calculate moles of its composition from Step 5.
- *Step 8*   Add moles/h of reflux to moles/h produced for each component in the overheads. Calculate the mole fraction for each of the summed components.
- *Step 9*   Calculate dew point temperature for the mole fractions obtained in Step 8 at the tower top pressure (Allow 5 psi above reflux drum pressure for tower top).
- *Step 10*   Calculate the bubble temperature for the isomerate at tower bottom pressure.
- *Step 11*   Calculate weight factor and a volume factor, for the overflow liquid from top tray and the vapour to the bottom tray. Use this to obtain mole weight and SG of the respective streams.
- *Step 12*   Develop the flash equilibrium separation for the feed at tower inlet conditions (or at least leaving the preheater). See Section 8.7.
- *Step 13*   Calculate the feed enthalpy based on the liquid and vapour from Step 12.
- *Step 14*   Carry out a heat balance over the condenser to obtain condenser duty.

- *Step 15* Carry out a heat balance over the top tray to arrive at rate of overflow (in lb/h) and also the vapour to top tray. The vapour is overflow plus product.
- *Step 16* An overall heat balance over the tower is calculated to arrive at a reboiler duty.
- *Step 17* Using the reboiler duty carry out a heat balance over the reboiler. The unknown in this case is the vapour leaving the reboiler. The feed to the reboiler is the liquid from the bottom tray which is the vapour and bottom product.
- *Step 18* Summarize the flows in terms of actual ft^3/s for vapour and liquid. Also calculate their densities at tower conditions.

An example calculation now follows.

Example Calculation

To predict C_4s leaving stab. offgas

Stab. overhead   (Lab result)

	Moles/h
H_2	26.48
C_1	4.28
C_2	2.22
C_3	0.62
iC_4	$3.86 + x$
nC_4	$0.54 + 0.123\,x$
	$38.0 + 1.123\,x$

Reflux drum condition $= 300$ psia and $100\,°$F

Carry out dew point calculation

	Y moles/h	K	$x = Y/K$	Vapour Moles/L	Moles	Liquid % mol.	Weight	Volume
H_2	26.48	490	0.054	26.46	0.054	0.13	0.26	—
C_1	4.28	8.4	0.054	4.28	0.510	0.13	18.08	—
C_2	2.22	2.1	1.057	2.22	1.057	2.35	70.50	0.198
C_3	0.62	0.7	0.888	0.62	0.886	1.97	86.68	0.171
iC_4	$3.86 + x$	0.34	$11.35 + 2.9\,x$	13.60	35.596	79.21	4594.18	8.160
nC_5	$0.54 + 0.123\,x$	0.25	$2.16 + 0.48\,x$	1.74	6.835	15.21	882.18	1.510
Total	$38 + 1.123\,x$		$16.022 = 3.38\,x$	48.94	44.943	100.00	5651.88	10.039

$38 + 1.123\,x = 16.022 + 3.38\,x$ Mol. wt $= 56.5$

$$x = \frac{21.978}{2.257} = 9.74 \text{ moles/h}$$ SG $= 0.563$

The material balance can now be written (isomerate component analysis from lab).

The material balance

	Feed			Overheads			Isomerate		
	lb/h	Moles/h	gal/h	lb/h	Moles/h	gal/h	lb/h	Moles/h	gal/h
H_2	53	26.48	—	53	26.48	—	—	—	—
C_1	68	4.28	—	68	4.28	—	—	—	—
C_2	31	2.02	27.3	67	2,22	22.6	14	0.4	4.7
C_3	200	4.52	47.4	27	0.62	6.4	173	3.9	41.0
iC_4	14 713	253.50	3143.8	789	13.60	168.6	13 924	239.9	2975.2
nC_4	14 069	242.44	2894.9	101	1.74	20.8	13 968	240.7	2874.1
iC_5	603	8.40	116.1				603	8.4	116.1
nC_5	170	2.40	32.4				170	2.4	32.4
Total	29 957	544.64	6261.9	1105	48.94	218.4	28 852	495.7	6043.5

Mol. wt	55.0		22.6	58.2
SG at 60	0.5736		—	0.5726
lb/h	4.78		—	4.77
°API	115.2		—	115.6

Tower conditions

Tower top pressure	= 305 psia
External reflux rate	= 11.6 m³/h (plant data)
	= 3065 gal/h
Tower bottom pressure	= 312.5 psia

Carry out dew point and bubble point calculations to fix tower top and bottom temperature respectively.

Tower top moles/h reflux = 254 moles/h

	Total overheads				1st trial		liquid		
	Prod.	Ref.	Total	Y Mol. frac.	K 305 psi 200	$X = Y/K$	Wt. factor	Vol. factor	
H_2	26.48	0.33	26.81	0.088	46.0	0.002	0.004	—	
C_1	4.28	2.87	7.15	0.024	8.8	0.003	0.048	—	Mol. wt = 57.4
C_2	2.22	5.97	8.19	0.027	3.8	0.007	0.210	0.071	lb/gal = 4.7
C_3	0.62	5.00	5.62	0.019	1.6	0.012	0.528	0.125	Sg = 0.564
iC_4	13.60	201.20	214.8	0.709	0.92	0.771	44.718	9.555	°API = 119.4
nC_4	1.74	38.63	40.37	0.133	0.72	0.185	10.73	2.208	
Total	48.94	254.0	302.94	1.000		0.980	56.238	11.959	

Actual temp. = 0.92 × 0.980
           = 198 °F

Tower bottom

	X_1	1st trial 230 K 313 psi	$Y = XK$	Vapour Wt. factor	Vol. factor	
C_2	0.001	4.0	0.004	0.12	0.04	
C_3	0.008	1.9	0.015	0.66	0.16	Mol. wt $= 57.8$
iC_4	0.484	1.15	0.557	32.31	6.90	lb/gal $= 4.75$
nC_4	0.485	0.90	0.437	25.35	5.22	SG $= 0.5708$
. . .	0.017	0.54	0.009	0.65	0.13	°API $= 116.4$
nC_5	0.005	0.46	0.002	0.14	0.03	
Total	1.000		1.024	59.23	12.48	

$$K_2 = \frac{1.15}{1.024} = 1.12$$

Actual temp. 226 °F

Feed condition and enthalpy

Feed pressure $= 309$ psia 160 °F (plant data)

	x_F Moles/h	K 309 psi 160 °F	V/LK	$V/L = 0.1$ $L = \dfrac{X_F}{1+(V/L)K}$	Liquid Wt. factor lb/h	Vapour Moles/h	lb/h
H_2	26.48	54.0	5.4	4.14	8		
C_1	4.28	8.5	0.85	2.31	37		
C_2	2.62	2.9	0.29	2.03	61		
C_3	4.52	1.2	0.12	4.04	178		
iC_4	253.5	0.64	0.064	238.25	13 819		
nC_4	242.44	0.48	0.048	231.34	13 415		
iC_5	8.40	0.25	0.025	8.20	590		
nC_5	2.40	0.205	0.0205	2.35	159		
Total	544.64			492.66	28 280	51.98	1677
Calculated V/L				0.105			

Liquid mol. wt $= 57.4$
Vapour mol. wt $= 32.3$

This is close enough.

Feed enthalpy—using Maxwell (datum—200 °F)

Liquid portion (use iC_4)

   $28\,280$ lb/h $\times 190$ BTU/lb $= 5.373 \times 10^6$ BTU/h

Vapour portion (use C_2)

   1677 lb/h $\times 342$ BTU/lb $= 0.574 \times 10^6$ BTU/h

   Total enthalpy $= 5.947 \times 10^6$ BTU/h

Carry out heat balances over top and bottom of tower.

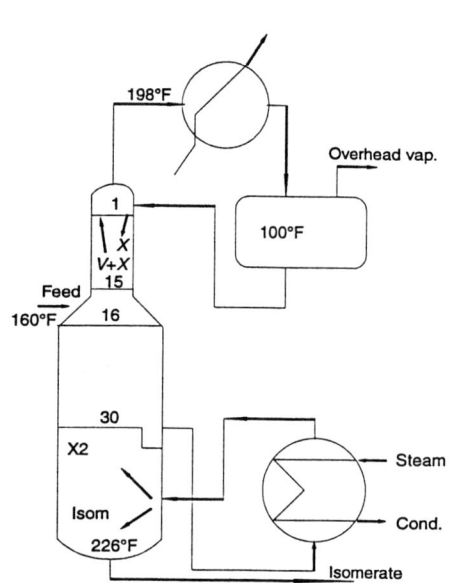

Stream	V/L	MW	°F	lb/h	BTU/lb	Enthalpy BTU/h × 10⁶
1. Calculate condenser duty						
In						
Prod. vap.	V	22.6	198	1105	382.5	0.423
Reflux vap.	V	56.5	198	14 351	315	4.521
Total in				15 456		4.944
Out						
Prod. vap	V	22.8	100	1105	337.5	0.373
Reflux vap.	L	56.5	100	14 351	154	2.210
Condenser				By difference		2.361
Total out				15 456		5.944
2. Calculate top tray overlow						
In						
Vap. ex tray 2	V	22.6	200	105	385	0.425
Ref. ex tray 2	V	57.4	200	X	316	316 X
Total in				1105 + X		0.425 + 316 X
Out						
Liquid ex 1	L	57.4	199	X	217	217 X
O/head vap.	V	22.6	100	1105	337.5	0.373
Condenser						2.361
Total out				1105 + X		2.734 + 217 X

$$X = \frac{2.309 \times 10^6}{99} = 23\ 323 \text{ lb/h}$$

$$= 406 \text{ moles/h}$$

Stream	V/L	MW	°F	lb/h	BTU/lb	Enthalpy BTU/h × 10⁶
3. Overall heat balance						
In						
Feed	V/L	—	160	29 957		5.947
Reboiler				By difference		3.654
Total in				29 957		9.601
Out						
O/head vap.	V	22.6	100	1105	337.5	0.373
Isomerate	L	58.2	226	28 852	238	6.867
Condenser						2.361
Total out				29 957		9.601
4. Heat balance over reboiler						
In						
Stripping vap.	L	57.8	223	X_2	231	231 X_2
Isomerate	L	58.2	224	28 852	232	6.694
Reboiler						3.654
Total in				28 852 + X_2		10.348 + 231 X_2
Out						
Vap. to tray 30	V	57.8	226	X_2	325	325 X
Isomerate	L	58.2	226	28 852	238	6.867
Total out				28 852 + X_2		6.867 + 325 X_2

$$X_2 = \frac{3.481 \times 10^6}{94} = 37\ 032 \text{ lb/h}$$

$$= 641 \text{ moles/h}$$

Tower loading summary

	Top tray—tray 1	Bottom tray—tray 30
Vapour		
Temp. (°F)	200	226
Press. (psia)	305	313
lb/h	24 428	37 032
Moles/h	455.26	641
ACFs	2.92	4.17
ρ_v (lb/ft³)	2.32	2.47
Liquid		
Temp. (°F)	199	224
Press. (psia)	305	313
lb/h	23 323	65 884
lb/gal at 60	4.7	4.74
lb/gal at temp.	3.84	3.74
gal min	101.2	293.6
ACFs	0.226	0.654
ρ_L (lb/ft³)	28.7	28.0

14.9 Evaluating Isomerate Stabilizer Performance in Terms of Downcomer Backup

This section describes the procedure for evaluating the tray performance in terms of its downcomer fill or liquid backup criteria. The calculation for the isomerate stabilizer uses data calculated in Section 14.8. The procedure to calculate the percentage of downcomer filled for the bottom tray is described by the following steps

● *Step 1* Calculate clear liquid height h_c using the expression

$$h_c = 0.5 \left[\frac{GPM}{L_O N_P} \right]^{2/3} + 0.5 \, h_w$$

where

h_c = the clear liquid height in inches
GPM = liquid flow (Section 14.8)
L_O = the outlet weir length in inches
N_P = number of passes
h_w = outlet weir height

● *Step 2* Find the hole area A_O from drawings if possible or use 12% of A_b

Hole area = $A_O/A_b = 0.12$

where

$$A_b = (A_s - A_{dc} - A_w)$$
A_s = total tray area
A_{dc} = total downcomer area
A_w = waste area

All in square feet.

- *Step 3* Hole velocity V_O is calculated using the expression CFS/A_O in ft/s.

- *Step 4* Calculate dry tray pressure drop at valves partially open using the expression

$$\Delta P_{PD} = (1.35\, t_M \cdot \rho_M/\rho_L) + K_1 (V_O)^2 \rho_V/\rho_L$$

where

ΔP_{PD} = dry tray pressure drop at valves partially open in inches of hot liquid
t_M = thickness of valve in inches
ρ_M = density of metal in lb/ft^3
K_1 = coefficient 0.2.

- *Step 5* Calculate dry tray pressure drop with valves fully open. This is given by

$$\Delta P_{FD} = K_2 (V_O)^2 \rho_V/\rho_L$$

where

ΔP_{PD} = is dry tray pressure drop with valves fully open in inches of hot liquid
K_2 = is a coefficient (0.92)

- *Step 6* Set h_{ed} which is the effective dry tray pressure drop. This is the greater of ΔP_{PD} and ΔP_{FD}.

- *Step 7* Determine the head loss under the downcomer using:

$$h_{vd} = 0.06 \left[\frac{GPM}{CL_i N_P} \right]^2$$

where

h_{vd} = head loss in inches hot liquid
C is calculated by $\dfrac{0.25\, GPM}{L_i \times N_P}$
L_i = length of inlet weir inches

- *Step 8* Calculate the inlet head by the expression:

$$h_i = 0.5 \left[\frac{GPM}{N_P L_i} \right]^{2/3} + h_{wi}$$

where

h_i = the inlet head in inches of hot liquid
h_{wi} = the inlet weir height in inches

● *Step 9* Determine the height of liquid in the downcomer in inches from

$$L_D = h_i + (h_t + h_{vd})\frac{\rho_L}{\rho_L - \rho_V} + 1.0$$

where

$$h_i = h_c + h_{ed}$$

● *Step 10* Divide L_D by tray spacing and multiply by 100 for per cent full. Fifty per cent and under is good. Seventy per cent and above would give cause for concern. Anything above seventy per cent usually leads to instability and tower flooding.

Example Calculation

This example is based on the tower loadings calculated in Section 14.8. The calculation concerns only the lower section of the tower and determines the amount of backup in the downcomer of the bottom tray.

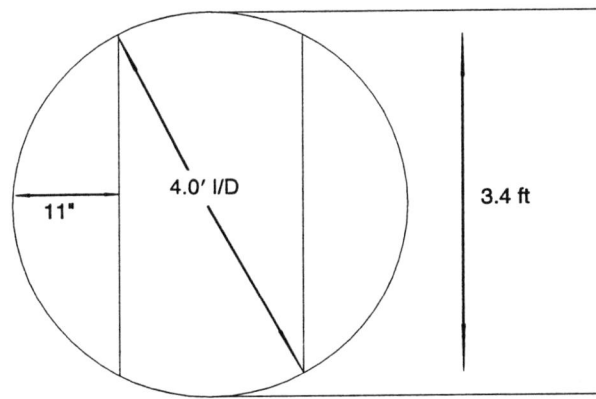

4.0' I/D

11"

3.4 ft

Downcomer area = 2.18 ft² (each)

Valve trays single pass

Tray spacing 24 in.

A_s = total tray area = 12.57 ft²
A_{dc} = inlet and outlet downcomer area = 2 × 2.18 ft² = 4.36 ft²
A_w = waste area estimate 15% = 1.9 ft²

1. Clear liquid height h_c

$$h_c = 0.5 \left[\frac{\text{GPM}}{N_P L_O}\right]^{2/3} + 0.5 h_w$$

where

h_c = clear liquid height inches of hot liquid
N_P = number of passes = 1.0
L_O = weir length in inches = 40.8 in.
h_W = outlet weir height (inches) = 24

$$h_c = 0.5 \left[\frac{293.6}{40.8} \right]^{2/3} + 0.5 \times 1$$

$$= 1.865 + 0.5 = 2.365 \text{ in}$$

2. Effective dry tray ΔP, h_{ed}

Hole area $= A_O/A_B = 0.12$
$A_B = (A_S - A_X - A_W)$
$A_B = (12.57 - 4.36 - 1.9)$
$\quad = 6.31$
$A_O = 0.757 \text{ ft}^2 \quad = \dfrac{\text{No. valves}}{78.5}$

Number of valves $\qquad = 59$

V_O = hole velocity ft/s $\qquad = \dfrac{\text{CFS}}{A_O}$

$$\frac{2.92}{0.757} = 3.86 \text{ ft/s}$$

$$\Delta P_P O = (1.35 \, t_M . \rho_M / \rho_L) + (K_1 (V_O)^2 . \rho_V / \rho_L)$$

where

$t_M \quad$ = valve metal thickness in inches (0.06)
$\rho_M \quad$ = valve metal density lb/cu ft
$K_1 \quad$ = ΔP_O coefficient (0.2)
ΔP_O = dry tray ΔP, valve partially open, inches of hot liquid

$$\Delta P p_O = \left(1.35 \times 0.06 \times \frac{480}{28.0} \right) + \left(0.2 \times (3.86)^2 \times \frac{2.47}{28.0} \right)$$

$$= 1.65 \text{ in.}$$

$$\Delta P p_O = K_2 (V_O)^2 \frac{\rho_V}{\rho_V}$$

where

ΔPp_O = dry tray Δp valve fully open, inches of hot liquid
K_2　　= ΔPp_O coefficient (0.92)

$$= 0.92 \times 14.9 \times \frac{2.47}{28.0}$$

$$= 1.21 \text{ in.}$$
$$h_{ed} = \text{greater of } \Delta Pp_O \text{ or } \Delta Pf_O$$
$$= 1.65 \text{ in.}$$

3. Head loss under downcomer (h_{VD})

$$h_{VD} = 0.06 \left[\frac{\text{GPM}}{CL_i N_P} \right]^2$$

where

C = 1.5 in. if $h_{VD} > 1.0$
L_i = inlet weir length

Set $h_{VD} = 1$ in. and calculate C

$$C = \frac{0.25 \text{ GPM}}{L_i N_P (h_{VD})^{0.5}} = \frac{0.25 \times 293.6}{40.8 \times 1.0 \times 1.0}$$

$$h_{VD} = 0.06 \left[\frac{293.6}{1.8 \times 40.8 \times 1.0} \right]^2$$

$$= 0.96 \text{ in. of hot liquid}$$

4. Inlet head h_i

$$h_6 = 0.05 \left(\frac{\text{GPM}}{N_p L_i} \right)^{2/3} + h_{wi}$$

where

h_{wi} = inlet weir height = 0.75 in.

$$h_i = 0.5 \left(\frac{293.6}{1 \times 40.8} \right)^{2/3} + 0.75$$

$$= 2.615 \text{ in.}$$

5. Downcomer filling L_D

$$L_D = h_i + (h_t + h_{vd}) \frac{\rho_L}{\rho_L - \rho_V} + 1.0$$

where

h_t = total tray $\Delta P = h_c + h_{ed}$

$$L_D = 2.615 + (4.015)\frac{28.0}{28 - 2.47} + 1.0$$

$$= 8.02 \text{ in. hot liquid}$$

Downcomer is 33.4% full—OK

Tower tray loading is good at 50% or less full.

14.10   Anhydrous Fluoric Acid (AHF)

Because of AHF's high toxic and corrosive nature this section is included to highlight its characteristics and the safe handling of the acid.

Anhydrous hydrofluoric acid (AHF) is a colourless, mobile liquid which boils at 67 °F and therefore requires pressure containers. The acid is so hydroscopic that its vapour combines with the moisture in air to form 'fumes'. This tendency to fume provides users with a built-in detector of leaks in AHF storage and transfer equipment. On the other hand, care is needed to avoid accidental spillage of water into tanks containing AHF, because its dilution is accompanied by a high release of heat.

Physical properties of anhydrous hydrofluoric acid

Boiling point (1 atm) °F	66.9
°C	19.4
Freezing point °F	− 117.4
°C	− 83
Specific gravity (32 °F, 0 °C)	1.00
Weight per gallon (32 °F, 0 °C), lb	8.35
Viscosity (32 °F, 0 °C), cp	0.31
Specific heat (32 °F, 0 °C) cal/g/°C	0.80

AHF vapour, even at very low concentrations in air, has a sharp penetrating odour that is an effective deterrent to wilful overexposure by operating personnel. Both the vapour and liquid forms of AHF cause severe and painful burns on contact with the skin, eyes or mucous membranes.

Hydrofluoric acid is very corrosive. It attacks glass, concrete and some metals, especially cast iron and alloys which contain silica (e.g. Bessemer steels). The acid attacks such organic materials as leather, natural rubber and wood, but does not promote their combustion.

Although AHF is non-flammable its corrosive action on metals, particularly in the presence of moisture, can result in hydrogen forming in containers and piping to create a fire and explosion hazard. Potential sources of ignition (sparks and flames) should be excluded from areas around equipment containing hydrofluoric acid.

Despite its corrosive nature, AHF can be handled with relative safety if the hazards are recognized and the necessary precautions taken. This section describes certain procedures for the safe handling of large bulk quantities of AHF.

SAFE HANDLING

The safe handling of AHF requires that well-designed equipment, be properly operated and maintained by well-trained, adequately protected, responsible personnel. Tanks and other containers of AHF should be protected from heat and the direct rays of the sun. Storage-area temperatures should preferably remain below 100 °F. If they reach or exceed 125 °F, means for cooling the containers must be applied.

Acid-transfer lines between the unloading station and the storage tank should tilt towards the latter to ensure free drainage. Relief valves should be installed in those sections of acid transfer lines where acid may be entrapped between two closed valves in the line, because expansion of the liquid might create excessive pressure and rupture the line.

No open fires, open lights or matches should be allowed in or around acid containers or lines. The possibility of acid acting on metal to produce hydrogen gas is ever present. Only non-sparking tools and spark-proof electrical equipment should be used in the AHF storage and handling areas.

Safety showers should be readily accessible at the unloading station, in the storage area and at other locations where acid is handled. The showers should be capable of supplying volume flows of 30 gal/min through quick-opening valves in 2-inch water lines. Handles at hip level should actuate the valves which, with a 0.25-inch weep hole directly above the valve, should be positioned below the frost line and surrounded by crushed rock or gravel to provide drainage. A water hydrant and hose should also be available in the unloading area to flush away spilled acid. Good drainage should be provided, also a supply of dry soda ash, ground limestone, or hydrated builder's lime. Accidental spills of acid on walkways or equipment should be washed off immediately with large volumes of water and, if necessary, neutralized with one of the agents mentioned.

PERSONAL PROTECTIVE EQUIPMENT

Personal protective equipment is not a substitute for good, safe working conditions. Its purpose is to protect the wearer in the event of an accident, major or minor. The extent of protection needed depends upon the degree of exposure attending the particular job at hand. Protective equipment should not be worn or carried beyond the operating area. It should be thoroughly washed with sodium bicarbonate solution immediately after each use.

The minimum protection required for operating and maintenance personnel includes the following items:

- Coveralls with sleeves to the wrists
- Face shield or chemical safety goggles
- Hard hat
- Poly(vinyl chloride) or neoprene-dipped gauntlets
- Poly(vinyl chloride) or neoprene-soled rubber shoes

When taking acid samples, opening equipment which may contain hydrofluoric acid or performing similar hazardous duties operators should wear the following:

- Poly(vinyl chloride) or neoprene overalls
- Poly(vinyl chloride) or neoprene boots
- Lightweight poly(vinyl chloride) or neoprene gloves under poly(vinyl chloride) or neoprene-dipped gauntlets
- Poly(vinyl chloride) or neoprene jumper
- Airline hood—air should be applied to the hood until the absence of fumes in the work area has been fully established.

UNLOADING AND TRANSFER OF AHF

AHF is shipped in tank cars having capacities ranging from approximately 5400 gallons to 25 000 gallons and in tank trucks of approximately 5250 gallons AHF capacity. Compressed dry gas (air, hydrocarbon or nitrogen) is the preferred means for transferring bulk quantities of AHF, but a centrifugal, rotary or positive-pressure pump can be used if necessary.

The unloading of AHF tank cars or tank trucks, with transfer of the acid to plant storage, consists of five steps:

(1) Spotting the tank car or tank truck at the unloading station.
(2) Connecting the plant compressed-gas (or vapour) and AHF-unloading lines to the proper valves on the carrier tank.
(3) Transferring the AHF from the carrier tank to the storage tank.
(4) Disconnecting the plant compressed-gas (or vapour) and AHF-unloading lines from the carrier tank valves.
(5) Releasing the tank car or tank truck for return to the shipper.

EQUIPMENT

Mild steel is satisfactory for storing and handling AHF at temperatures up to 150 °F maximum. Type 300 stainless steels are useful up to 200 °F. Monel nickel–copper alloy and Hastelloy C nickel steel are suitable for higher temperatures. Teflon TFE fluorocarbon resin is completely resistant to all concentrations of hydrofluoric acid at temperatures up to 500 °F.

Steel should not be used for movable parts because the corrosion-product film will cause such parts to freeze. Cast iron, type 400 stainless steel and hardened steels are unsatisfactory for AHF handling. Copper is velocity-sensitive. Stressed Monel may stress crack if exposed to moist vapours or aerated acid containing water. Welds in Monel corrode rapidly.

The selection of construction materials used for AHF equipment depends very much on such corrosion-affecting variables as moisture, temperature, aeration, fluid velocity, and impurities. Each storage and handling situation requires separate study to evaluate these factors before selecting materials which must meet the requirements of the installation.

STORAGE TANKS

The capacity of the storage system should be approximately 3 times the maximum quantity normally ordered to ensure against running out of acid between receipt of shipments. As a rule, too large a storage system is preferable to too small a system. The additional investment required for the larger installation is not great. The larger installation permits further expansion, less precise scheduling of shipments and larger inventories when desired.

The horizontal cylindrical storage tank should be manufactured according to the current ASME Code for Unfired Pressure Vessels or other equivalent codes which meet state or local mandatory requirements. In addition, it is recommended that the wall thickness of the tank be at least ⅛-inch in excess of the ASME Code requirements. The tank should be double-welded, butt-joint construction, the welds slag-free (conforming to ASME Code, Section 8) and ground smooth inside to facilitate inspection, X-ray inspection of welds is recommended.

The storage tank should be suitably supported above ground level. Structural steel supports or concrete saddles (protected with an acid-resistant paint) are satisfactory.

Safety devices for relieving abnormal internal tank pressures should be obtained from qualified manufacturers who are familiar with AHF. The maximum working pressure of the storage system should not exceed ⅔ the rated relief or bursting pressure of the safety devices. The dual relief system is recommended which has a two-way valve and rupture disks ahead of the relief valves and also a separate rupture-disk line in case of relief valve failure.

PIPING

All pipelines should be installed so that they drain toward the storage tank, or toward the point of consumption. This will prevent the accumulation of acid in low points, thereby eliminating possible safety hazards when repairs are necessary. Relief valves should be installed in the various sections of the lines in case acid becomes confined between two closed valves in the line. All flanges in the lines should preferably be coated with an acid-indicating paint, which changes in colour from orange to yellow in the presence of AHF liquid or vapour.

The line from the unloading station to the storage tank should be equipped with a gate valve so acid flow can be stopped at any time. The line should also be securely anchored to the storage tank as considerable vibration may occur, especially when unloading by means of compressed gas.

Extra-heavy (Schedule 80) or, better, triple extra-heavy black seamless or welded steel pipe which is free from non-metallic inclusions is satisfactory.

PERSONAL SAFETY

Liquid anhydrous hydrofluoric acid causes immediate and serious burns to any part of the body on contact. Dilute solutions of hydrofluoric acid often do not cause an immediate burning sensation where they came in contact with skin. Several hours may pass before the solution penetrates the skin sufficiently to cause redness or a burning sensation.

Wearing clothing which may have absorbed small amounts of hydrofluoric acid (such as leather shoes or gloves) can result in painful delayed effects similar to those caused by dilute acid solutions.

Hydrofluoric acid vapour causes skin irritation and inflammation of the mucous membranes; the burns become apparent a few hours after exposure. Inhaling the vapour in high concentrations may cause lung damage (pulmonary oedema).

The American Conference of Governmental Industrial Hygienists recommends a Threshold Limit Value of 3 parts (by volume) anhydrous hydrofluoric acid vapour (hydrogen fluoride) per million parts air. This value refers to a time-weighted concentration for a seven- or eight-hour workday and 40-hour work week.

The 3 ppm figure is based on both experimental and occupational evidence. However, nosebleeds and sinus troubles have reportedly occurred among metal workers exposed to even lower concentrations of a fluoride or fluorine in air. Therefore, for protection against acute irritation, 3 ppm should be considered a ceiling limit. Anyone who knows or even suspects that he or she has come in contact with hydrofluoric acid should immediately seek first aid.

In the event of an accident, the plant nurse or physician should be called as soon as possible. However, all plant supervisors should be aware of first-aid procedures for HF burns. All affected persons should be referred to a doctor even when the injury seems slight.

15 GAS-TREATING PROCESSES

15.1 Process Description of a Gas-treating Unit

Gas-treating processes in a refinery are usually used for cleaning up fuel gas and recycle streams. On rare occasions gas treating is incorporated to meet saleable specifications for the gas streams. There are general proprietary processes available for gas treating and almost all have the same process configuration. By far the most common processes are non-proprietary using amines as absorbants. The two types of amine used are:

Monethanol amine (MEA)
Diethanol amine (DEA)

For product gas streams which must meet lower than 1 grain per scf of H_2S, MEA must be used. This amine, however, is degenerated by certain sulphide compounds found in gas from thermal crackers. The most common compound that degenerates MEA is carbonyl sulphide (COS). MEA can, however, be regenerated by batch vaporization of the MEA and the disposal of the sludge formed by the degeneration.

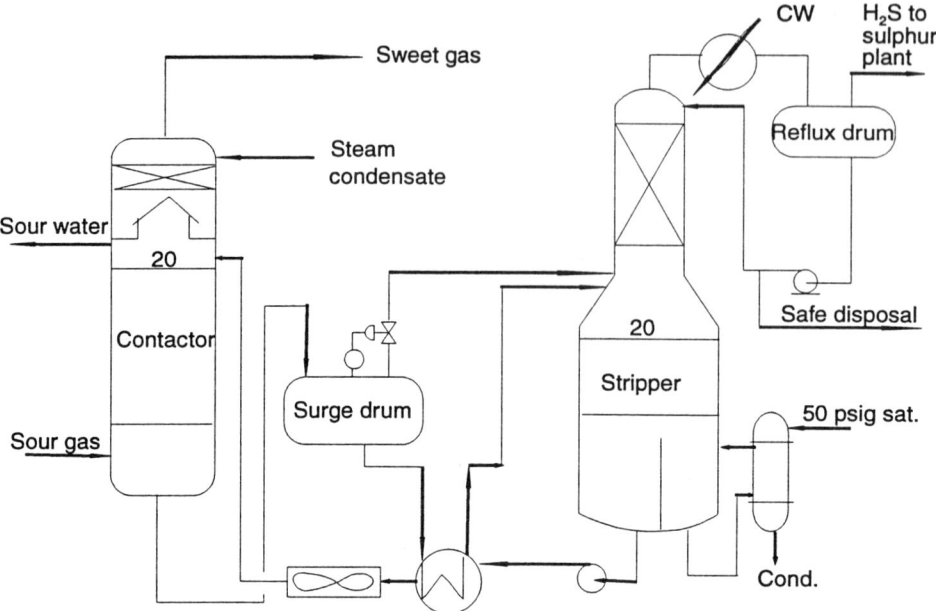

Figure 15.1. Amine-treating unit

Referring to Figure 15.1 sour gas (rich in H_2S) enters the bottom of the trayed absorber (or contactor). Lean amine is introduced at the top tray of the absorber section to move down the column. Contact between the gas and amine liquid on the trays results in the H_2S in the gas being absorbed into the amine. The sweet gas is water washed to remove any entrained amine before leaving the top of the contactor.

Rich amine leaves the bottom of the contactor to enter a surge drum. If the contactor pressure is high enough a flash stream of H_2S can be routed from the drum to a trayed stripping column. The liquid from the drum is preheated before entering a 20-tray stripping column on the top tray. This stripper is reboiled with 50 psig saturated steam. Saturated steam is used because high temperatures usually associated with superheated steam will cause amines to break down. The H_2S is stripped off and leaves the reflux drum usually to a sulphur production plant. Sulphur is produced in this plant by burning H_2S with a controlled air stream.

The lean amine leaves the stripper and is cooled. The cooled stream is routed to the contactor.

15.2   Review of Liquid Solvent Treating of Gases

There are several solvents in commercial use for the removal of H_2S and CO_2 from industrial gases. Among the more common are the amines. These include:

    MEA   monoethanol amine
    DEA   diethanol amine
    DGA   diglycol amine

In addition to these amine-based solvents there are also the hot potassium carbonate process, sulfinol and ADIP. The latter processes are marketed by the Shell Company and are quite common in worldwide usage. It is not the intent here to detail these processes with their various properties: only the three amine solvents will be described and discussed. Some mention will, however, be made of the other processes.

All amine systems circulate alkaline liquids. They chemically react with any sulphur or CO_2 in the gas and remove them from the gas stream. The rich liquid is heated to drive off the acid gases and then returned to the acid gas contactor in a continuous cycle.

MONOETHANOL AMINE (MEA)

This is the most common acid gas absorption process. Normally 15–20 wt% MEA in water is circulated down through a trayed absorber to provide intimate contact with the sour gas. The rich solution is sent to a steam stripping column where it is heated to about 250 °F (120 °C) at 10 psig (70 kPa) to strip out the acid gases. The lean MEA solution is pumped back to the absorber.

MEA is the most basic (and thus reactive) of the ethanol amines. It will completely sweeten natural gases, removing nearly all acid gases if desired. The process is well proven in refinery operations.

Like all the amine solvents used for acid gas removal, MEA depends upon its amino nitrogen group to react with the acidic CO_2 and H_2S in performing its absorption. The particular amines are selected with a hydroxyl group which increases their molecular weight and lowers their vapour pressures, yielding minimum solvent losses to the gas stream. MEA is considered a chemically stable compound. If there are no other chemicals present it will not suffer degradation or decomposition at temperatures up to its normal boiling point.

The process reactions are given below.

$$HOCH_2CH_2NH_2 = RNH_2$$

$$\text{Low temp.}$$
$$2\ RNH_2 + H_2S = (RNH_3)_2S$$
$$\text{High temp.}$$

$$\text{Low temp.}$$
$$(RNH_3)_2S + H_2S = 2\ RNH_3HS$$
$$\text{High temp.}$$

Some of the degradation products formed in these systems are highly corrosive. They are usually removed by filtration or reclaimer operations. Filtration will remove corrosive byproducts such as iron sulphide. Reclaiming is designed to remove heat-stable salts formed by the irreversible reaction of MEA with COS, CS_2 (carbonyl sulphide and carbon disulphide). The reclaimer operates on a sidestream of 1–3% of the total MEA circulation. It is operated as a stream-stripping kettle to boil water and MEA overhead while retaining the higher boiling point heat-stable salts. When the kettle liquids become saturated at a constant boiling point with the degradation products it is shut in and dumped to the drain.

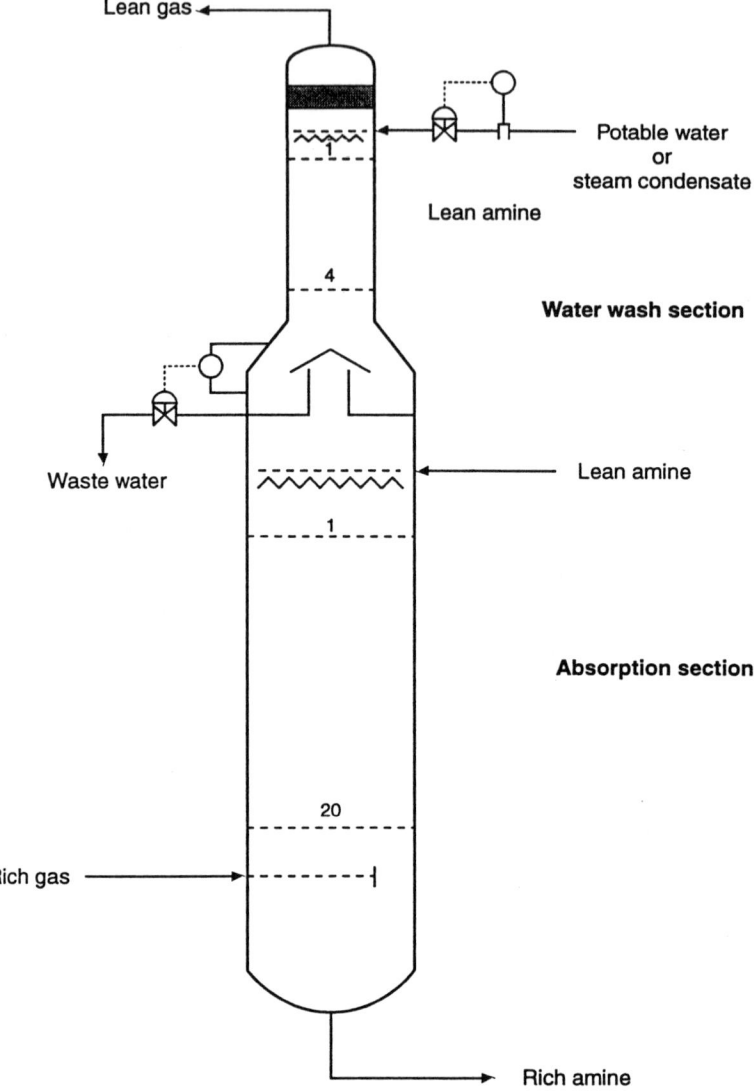

Figure 15.2. A typical DEA or MEA contactor

In addition to the chemical degradation mentioned above, MEA oxidizes when exposed to air. Storage and surge tanks must therefore be provided with inert blanket gases such as N_2 or sweet natural gas to avoid this degradation.

Amine systems foam rather easily, resulting in excessive amine carry-over in the contactor. Foaming can be caused by solids such as carbon or iron sulphide; condensed hydrocarbon liquids from the gas stream; degradation products; almost any foreign material introduced to the system such as valve grease, excess corrosion inhibitor, etc. Some of these items such as iron sulphide or carbon particles are removed by cartridge filters. Hydrocarbon liquids are usually removed by the use of a carbon bed filter on a lean amine sidestream (about 10% of total flow). Corrosion byproducts are removed by reclaiming as noted above.

High skin temperatures on the reboiler/reclaimer tubes promote amine degradation. Steam or hot oil used for the reboiler should be limited to a maximum of 285 °F (140 °C) to avoid excessive temperatures. The reclaimer should not see hot oil or steam above 415 °F (213 °C). MEA is non-selective in absorbing acid gases. It will absorb H_2S faster than CO_2 but the difference is not significant enough to allow its use to separate them. With the lowest molecular weight of the common amines, it has a greater carrying capacity for acid gases on a unit weight or volume basis. This generally means less pure amine circulation to remove a given amount of acid gases. Because the solvent is in solution with water, the gas with which it comes in intimate contact will leave the contactor at its water-saturation point. If dehydration is necessary it must be done after the MEA system.

MEA has a vapour pressure 30 times that of DEA (300 times that of TEA and DGA) at the same temperature. This causes MEA losses of 1–3 lb/MMscf (16–48 kg/MMNm^3) compared to 1/4 to 1/2 lb/MMscf (4–8 kg/MMNm^3) for the other amine systems using the same system design parameters. Entrainment and leakage losses prevent low vapour pressure amines from attaining the predicted losses.

DIETHANOL AMINE (DEA)

DEA does not degrade when contacted with CS_2, COS and mercaptans as MEA does. Because of this, DEA has been developed as a preferred solvent when these chemicals are present in the stream to be treated. The reaction with acid gas for any of the amines is a mole to mole reaction. As shown in Figure 15.2 the molecular weight of DEA is 1.7 times that of MEA. Even after correcting for density it requires 1.6 lb (kg) of DEA to react with the same amount of acid gas as 1 lb (kg) of MEA.

DEA is a weaker base (less reactive) than MEA. This has allowed DEA to be circulated at about twice the solution strength of MEA without corrosion problems. DEA systems are commonly operated at strengths up to 30 wt% in water and it is not unusual to see them as high as 35 wt%. This results in the DEA solution circulation rate usually being a little less than MEA for the same system design parameters.

The process reactions are shown below.

$$HOCH_2CH_2NHCH_2OH = R_2NH = DEA$$

$$2\ R_2NH + H_2S \underset{\text{High temp.}}{\overset{\text{Low temp.}}{\rightleftharpoons}} (R_2NH_2)_2S$$

$$(R_2NH_2)_2S + H_2S \underset{\text{High temp.}}{\overset{\text{Low temp.}}{\rightleftharpoons}} 2\ R_2NH_2HS$$

Because the system has much fewer corrosion problems and removes acid gases to nearly pipeline specifications it has been installed as the predominant system in recent years.

DIGLYCOL AMINE (DGA)

This process originally began as a combination of 15% MEA, 80% triethylene glycol, 5% water. The system would both sweeten and dehydrate (to the same level as 95% TEG) the gas in a single step. The high vapour release during regeneration (both water vapours and acid gases) causes severe erosion/corrosion problems in the amine/amine exchanger and in the regeneration column. This system has generally been abandoned. The present system uses 2-(2-amino ethoxy) ethanol at a recommended solution strength of 60 wt% in water. DGA has almost the same molecular weight as DEA and reacts mole for mole with acid gases. DGA seems to tie up acid gases more effectively so that the higher concentration of acid gas per gallon of solution does not cause corrosion problems as experienced with the usual amine systems.

The system reactions are given below:

$$HOCH_2CH_2OCH_2CH_2NH_2 = RNH_2 = DGA$$

$$2\ RNH_2 + H_2S \underset{\text{High temp.}}{\overset{\text{Low temp.}}{\rightleftharpoons}} (RNH_3)_2S$$

$$(RNH_3)_2S + H_2S \underset{\text{High temp.}}{\overset{\text{Low temp.}}{\rightleftharpoons}} 2\ RNH_3HS$$

DGA does react with COS and mercaptans similarly to MEA but forms N, N1, *bis* (hydroxy, ethoxy ethyl) urea, BHEEU. BHEEU can only be detected using an infra-red test rather than chromatography. Normal operating levels of 2–4% BHEEU are carried in the DGA without corrosion problems. BHEEU is removed by the use of a reclaimer identical to that for an MEA system but operated at 385 °F (196 °C). Materials of construction are the same as those for MEA systems.

There has been a concern that DGA might be a good solvent for unsaturated hydrocarbons. A survey of the DGA users indicates that many of the systems are operated on gas containing concentrations of C_5^+ above 2% without any indication of hydrocarbon loading of the system. Those systems near their hydrocarbon dew point are usually installed with a flash tank on the rich amine from the absorber. The flash tank is operated at a reduced pressure just high enough to get into the plant fuel gas system. It reduces the vapour load on the regenerator column. (A similar system is recommended on MEA systems operating near the hydrocarbon dew point.)

DGA allows H_2S removal to less than 1/4 grain per 100 scf (about 0.006 kg per 1000 cubic metres) and removes CO_2 to levels of about 200 ppm using normal absorber design parameters.

OTHER GAS-TREATING PROCESSES

Hot Potassium Carbonate (PC)

The basic process concept has been known since the early 1900s but it was not an economical, practically demonstrated process until the mid-1950s. Since that time the process has been used for bulk removal of acid gases where residual CO_2 content was not needed in the ppm range.

The process is very similar to the amine processes. High temperatures favour high solubility of PC in water leading to high concentrations of PC. High PC concentrations mean higher carrying capacity of acid gases in the system. The system is ideal for streams having CO_2 partial pressures of 30 to 90 psi (205–620 kPa). It has a high affinity for H_2S so that pipeline specification for H_2S can easily be reached at about 4 ppm. A stream with little or no CO_2 is not suited for the process due to making regeneration of the lean PC extremely difficult. This process is usually found as part of the hydrogen plant in those refineries that need to produce hydrogen.

Sulfinol

This is a proprietary system developed by Shell. The process uses a solution containing both a chemically reactive component, di-isopropanolamine (DIPA) and a physical solvent, tetrahydrothiophene 1-1 dioxide (sulfolane).

The sulfolane is a very active solvent for H_2S, COS and the mercaptans. CO_2 is also soluble in it, but not nearly as much as the S compounds. Because of this, sulfinol systems are most economically attractive (compared to amine systems) for H_2S/CO_2 ratios greater than 1:1. If the bulk of the acid gas can be dissolved the system is much cheaper to operate than amine.

The acid gases are picked up and released with very little heat increase or heat required. The solubility of acid gases is much higher in sulfinol than for the amines. Sulfinol loadings are limited to 4–6 scf (30–45 Nm_3) acid gas/gal (m^3) solvent versus 2.5 scf acid gas/gal (19 m^3 AG/m^3) amine solution. In addition, the heat capacity of sulfolane is about half that of the amines, further reducing the regeneration heat required.

Other, sweetening liquid processes such as Vetracoke, Stretford and Rectisol have found high usage in the coal gasification and natural gas industries. They have not reached the prominence of the amines or the PC processes in oil refining. Table 15.1 summarizes a comparison of the various solvents.

15.3 Calculating the Amine Circulation Rate

The circulation rate for amine solvents is important to ensure effective treatment of the sour gas. It is important also because it is a major contribution to the operating cost of the plant. These costs are incurred by pumping cost, steam to reboiler and air cooler/condenser costs. This section provides a calculation method to establish this circulating rate. It is based on some fixed parameters which have been accepted as the optimum. The steps are:

- *Step 1*   In DEA and MEA treating processes a ratio of the amine to H_2S is about 3:1 mol. Fix this figure.
- *Step 2*   Obtain gas feed rate and its composition (from plant and lab tests). Determine its mol wt and the volume per cent H_2S.
- *Step 3*   Fix the weight per cent of amine in the recirculating amine solution. This will be between 15% and 20% weight of DEA or MEA in water.
- *Step 4*   Calculate the H_2S in moles/h that is in the feed. This is done by resolving volume flow of gas to moles and using lab data to provide moles/h of H_2S.

Table 15.1. Comparison of H_2S and CO_2 solvents (metric)

Formula	DEA	MEA	DGA	DIPA	Sulfinol	Sulfolane
Molecular wt	61.1	105.1	105.14	133.19		120.17
Boiling point (°C)	170.3	268.4	207.0	248.7		285
Boiling range, 5–95% (°C)	169.3–171.7	111.1–169.3 (20 mm Hg)	96.1–110			27.6
Freezing point (°C)	10.3	25.1	−12.5	42.0		27.6
SG 25 °C	1.0113	1.0881 (30 °C)	1.0572	—		1.256 (30 °C)
60°C	0.9844	1.0693	1.022	0.981 (53.9 °C)		1.235
Pounds per gallon, 25 °C (kg/m³)	1012	1089 (30 °C)	1057	994 (30 °C)		1253 (30 °C)
Abs. visc., CP						
25°C	18.95	351.9 (30 °C)	40	870 (30 °C)		12.1 (30 °C)
60°C	5.03	53.85	6.8	86 (53.9 °C)		4.9
Flash point (°C)	93	146	127	124		177
Fire point (°C)	96	166	141	135		193
Sp. ht joules/kg °C	2774	2531	2389	3410		1464
Critical—temp. (°C)	341.3	442.1	407.6	—		528.0
Critical—press. (kPa)	5568	3273	3771	—		5289
Ht of vaporiz. (kJ/kg)	833	621	510	472		525
Ht of reaction (CO_2) kJ/kg (approx)	1919	1442	1977		1349	
Ht of reaction (H_2S) kJ/kg (approx)	1512	1280	1568	1163		
Molecular wt	61.1	105.1	105.14	133.19		120.17
Boiling point (°F)	338.6	515.1	404.6	479.7		545
Boiling range, 5–95% (°F)	336.7–341.0	231.9–336.8 (20 mm Hg)	205–230	—		—
Freezing point (°F)	50.5	77.2	9.5	107.6		81.7
SG 77 °F	1.0113	1.0881 (86 °F)	1.0572	—		1.256 (86 °F)
140 °F	0.9844	1.0693	1.022	0.981 (129 °F)		1.235
Pounds per gallon (77 °F)	8.45	9.09 (86 °F)	8.82	8.3 (86 °F)		10.46 (86 °F)
Abs. visc. CP						
77 °F	18.95	351.9 (86 °F)	40	870 (86 °F)		12.1 (86 °F)
140 °F	5.03	53.85	6.8	86 (129 °F)		4.9
Flash point (°F)	200	295	260	255		350
Fire point (°F)	205	330	285	275		380
Sp. ht BTU/lb (°F)	0.663	0.605	0.571	0.815		0.35
Critical temp. (°F)	646.3	827.8	765.6	—		982.4
Critical press. (atm)	44.1	32.3	37.22	—		52.2
Ht of vaporiz. (BTU/lb)	357.94	267.00	219.14	202.72		225.7
Ht of reaction (CO_2) BTU/lb (approx)	825	620	850		580	
Ht of reaction (H_2S) BTU/lb (approx)	650	550	674		500	

- *Step 5* Calculate the amount of H₂S to be left in the lean gas. This is usually 10 grains/100 SCF with DEA as absorbent and < 1.0 grains/100 SCF with MEA at absorbent (1 grain = 0.648 grams or 0.0022857 ounces).
- *Step 6* Calculate H₂S absorbed. This is the difference between Steps 4 and 5.
- *Step 7* Using the ratio 3:1 fixed in Step 1 calculate the rate of DEA (or MEA) that will be required.
- *Step 8* Mol wt of DEA is 105.1 and the mol wt of MEA is 61.1. Calculate weight of the amine using percentage weight of amine in solution. Calculate weight per unit time of solution.
- *Step 9* Using the data from Section 15.2 calculate the solution's gallons per hour or per minute.

An example calculation now follows.

Example Calculation

To calculate circulation rate of the following system:

Feed gas rate	$= 17.359 \times 10^6$ SCF D
Feed gas mol wt	$= 10.84$
H₂S vol.%	$= 6.15$
Treated gas spec.	$= 10$ grains H₂S/100 SCF
DEA/H₂S ratio	$= 3.0$
DEA solution	$= 20\%$ wt

$$\text{Feed gas} \quad = \frac{17.359 \times 10^6}{24 \times 378} \quad = 1910 \text{ moles/h}$$

$$= 1910 \times 10.84 \quad = 20\,700 \text{ lb/h}$$

$$\text{Acid gas in feed} \quad = 1910 \times 0.0615 \quad = 117.4 \text{ moles/h } H_2S$$

Acid gas in lean gas
$$= 10 \text{ grains/100 SCF}$$
$$= 160 \times 10^{-6} \text{ mols } H_2S/\text{mol. lean gas}$$
$$= 160 \times 10^{-6} (1910{-}117.4)$$
$$= 0.29 \text{ mols } H_2S/\text{h}$$

Acid gas absorbed
$$= 117.4{-}0.29$$
$$= 117.11 \text{ mols } H_2S/\text{h}$$

Amine circulation:

3.0 mol. DEA/mol. H₂S × 117.4	$= 352$ mol. DEA
MW DEA	$= 105.1$
lb DEA	$= 352 \times 105.1 = 37\,000$ lb DEA/h
20% wt DEA solution	$= \dfrac{37\,000}{0.2} = 185\,000$ lb/h solution

lb/gal DEA	$= 9.09$	$= 4070$ gal/h
lb/gal Water	$= 8.328$	$= 17771$ gal/h

Solution rate at 60 °F $= 21841$ GPH
$$= 364 \text{ GPM}$$

15.4   Evaluating the Tray Efficiency of an Amine Contactor

Amine absorbers do not have a high tray efficiency. Generally, the efficiency of a contactor will range between 10% and 20%. This can be determined on an operating plant using plant data to determine the number of theoretical trays required to achieve the plant's operating performance. There are several accepted methods to calculate theoretical trays in this service. Among these are the McCabe Thiele Graphical Method and the calculation method described by the following steps. This calculation method is considered by many to be the sounder and more accurate of the methods available. The following steps describe this calculation procedure:

● *Step 1*   The equation to be solved is:

$$N = \left[\frac{\text{Log } 1/q \ (A-1)}{\text{Log } A} \right] - 1$$

where

N = number of theoretical trays

$q = \dfrac{\text{mol. } H_2S \text{ in lean gas}}{\text{mol. } H_2S \text{ in feed gas}}$

A = the absorption factor $\dfrac{L}{V} K$

● *Step 2*   Calculate H_2S in lean gas in mol./h and mol. H_2S in the rich gas. Divide H_2S in lean gas by that in the feed. This is q.
● *Step 3*   Calculate the amine circulation rate in mol/h amine, lb/h amine and lb/h solution (20%) as shown in Section 15.3. Resolve the gal/h of solution.
● *Step 4*   Using the grains/h of H_2S leaving in the richer amine, (that is, grains in feed gas less grains in product gas), calculate grains per gallon absorbed by amine solution. Add to that value the H_2S residual in the lean amine. The sum of these is the grains of H_2S per gallon of amine, and is the value used in Figure A1.11 to determine the partial pressure of H_2S in MEA.
● *Step 5*   The absorption factor is obtained from the equation:

$$A = \frac{a \ (1 + R \cdot r) \ (1 - q)}{pp/P}$$

where

A = the absorption factor
a = mole fraction of H_2S in gas feed
R = moles MEA/moles H_2S absorbed
r = Residual H_2S in lean MEA solution (mole H_2S per mole MEA)
pp = partial pressure of H_2S in rich solution (psia)
P = systems pressure (tower pressure) (psia)

● *Step 7*   Solve for A. Then solve equation given in Step 1 for N. Divide N by actual number of trays, multiply by 100 to give efficiency as a percentage

An example calculation now follows:

Example Calculation

H_2S is removed from a hytrotreater recycle gas stream to meet a lean gas H_2S content of 0.1 grain/100 SCF of gas. The H_2S content of the rich gas is 4048 grains/100 SCF·MEA in a 20 wt% solution will be used as the absorbant in a 15-valve tray tower operating at 315 psia and 120 °F. The amine circulation rate is fixed at 3.0 moles MEA/mol. H_2S absorbed and the recycle gas rate is 17.5 mm SCF/day. These test run data are used to determine the absorber tray efficiency as follows:

Calculate number of theoretical trays in the absorber.

$$q = \frac{\text{mol. } H_2S \text{ in lean gas}}{\text{mol. } H_2S \text{ in feed gas}}$$

H_2S in lean gas   = 0.1 grains/100 SCF
                              = 0.003 moles H_2S/h (See Section 15.3)
H_2S in rich gas   = 124 moles/h (4048 grains/100 SCF)

$$q \quad = \frac{0.003}{124} = 0.24 \times 10^{-4}$$

20% wt MEA solution again at 3.0 moles/mol. H_2S

3.0 × 124 = 372 mol. MEA
MW MEA = 61.1
lb MEA/h = 22 729
lb MEA Sol = $\dfrac{22\,729}{0.2}$ = 113 646 lb/h

	lb/h	lb/gal	gal/h
MEA	22 729	8.45	2 690
WATER	90 917	8.38	10 849
MEA solution	113 646		13 539

Grains H_2S in rich solution:

Grains H_2S absorbed/h = 4048 × 0.729 × 10^4 = 29.51 × 10^6/h

Grains H_2S residual in MEA = 0.09 moles H_2S/mol. MEA
moles MEA/h = 372
moles H_2S/h = 33.48

Grains H_2S/h = 8.13 × 10^6

total grains/h of H_2S in MEA sol = 37.48 × 10^6

Total grains H$_2$S in rich solution =

$$\text{grains } H_2S/\text{gal MEA sol} = \frac{37.64 \times 10^6}{13\ 539} = 2780 \text{ grains/gal}$$

Based on this H$_2$S content the partial pressure is read from Figure A1.11 as 0.35 psia.

Calculate the absorption factor A from the equation:

$$A = \frac{a'(1 + R \cdot r)(1 - q)}{pp/P}$$

where

$$a = \text{mol. fract. of } H_2S \text{ in gas feed}$$

$$= \frac{\text{moles } H_2S/h}{\text{moles of recycle gas}}$$

$$\text{moles recycle gas} = \frac{0.729 \times 10^6}{378} = 1929$$

$$a = \frac{124}{1929} = 0.0643$$

$$A = \frac{0.0643 \times (1 + (3 \times 0.09)) \times (1 - (0.24 \times 10^{-4}))}{0.35 \div 315} = 73.5$$

$$\text{Then } N = \frac{\log\ [1 \div (0.24 \times 10^{-4}) \times 73.5]}{\log 73.5} - 1 = 2.5$$

$$\text{Then tray efficiency is } \frac{2.5}{15} = 17.0\%$$

15.5 Downstream Plant Protection Against Amine Contamination

Amines generally are poisons to most hydrotreaters and hydrocracking catalysts. Consequently where amine contactors are used in the recycle gas systems of these plants a guard must be installed to prevent amine carry-over into the treated gas streams. The method generally used as a guard is the installation of a water wash section above the amine absorber section. Figure 15.2 shows this feature.

The lean gas from the top absorber tray is allowed to flow through a chimney tray. This is a tray with a chimney usually about 30 inches in height. The tray is hooded to prevent water from the bottom wash tray flowing into the absorber section. The water gathers in the space between the chimney and the column wall on the sealed tray from where it is discharged under level control to the sour water system.

The wash section consists of four to six trays located above the chimney tray. Potable water or condensate is introduced over the top tray of the section through a suitably designed distributor. The water moves downwards from tray to tray, washing the lean gas moving upwards from the chimney tray. Any entrained amine is removed by this washing, allowing the gas to emerge from the top of the wash section into the gas system.

The water wash rate is usually about 25% of the amine circulation rate. The water is introduced under flow control and is removed from the tower under level control. The water will absorb some of the heavier hydrocarbons from the gas stream and, of course, any amine that may be present. The effluent water should therefore be steam stripped before disposal to the effluent system.

15.6 Calculating the Overall Heat Transfer Coefficient for the Rich/lean Amine Exchanger

The amine in the contactor picks up the heat of reaction which occurs with the absorption of H_2S. On leaving the amine contactor or absorber the rich amine is heat exchanged against hot lean amine leaving the bottom of the stripper. The performance of this exchanger is critical to the process as a whole. Usually the rich amine receives no other heat before it enters the stripper, and as discussed in Chapter 8, the feed condition is vital to the proper operation of the tower. The purpose of this section is to provide a calculation procedure to evaluate the heat transfer coefficient of this exchanger. The following steps gives the procedure:

- *Step 1* Using lab data determine the quantity of H_2S absorbed in the contactor.
- *Step 2* From the data given in Table 15.1 calculate the total heat of reaction in BTU/h.
- *Step 3* Set the temperature of the lean gas leaving the absorber to be the same as the amine entering.
- *Step 4* Carry out a heat balance over the contactor with the rich amine temperature being the unknown. Equate and resolve for this.
- *Step 5* From plant data and the temperatures and flow in and out of the exchanger calculate its heat duty in BTU/h.
- *Step 6* Calculate the LMTD (Log Mean Temperature Difference) and from equipment data obtain the heat exchange surface area. (See also Sections 6.15 and 7.8).
- *Step 7* Using the energy equation

$$Q = UA\Delta TM$$

where

$$Q = \text{the duty in BTU/h}$$
$$A = \text{the area in ft}^2$$
$$\Delta TM = \text{the log mean temperature difference in °F}$$
$$U = \text{the overall heat transfer coefficient in BTU/h ft}^2 \text{ °F}$$

solve for U.

- *Step 8* Compare U to those acceptable values given in Table 6.2.

An example calculation now follows.

Example Calculation

Calculate the temperature of rich amine leaving the contactor

Rich gas flow = 1913 moles/h
MW   = 10.84
lb/h   = 20 737

Temperature of gas into contactor	= 100 °F
Lean MEA flow	= 107 536 lb/h (Section 15.4)
Temperature of lean MEA	= 105 °F
Moles/h of H_2S to be removed	= 117.4−0.044 (Section 15.4)
	= 117.356 moles/h
	= 3990 lb/h

Heat of reaction H_2S/MEA	= 650 BTU/lb H_2S
Total heat by reaction	= 650 × 3990 BTU/h
	= 2.594 × 10⁶ BTU/h

Temp. of gas leaving contactor	= 105 °F (same as lean MEA)
Enthalpy in gas entering	= 20 737 × 350 = 7.258 × 10⁶
Enthalpy in gas leaving	= 20 737 × 355 = 7.362 × 10⁶

Heat in with lean MEA

$$\left. \begin{array}{l} = 21\,507 \times 0.663 \times 105 \\ + \\ = 86\,028 \times 1.0 \times 105 \end{array} \right\} 10.530$$

Enthalpy balance

Heat in
With gas	= 7.258 × 10⁶
With lean MEA	= 10.530 × 10⁶
Reaction	= 2.594 × 10⁶
Total in	= 20.382 × 10⁶ BTU/h

Heat out
With gas	= 7.362 × 10⁶
With rich MEA	= 13.021 × 10⁶ (by difference)
Total out	= 20.382 × 10⁶ BTU/h

$$\text{MEA outlet temp.} = \frac{13.021 \times 10^6}{(21\,507 \times 0.663) + 86\,028}$$

$$T_1 = 130\,°F$$

The stripper will operate at about 7 psig overhead and 10 psig bottom

Overhead temp. 210 °F
Bottoms temp. 240 °F

Rich amine is exchanged against bottoms from the stripper

The lean amine is cooled to 178 °F

Duty of exchanger is

$$(21\,507 \times 0.663 \times (240 - 178)) + (86\,029 \times (240 - 78)) = 6.218 \times 10^6 \text{ BTU/h}$$

Temperature of rich amine leaving

$$(13.021 + 6.218) \times 10^6 \quad = (21\,507 \times 0.663 \times T_1) + (86\,029 \times T_1)$$

$$192 \text{ °F} \qquad = T_1 \text{ (DATUM} = 0 \text{ BTU/lb @ 0 °F)}.$$

The heat exchanger has 1300 ft² of surface area

The $\dfrac{6.218 \times 10^6}{A \times \text{LMTD}} = U_0$

$$\text{LMTD} \quad = 240 \rightarrow 178$$
$$\underline{192 \leftarrow 130}$$
$$48 \qquad 48 \quad \text{LMTD} = 48 \text{ °F}$$

$$U_0 \qquad = \frac{6.218 \times 10^6}{1300 \times 48} = 99.6 \text{ BTU/h ft}^2 \text{ °F}$$

Table 6.2 gives 110 for MEA/cooling water

Calculated U_0 is therefore acceptable

15.7 Evaluating the Amine Stripper Performance

This evaluation is based on reconciling the steam used for stripping the rich amine. The quantity of steam is a major operating cost in this type of plant and therefore deserves attention. An acceptable level of steam usage is about 0.8–1.1 lb steam/gallon of circulating solution for MEA and 1.1–1.3 lb steam/gallon in the case of DEA. This section describes a procedure to calculate this steam rate.

● *Step 1* From plant data ascertain the feed rate of rich amine to the stripper. Obtain from lab data its composition in terms of H_2S, water, MEA (or DEA), and hydrocarbons.

- *Step 2* Again from lab and plant data obtain flow rate of lean amine leaving the bottom of the tower. Obtain from the lab data its composition in terms of residual H_2S, water, MEA (or DEA) and hydrocarbon (if any).
- *Step 3* Develop the material balance over the tower. Check flow rate of overhead gas against plant readings.
- *Step 4* Using the data in Step 3 and plant reading for the external tower top reflux calculate the heat balance over the tower to find the condenser duty.
- *Step 5* Calculate the heat in with the rich amine feed before the preheat exchangers (see Section 15.6).
- *Step 6* Using plant data for lean amine temperature in and out of the preheat exchangers calculate its duty. Add the duty to the enthalpy from Step 5 to give feed enthalpy into the tower.
- *Step 7* Calculate the overall heat balance over the stripper to find the reboiler duty. Remember to add in heat of dissociation which is equal to the heat of reaction in the contactor (see Section 15.6).
- *Step 8* Saturated 50 psig steam is usually used as the heating medium. Calculate the amount of steam from its enthalpy data (steam tables) and the reboiler duty. Check plant reading.
- *Step 9* If the steam usage is excessive check overhead stream and the lean amine concentration. It is possible that a high volume of water is being evaporated.

An example calculation now follows.

Example Calculation

Calculate the material balance over the rich amine stripper

In

Hydrocarbons	1075 lb/h (estimated)
H_2S	5055 lb/h
Water	86 029
MEA	21 507
Total in	113 666 lb/h

Out

		lb/h	Moles
Overhead	Hydrocarbons	1075	10.75 (C_7)
	H_2S	3982	117.1
	Water	344	19.1
	MEA	86	1.4
Total overheads		5487	148.35

Bottoms	lb/h	
H_2S	1073	(0.09 mol/mol MEA)
MEA	21 421	
Water	85 685	

Calculate heat balance over tower top

Liquid reflux mol. wt = 46
lb/h reflux = 13 648

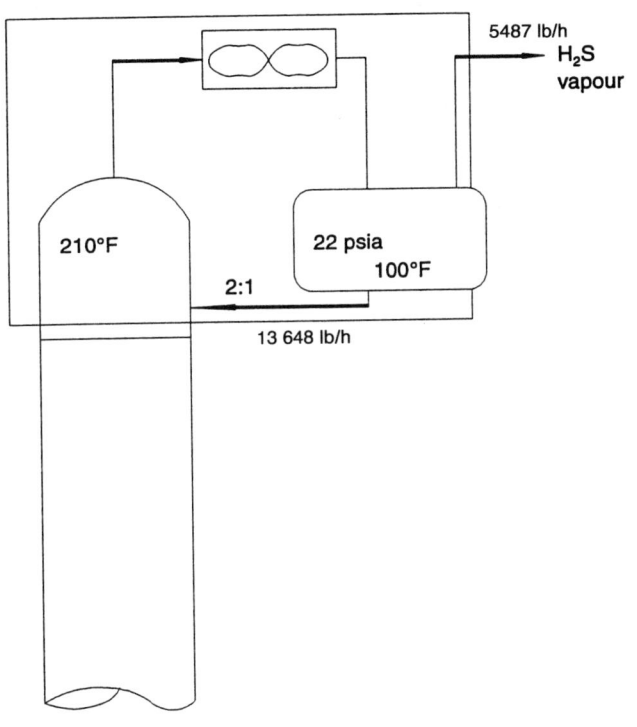

					Enthalpy	
	V or L	MW	°F	lb/h	BTU/lb	BTU/h × 10⁶
In						
Prod Vapour	V	—	210	5487	(1)	1.181
Reflux Vapour	V	46	210	13 648	265	3.617
Total in				19 135		4.798
Out						
Prod Vapour	V	—	100	5487	(2)	0.943
Reflux Liquid	L	46	100	13 648	65	0.887
Condenser				(by difference)		2.968
Total out				19 135		4.798

(1) Enthalpy in product vapour stream at 210 °C.

$$BTU/h \times 10^6$$

HC	1075×345	$= 0.371$
H_2S	3982×90	$= 0.358$
Water	344×1188	$= 0.409$
MEA	$(86 \times 0.663 \times 210) + (86 \times 357.9)$	$= 0.043$
Total Enthalpy in prod vapour		1.181

(2) Enthalpy in product vapour stream at 100 °F:

HC	1075×295	$= 0.317$
H_2S	3982×50	$= 0.199$
Water	344×1137	$= 0.391$
MEA	$(86 \times 0.663 \times 100) + (86 \times 357.9)$	$= 0.036$
		$= 0.943$

Calculate overall heat balance over stripper

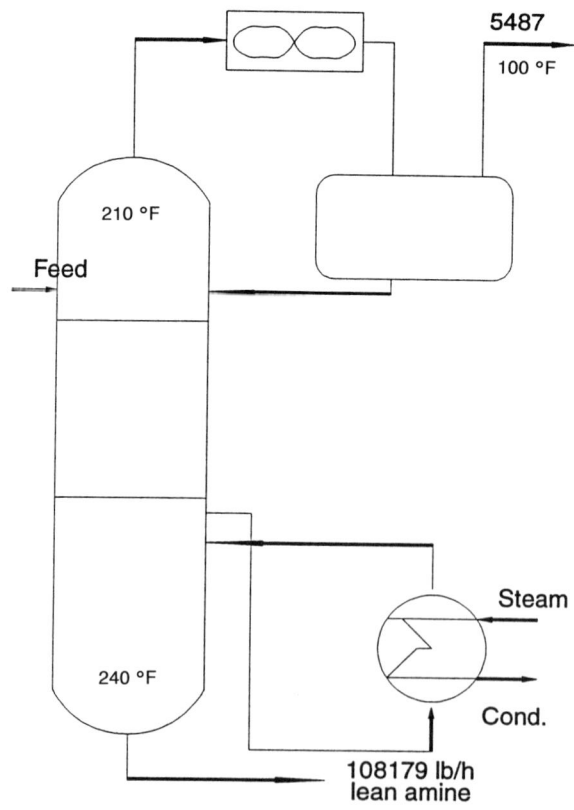

	V or L	MW	°F	lb/h	BTU/lb	Enthalpy BTU/h $\times 10^6$
In						
Feed				11 366		20.336
Reboiler		By difference				10.360
Total in				11 366		30.696
Out						
Out heat of dissoc.						2.594
Overheads	V	—	100	5487		0.943
Lean amine	L	—	24C	108 179		24.191*
Condenser						2.968
Total out						30.696

*Lean amine enthalpy:

	lb/h	BTU/h $\times 10^6$
H_2S	1073×203	0.218
MEA	21 421	3.409
Water	85 685	20.564

Reboiler duty	$= 10.360 \times 10^6$
50 psig sat. steam	$= 1210.6$ BTU/lb
50 psig cond.	$= 296$ BTU/lb

$$\text{Steam used} = \frac{10.360 \times 10^6}{(1210.6 - 296)}$$

$$= 11327 \text{ lb/h}$$

Which is 0.9 lb steam/gallon MEA Sol^N

The norm is 0.8 to 1.1 lb steam/gal

15.8 Procedure for Determining Tray Performance in Amine Towers

This procedure is similar to those described in other sections of this manual. Emphasis is made here, however, on the foaming tendencies of MEA and DEA. Most of these units contain facilities to inject anti-foaming chemicals. Under normal operation, these chemicals subdue foaming to a large extent particularly in the contactors. The addition of a filter in the rich amine line also helps to combat severe foaming.

It is believed that foaming is enhanced by the absorption of large quantities of hydrocarbons into the amine stream. This can occur quite easily if the temperature or pressure change forces the hydrocarbon gas into a condition below its dew point. The presence of large quantities of impurities such as COS (carbonyl sulphide) or the tarry derivative of COS may also induce foaming. The addition of a filter and a reclaimer usually solves this problem to a large extent.

Loading of amine towers, nevertheless, are critical due to the foaming nature. Consequently the towers are designed with more latitude than most other towers. In evaluating its performance the acceptable level of flooding is lower by 30 to 40% than that for normal towers.

The calculation to evaluate the towers flooding follows the same steps as those described in Sections 6.14, 8.8, etc. In this case, however, the calculated value for Gf is multiplied by a system factor between 60 to 70%. If the unit is being used in service where high concentration of impurities are present a figure of 60% should be used. An example calculation now follows

Example Calculation

Feed gas rate to contactor	$= 34.71 \times 10^6$ SCFD
	$= 3820$ moles/h
	$= 41\,400$ lb/h

lb/h DEA $= 370\,000$

gal/h of 20% solution $= 43\,682$	$= 728$ GPM
Say inlet temp. of rich gas	$= 100\,°F$
Temp. gas leaving	$= 105\,°F$
Temp. of rich amine leaving	$= 135\,°F$ (estimated)
lb/gal 20% DEA solution	$= \dfrac{370\,000}{43\,682} = 8.47$ at 60
	$= 1.016$ SG at $60\,°F$
SG at $135\,°F$	$= 1.000$
gal/min at $135\,°F$	$= 7940$ gal/min
ACFs of rich gas	$= \dfrac{3820 \times 378 \times 14.7 \times 560}{3600 \times 520 \times 315}$
	$= 20.16$ cfs

ρ_V $\qquad = 0.570 \, \text{lb/ft}^3$
ρ_L $\qquad = 62.52 \, \text{lb/ft}^3$

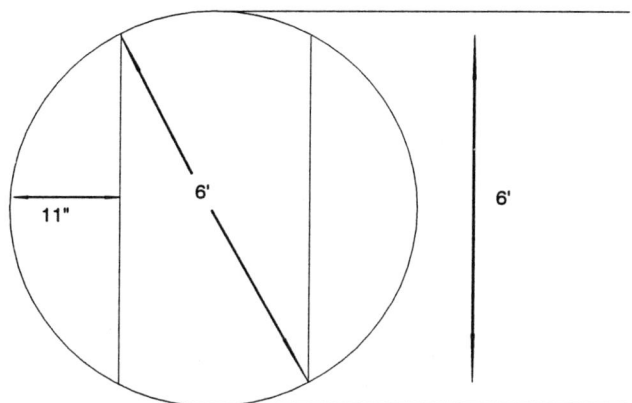

use a foaming factor of 60%
Tray details are:
Diameter = 6 ft
A_s $\qquad = 28.27 \, \text{ft}^2$
A_{dc} $\qquad = 2 \times 2.7 = 5.4 \, \text{sq ft}$
A_w $\qquad = 4.24$
A_b $\qquad = 18.63 \, \text{sq ft}$
Tray spacing = 24 in

Actual vapour loading:

$$Ga = \frac{41\,400}{18.63}$$

$$= 2222 \, \text{lb/h.sqft}$$

using a factor of 0.6 K then vapour load allowable at flood with 24 in tray spacing:

$$0.6 \times 1110 \, \text{(from figure 6.9)}$$
$$= 666$$

Then $Gf = 666\sqrt{(0.57 \times 62.52 - 0.57)}$

$$= 3957 \, \text{lb/h sq ft}$$

Percentage of flood $= \dfrac{2222}{3957} \times 100$

$$= 56.2\%$$

15.9 Combating Degradable Impurities in MEA Gas Treating

Although MEA (mono-ethanolamine) is the most efficient absorbent in the amine family, it has one major shortcoming. It is readily degraded by certain sulphur compounds such as carbonyl sulphide (COS) and by carbon disulphides. Both these compounds are found in significant quantities in gases from cracking processes such as thermal crackers and catalytic crackers.

It is difficult (if not impossible) under normal refinery conditions to remove these sulphides from the gas. What can be done and is the normal practice is to remove these products of degradation and return the 'clean' amine to the system. Two items of equipment are added to the process to achieve this: a filter and a reclaimer. The filter is a normal leaf type filter contained in two filter casings. These are piped up in parallel with one onstream and the other shut down for cleaning and as spare.

Reclaimers are really kettle-type reboilers. It takes as feed a portion of the lean amine leaving the stripper. This stream is vaporized and the vapour returned to the stripping tower. The residue or sludge is the product of degradation and is dumped to waste. Reclaimers can be designed to operate continuously or on a batch basis. Steam is used as the heating medium, and, as in the case of the stripper reboilers, the heating steam temperature to the reclaimer is carefully controlled. Thus the steam medium is saturated 50 psig steam.

The example calculation which follows demonstrates the operation of a batch reclaimer process. It must be remembered that the duty to condense the vaporized amine from the reclaimer must be added to the stripper overhead condenser duty. It must also be included in any tower loading exercises that may be done.

Example Calculation

MEA circulating $= 108\,179\,\text{lb/h}$ of solution.

The reclaimer will be sized for batches of 10 000 lb.

$$= 2000\,\text{lb/h MEA} \qquad = 236.5 \text{ gallons at } 60$$
$$ 8000\,\text{lb/h } H_2O \qquad = \underline{954.5} \text{ gallons at } 60$$
$$\qquad\qquad\qquad\qquad\quad = 1191 \text{ gallons at } 60$$

Expansion factor to 250 °F $= 1.1$

Capacity of reclaimer $= 1310.1$ gallon

To calculate one cycle time

Slip stream at 10% normal pump rate $= 14172.5 \times 0.1$
$$= 1417.25\,\text{gal/h} \qquad\qquad = 23.63\,\text{gal/min}$$

Filling time $= \dfrac{1310.1}{23.62} = 55.5\,\text{min}$

Reclaimer to be heated by 50 psig saturated.

$$\text{Steam} = 298\,^\circ\text{F} = 1179.15 + 32\ \text{BTU/lb}$$

Heat flux for reclaimer (use kettle reboiler)

$$= 26\,500\ \text{BTU/h ft}^2\ \text{at}\ 35\ \text{psia}$$

and 260 °F enthalpy $= 1167.4 + 32\ \text{BTU/lb}$ (datum $= 0$ BTU/lb at 0 °F)

Total $\qquad\qquad = 1199.4 \times 9500 = 11.394$

$(9500 = 10\,000$ less 5% product of degradation)

Bundle consists of 450 tubes ¾″ O/D on 1″ □ pitch each 0.1963 ft²/ft length

Tube length	$= 10\ \text{ft}$
Heating surface area	$= 0.1963\ \text{ft}^2 \times 10 \times 450$
	$= 883.4\ \text{ft}^2$
Heat transfer rate	$= 26\,500 \times 883.4\ \text{BTU/h}$
	$= 23.41\ \text{BTU/h} \times 10^6$
Vaporizing time	$= \dfrac{11.394 \times 10^6\ \text{h}}{23.41\ \times 10^6\ \text{h}}$
	$= 0.487$ say ½ h
Sludge emptying time	$= 15\ \text{min}$

Total cycle time:
    Filling   1 h
    Vaporizing   30 min
    Discharging   15 min
    Total   1 h 45 min

    Say 2 h

Reclaiming to be done three times a shift if degradation requires it.

Note: Amine stripper overhead condenser to be large enough to accommodate reclaimer heat load.

15.10   Effect of High Temperature on MEA and DEA

Both MEA and DEA deteriorate rapidly when exposed to temperatures in excess of 300 °F, and, because of this, the amine stripper reboiler and reclaimers use 50 psig *saturated steam* as the heating medium. The control of the amine stripper reboiler is usually a conventional flow control on the steam inlet side. In many units a de-superheater is added downstream of the control valve as protection against hot temperature surges. This de-superheater feeds boiler feedwater to retain the inlet temperature to the reboiler at below 300 °F. The BFW is controlled on temperature and resets the main steam flow controller.

A similar feature is included on the reclaimer. These controls against high skin temperatures are essential where surges, particularly in low-pressure steam systems, are experienced. Many refineries, of course, have de-superheaters as an integral part of the steam distribution system. In these cases there would be no need for the high sophistication of stripper reboiler skin temperature control. High steam temperature to these items of equipment should nevertheless be properly alarmed for remedial action. DEA and the potassium carbonate processes are not greatly affected by high temperature.

APPENDIX 1: GENERAL
DATA AND CORRELATIONS

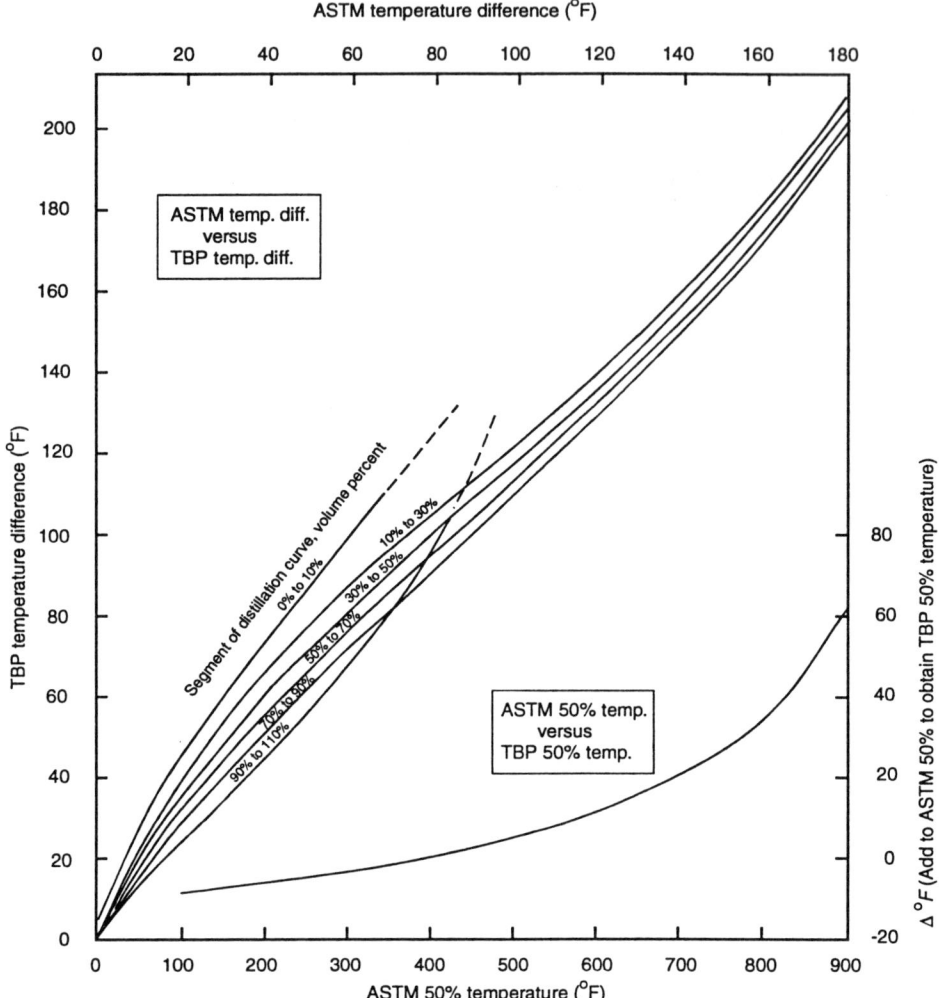

Figure A1.1 ASTM–TBP Correlation
From: Applied Hydrocarbon Thermodynamics Volume 2, by Wayne C. Edmister. Copyright
© 1988 by Gulf Publishing Company, Houston, TX, used with permission. All rights reserved

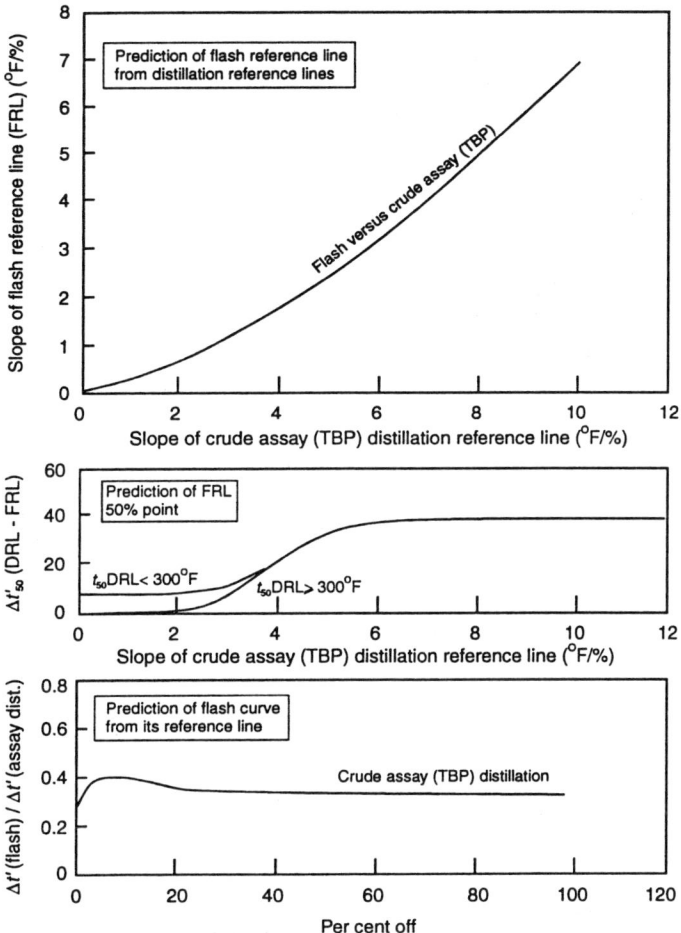

Figure A1.2 TBP–EFV Correlation
From: Maxwell, Data Book on Hydrocarbons. Copyright by A. K. Krieger Publishing
Company, used with permission. All rights reserved

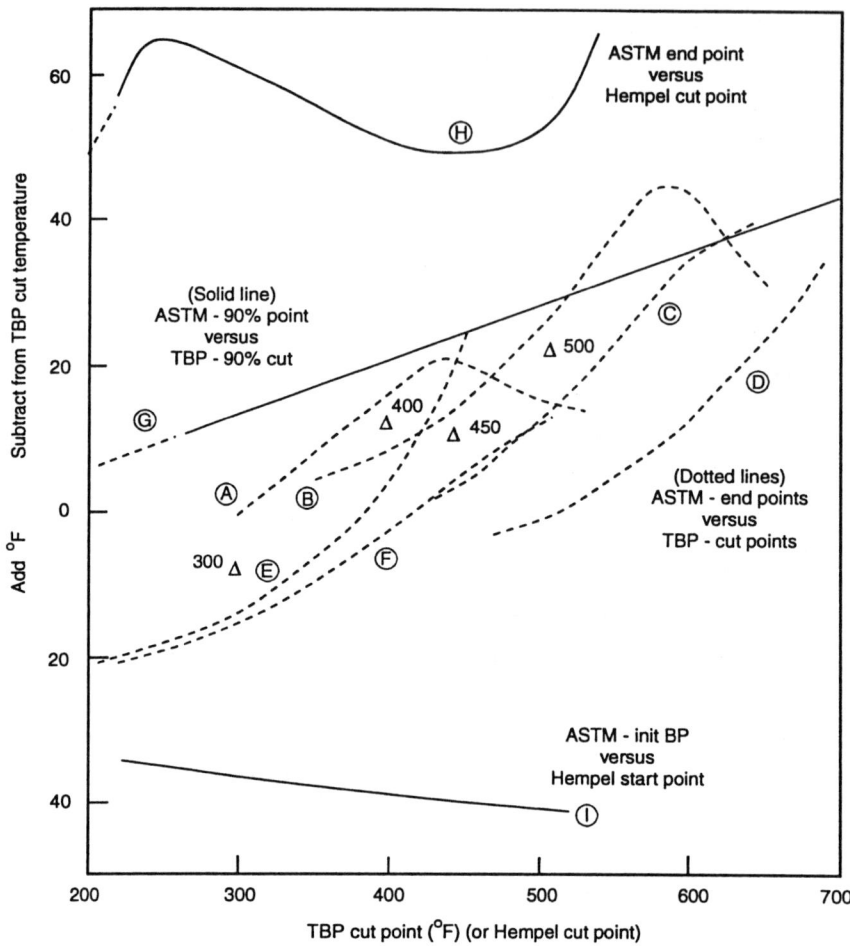

Curves:
A. ASTM end points vs TBP cut point for fractions starting at 200 °F TBP or lower
B. ASTM end points vs TBP cut point for fractions starting at 300 °F
C. ASTM end points vs TBP cut point for fractions starting at 400 °F
D. ASTM end points vs TBP cut point for fractions starting at 500 °F
E and F. ASTM end points vs TBP cut point for 300 ml STD col and 5′ packed towers
G. ASTM 90% vol temp vs temp @ 90% vol TBP cut (all fractions)
H. ASTM end point vs Hempel cut point temp

Figure A1.3 TBP cut point vs ASTM end points

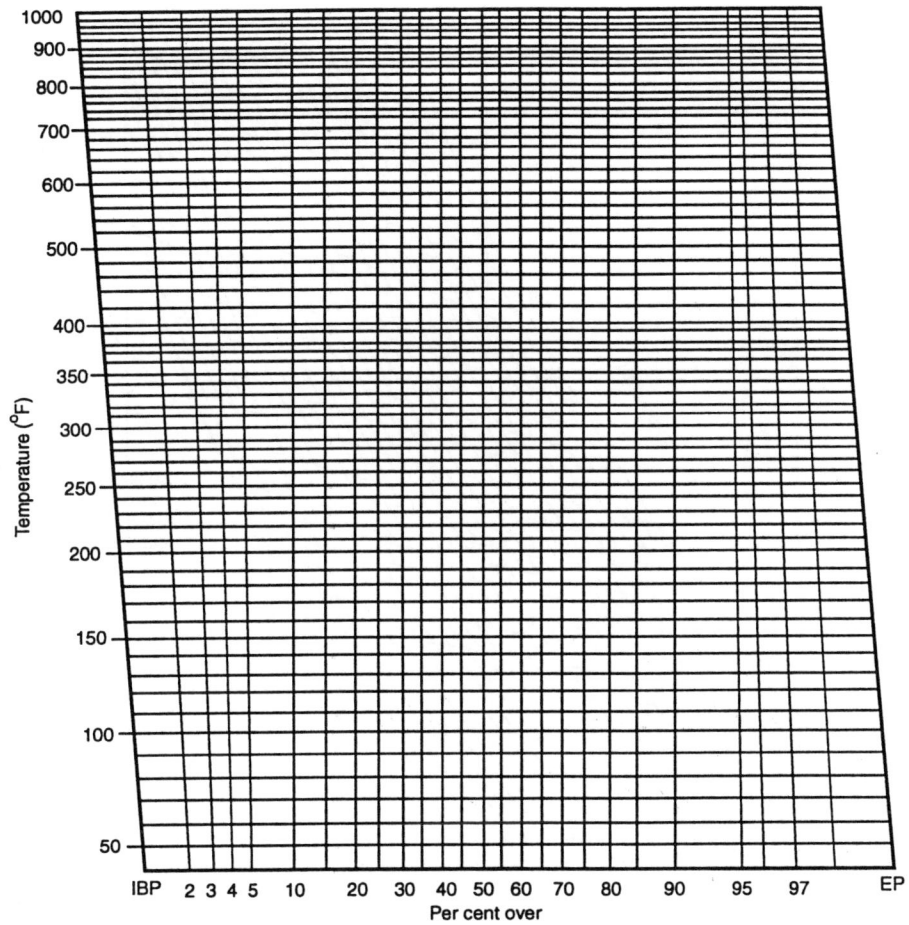

Figure A1.4 Probability graph for ASTM distillations
Reproduced by permission of Pennwell Publishing Company from the 'Oil and Gas' Journal
4 September 1981

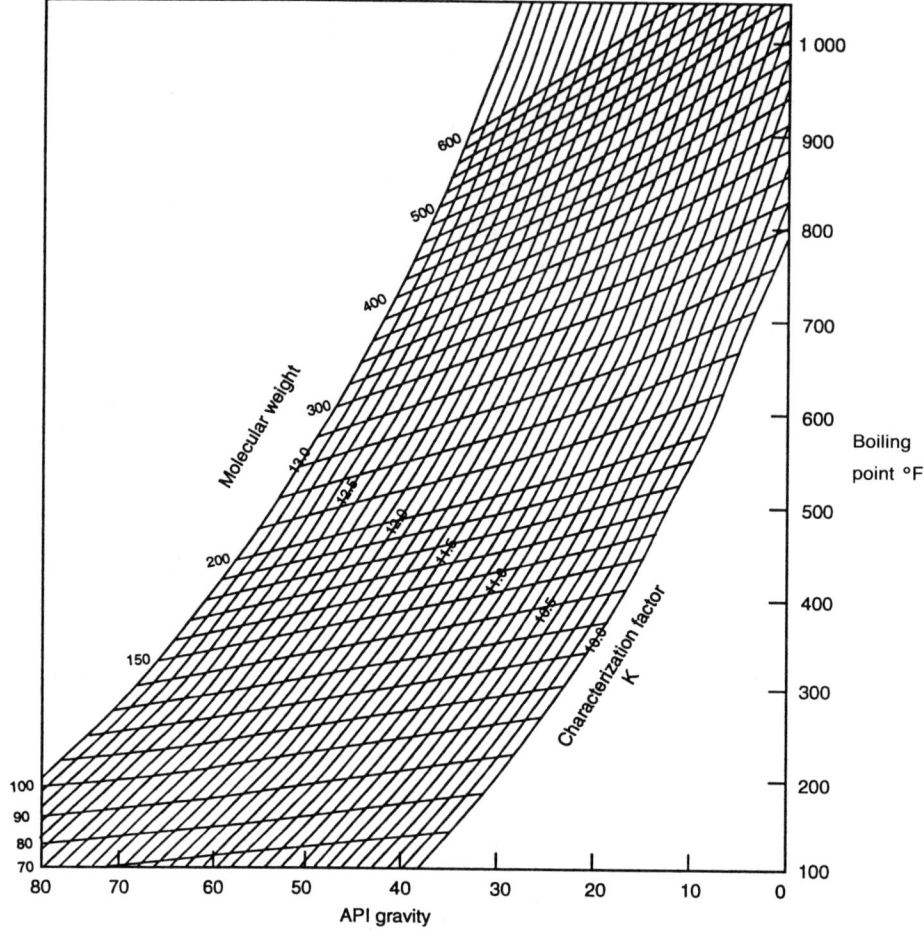

Figure A1.5 Correlation between boiling point, characterization factor K, °API and Mol wt
for hydrocarbons
Reproduced by permission of Fluor Daniel Inc

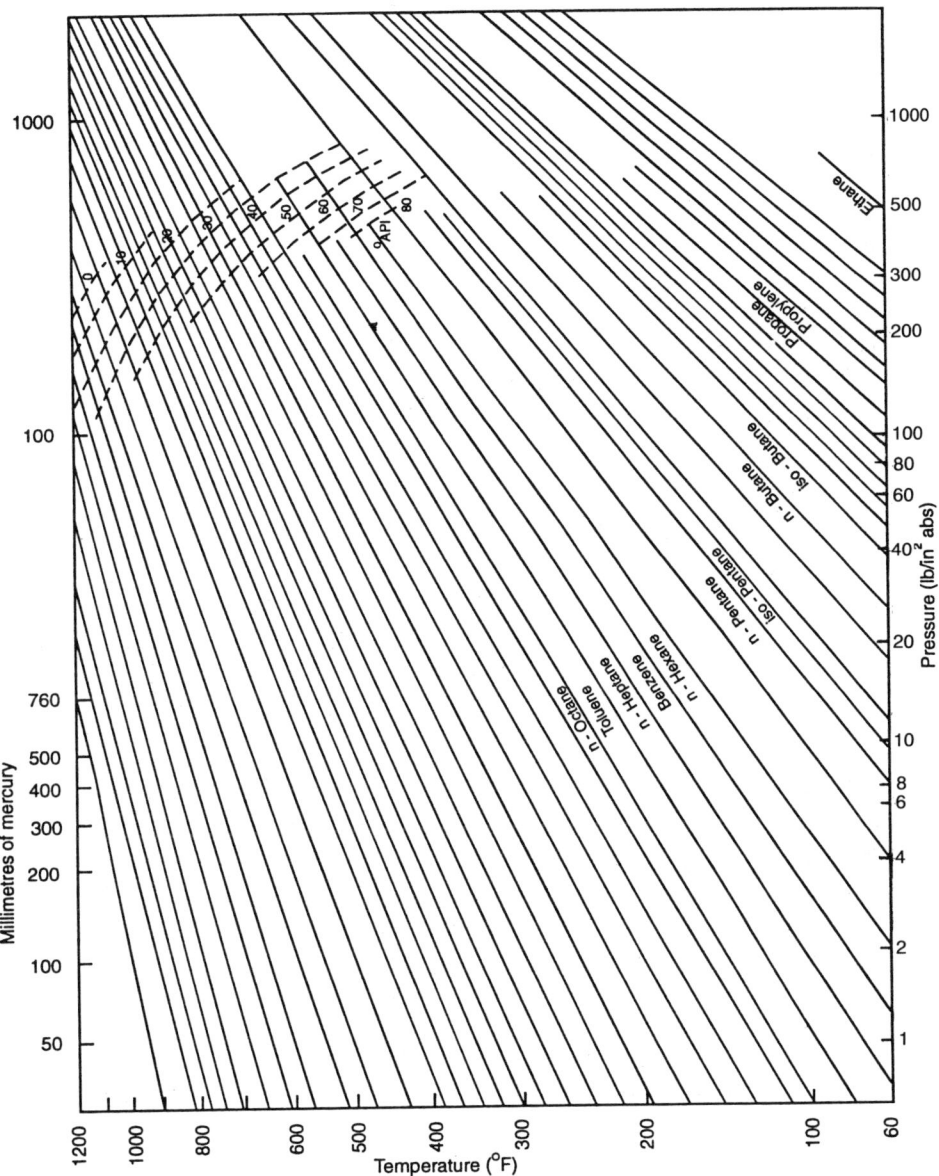

Figure A1.6 Vapour pressure chart for hydrocarbons (high temperature range)
Reproduced by permission of Fluor Daniel Inc

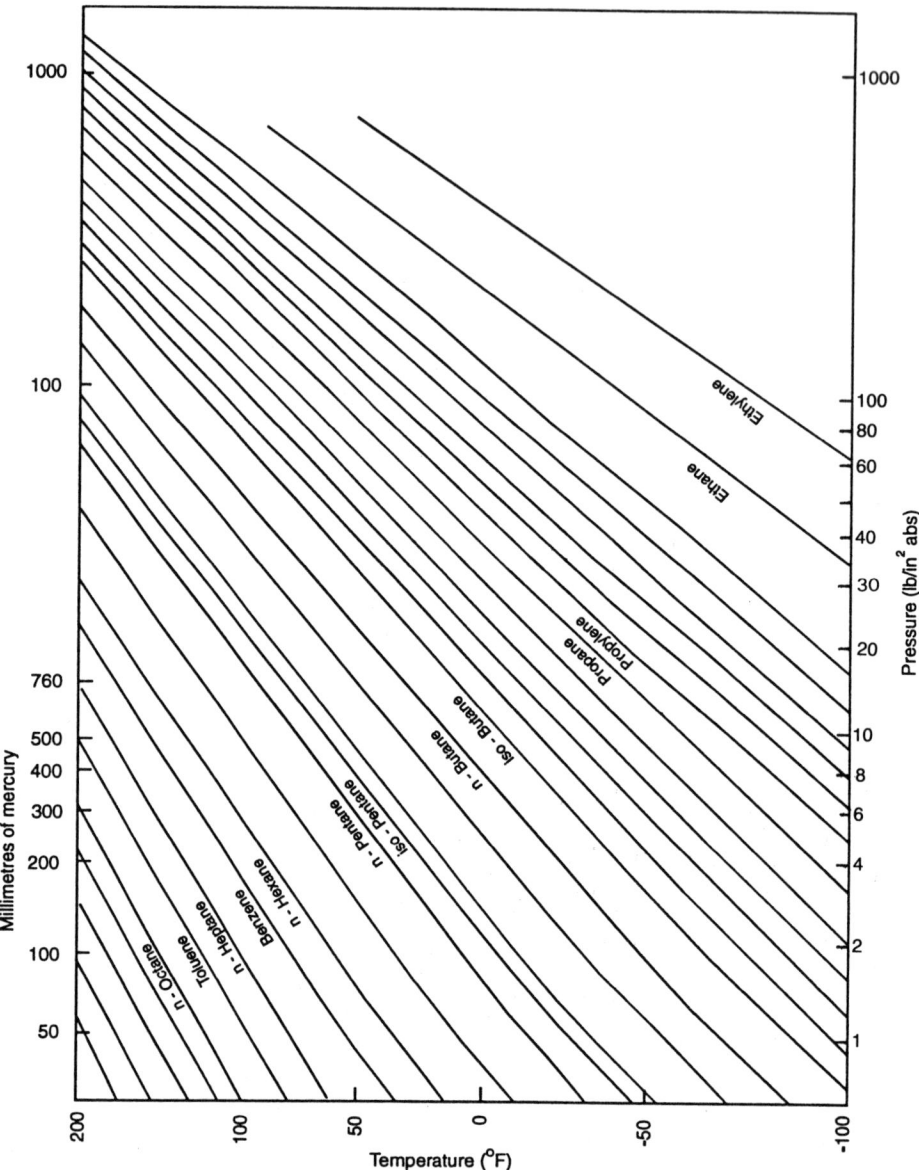

Figure A1.7 Vapour pressure for hydrocarbons (low temperature range)
Reproduced by permission of Fluor Daniel Inc

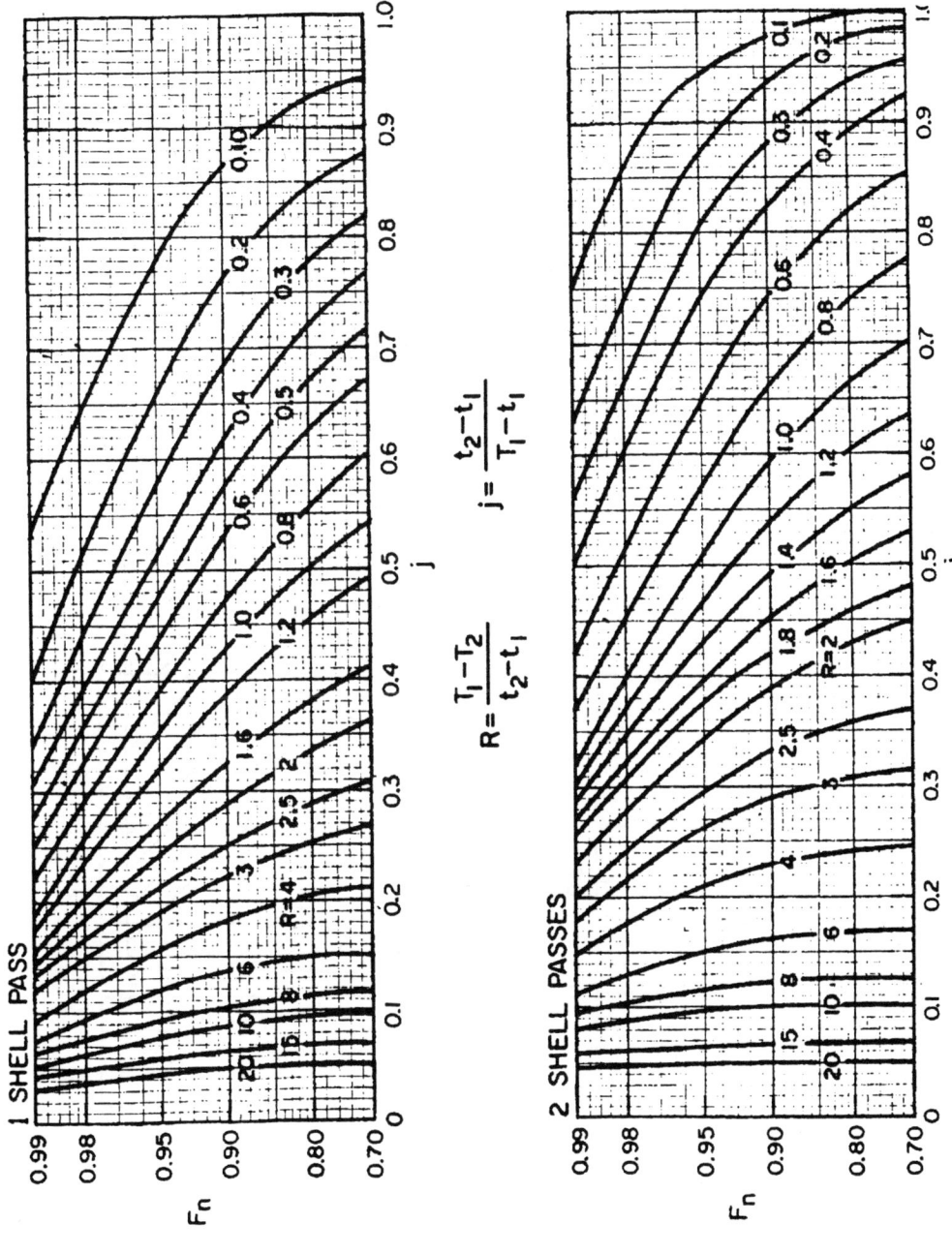

Where T_1 = Shell inlet temp T_2 = Shell outlet temp F_N = Correction factor t_1 = Tube inlet temp t_2 = Tube outlet temp

Figure A1.8 LMTD correction factors

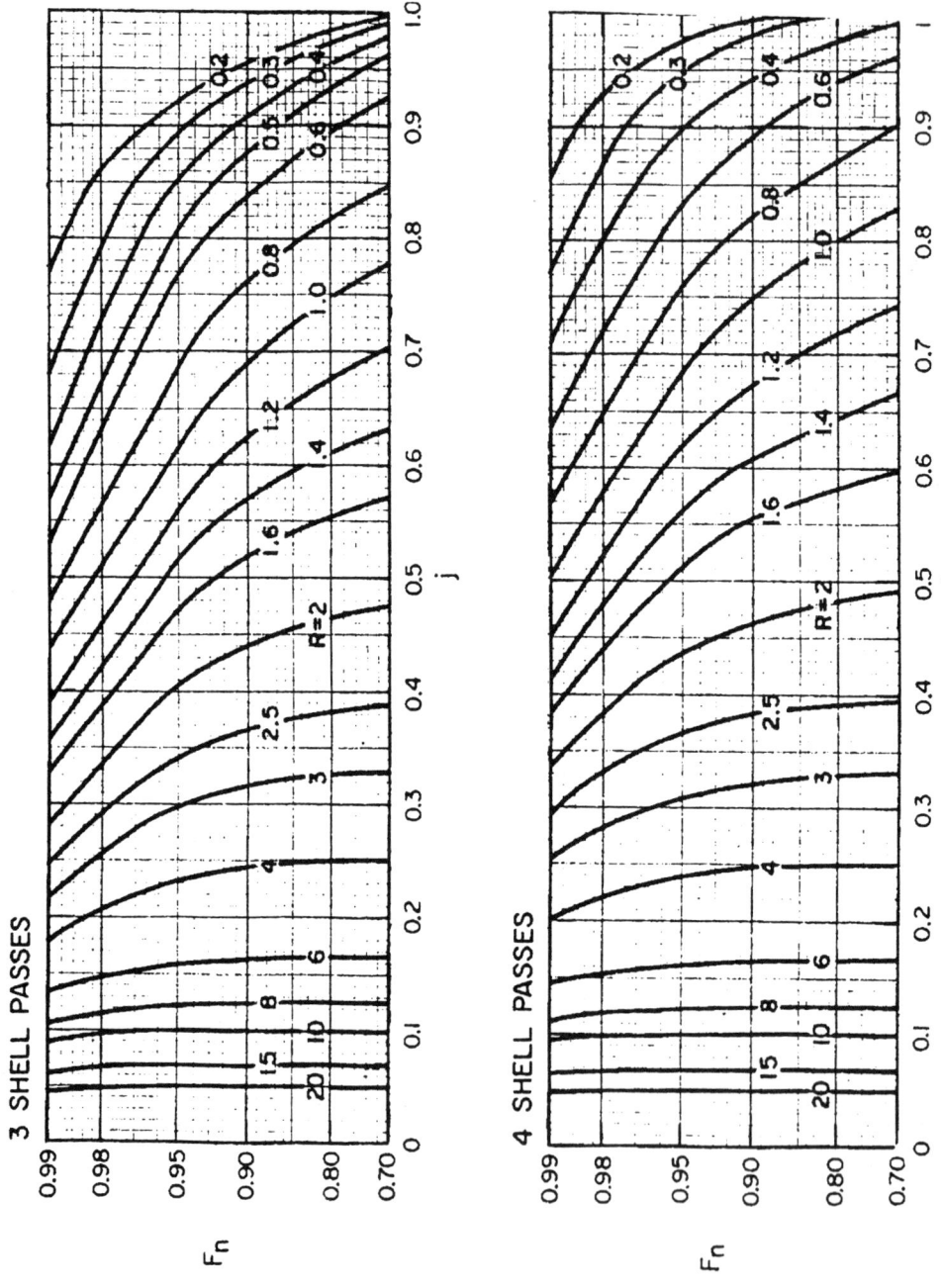

Figure A1.8 (continued)

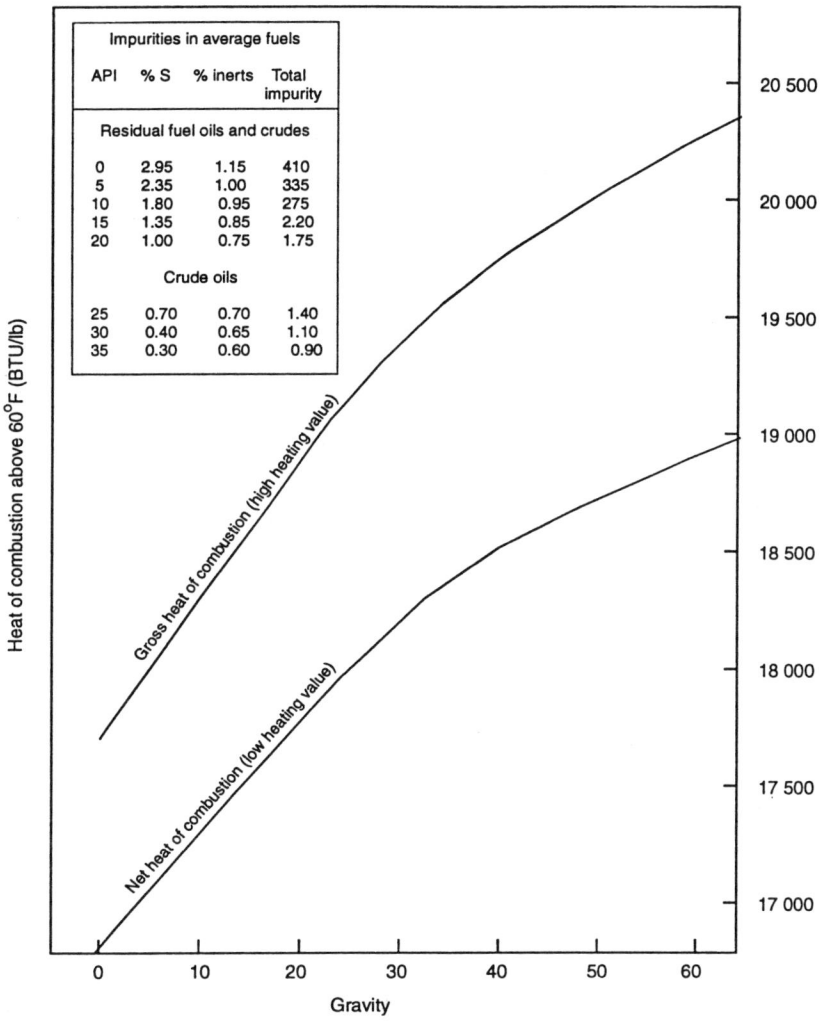

Figure A1.9 Heat of combustion of fuel oils and petroleum fractions
From: Maxwell, Data Book on Hydrocarbons. Copyright by A. K. Krieger Publishing
Company, used with permission. All rights reserved

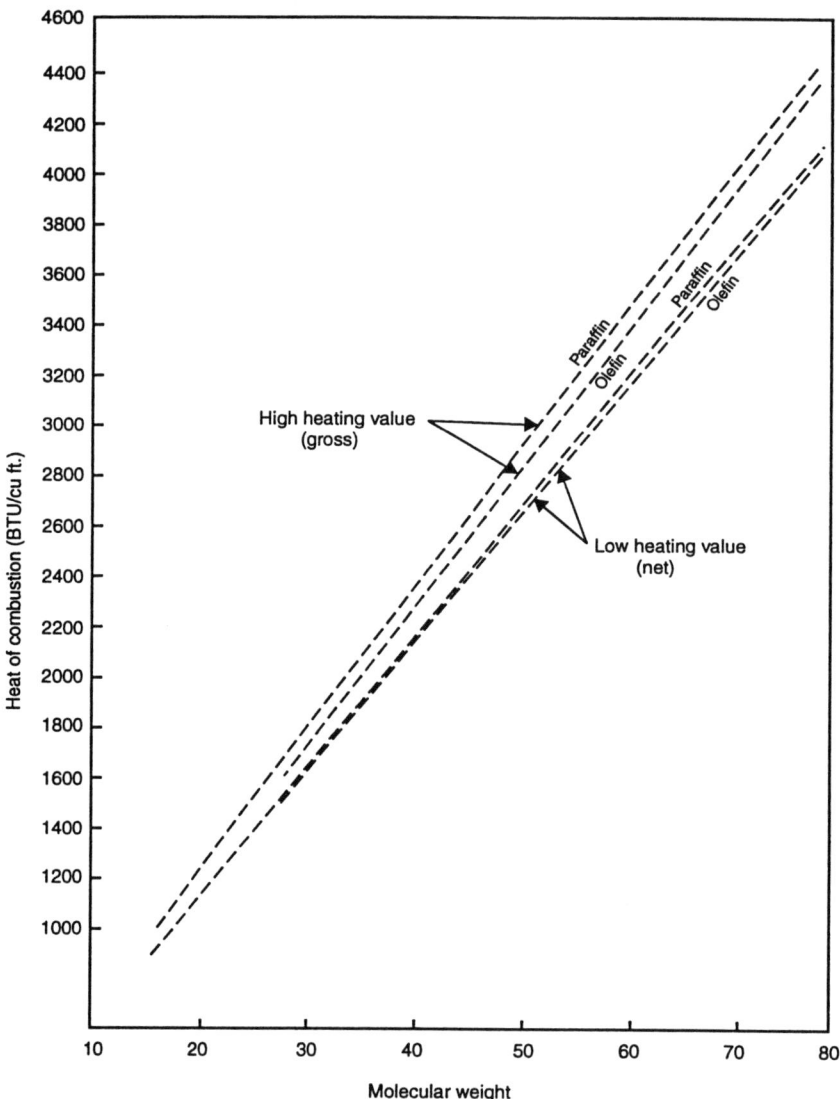

Figure A1.10 Heat of combustion of paraffin and olefin gases
From: Maxwell, Data Book on Hydrocarbons. Copyright by A. K. Krieger Publishing
Company, used with permission. All rights reserved

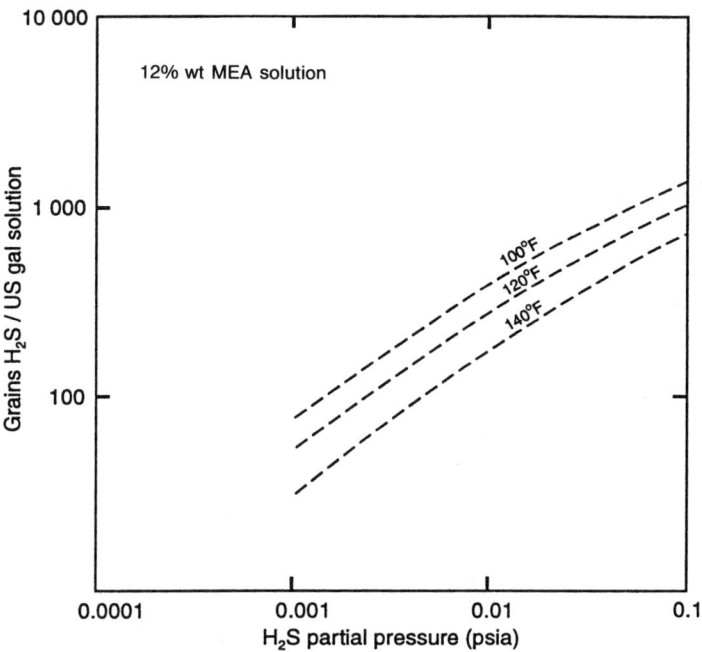

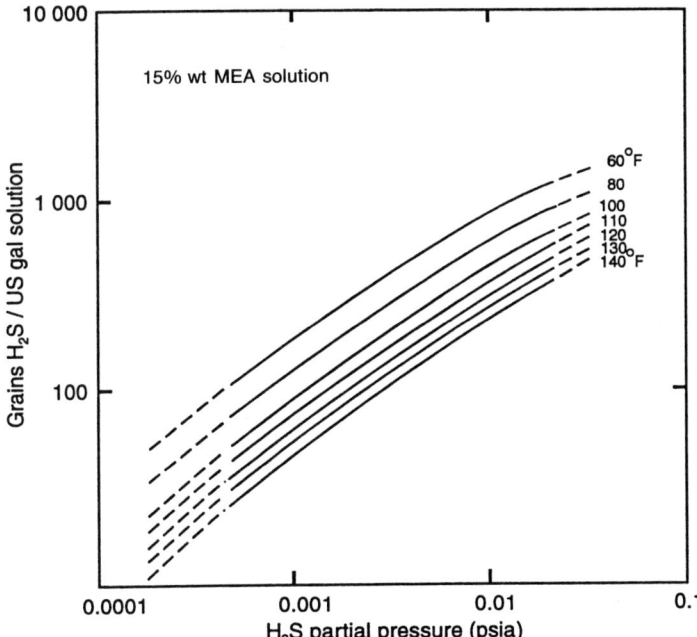

Figure A1.11 H_2S partial pressure in MEA solution

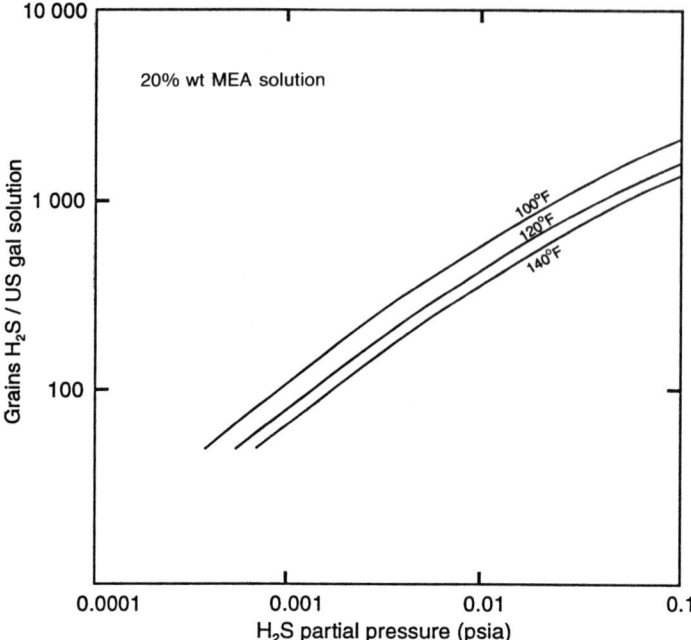

Note: $\left(\dfrac{\text{Moles } H_2S}{\text{Mole MEA}}\right) (325) \, (\text{wt \% MEA}) = \dfrac{\text{Grains } H_2S}{\text{U.S. gals}}$

Figure A1.11 (continued)

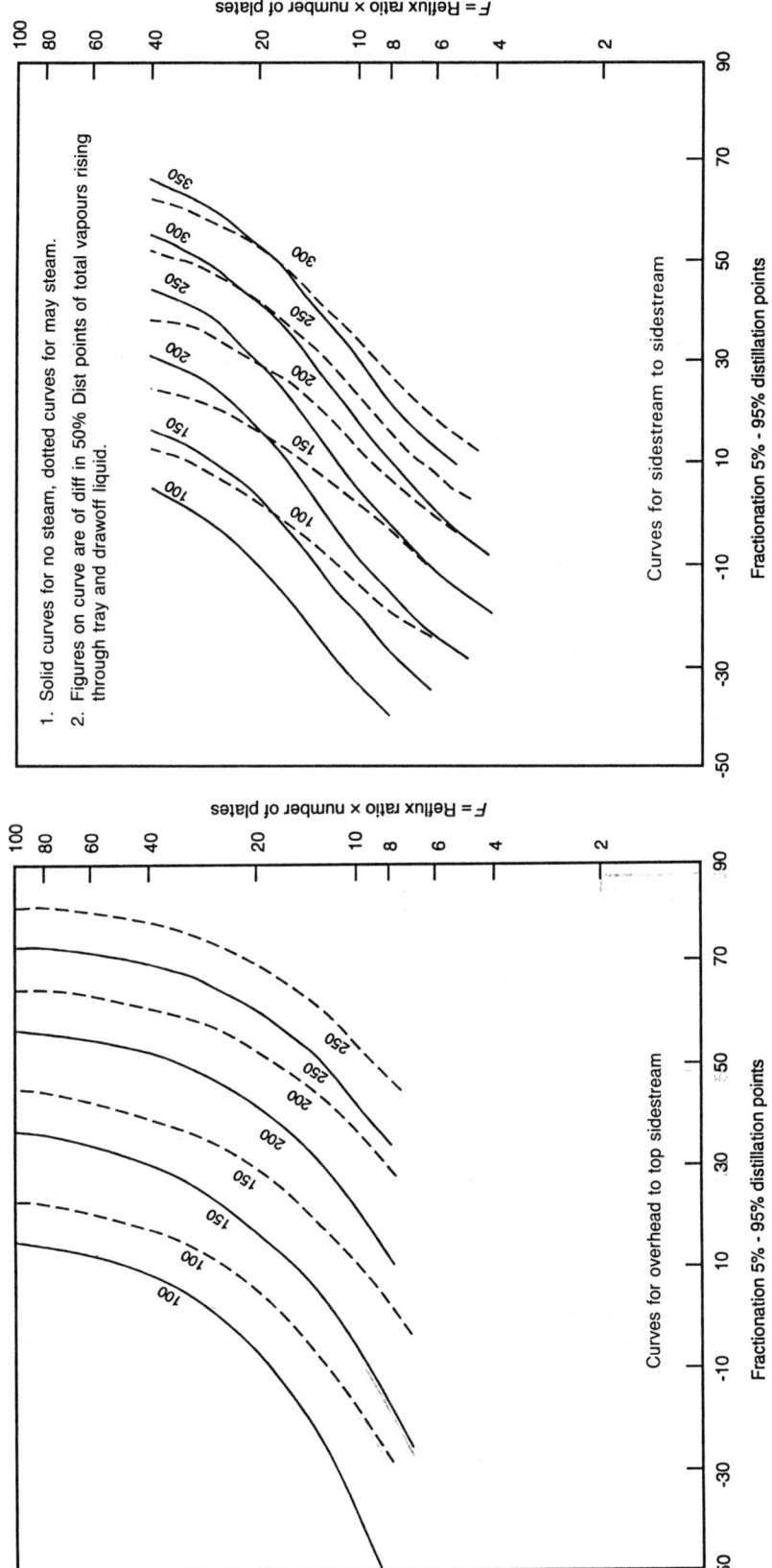

Figure A1.12 Fractionation curves

From: Packie Distillation Equipment in the Oil Refining Industry. *Am Inst of Chem Eng Transactions* **Vol 317** 1941. By permission of American Institute of Chemical Engineers. All rights reserved

Table A1.1 Weir lengths and downcomer area*

R	L	A	R	L	A	R	L	A	R	L	A	R	L	A	R	L	A
0.070	0.511	0.0308	0.120	0.650	0.0680	0.170	0.751	0.113	0.220	0.828	0.163	0.280	0.898	0.230	0.390	0.977	0.361
1	0.514	0.0315	1	0.652	0.0688	1	0.753	0.114	1	0.829	0.164	5	0.903	0.236	5	0.979	0.367
2	0.517	0.0321	2	0.654	0.0697	2	0.755	0.115	2	0.831	0.165	0.290	0.908	0.241	0.400	0.980	0.374
3	0.521	0.0328	3	0.657	0.0705	3	0.756	0.116	3	0.832	0.166	5	0.913	0.247	5	0.983	0.380
4	0.524	0.0335	4	0.659	0.0714	4	0.758	0.117	4	0.834	0.167	0.300	0.917	0.252	0.410	0.984	0.386
5	0.527	0.0342	5	0.661	0.0722	5	0.760	0.117	5	0.835	0.169	5	0.921	0.258	5	0.986	0.392
6	0.530	0.0348	6	0.663	0.0731	6	0.762	0.118	6	0.836	0.170	0.310	0.925	0.264	0.420	0.987	0.398
7	0.533	0.0355	7	0.665	0.0739	7	0.763	0.119	7	0.838	0.171	5	0.930	0.270	5	0.989	0.405
8	0.536	0.0362	8	0.668	0.0748	8	0.765	0.120	8	0.839	0.172	0.320	0.933	0.276	0.430	0.991	0.412
9	0.539	0.0368	9	0.670	0.0756	9	0.766	0.121	9	0.841	0.173	5	0.937	0.282	5	0.993	0.418
0.080	0.542	0.0375	0.130	0.672	0.0765	0.180	0.768	0.122	0.230	0.842	0.174	0.330	0.941	0.288	0.440	0.994	0.424
1	0.545	0.0382	1	0.674	0.0774	1	0.770	0.123	1	0.843	0.175	5	0.945	0.294	5	0.995	0.430
2	0.548	0.0389	2	0.677	0.0782	2	0.772	0.124	2	0.845	0.176	0.340	0.948	0.300	0.450	0.996	0.437
3	0.552	0.0396	3	0.679	0.0791	3	0.773	0.125	3	0.846	0.177	5	0.951	0.306	0.460	0.997	0.450
4	0.555	0.0403	4	0.682	0.0799	4	0.775	0.126	4	0.848	0.178	0.350	0.955	0.312	0.470	0.998	0.462
5	0.558	0.0410	5	0.684	0.0808	5	0.777	0.127	5	0.849	0.179	5	0.958	0.318	0.480	0.998	0.475
6	0.561	0.0418	6	0.686	0.0817	6	0.778	0.128	6	0.850	0.180	0.360	0.961	0.324	0.490	0.999	0.488
7	0.564	0.0425	7	0.688	0.0825	7	0.780	0.129	7	0.851	0.181	5	0.964	0.330	0.500	1.0	0.50
8	0.567	0.0432	8	0.691	0.0834	8	0.781	0.130	8	0.853	0.182	0.370	0.967	0.337			
9	0.570	0.0439	9	0.693	0.0842	9	0.783	0.131	9	0.854	0.183	5	0.969	0.343			
0.090	0.573	0.0446	0.140	0.695	0.0851	0.190	0.784	0.132	0.240	0.855	0.184	0.380	0.971	0.348			
1	0.576	0.0454	1	0.697	0.0860	1	0.786	0.133	5	0.860	0.190	5	0.977	0.354			
2	0.578	0.0461	2	0.699	0.0869	2	0.787	0.134	0.250	0.866	0.196						
3	0.581	0.0469	3	0.700	0.0878	3	0.789	0.135	5	0.872	0.202						
4	0.583	0.0476	4	0.702	0.0887	4	0.790	0.136	0.260	0.878	0.207						
5	0.586	0.0484	5	0.704	0.0896	5	0.792	0.137	5	0.883	0.213						
6	0.589	0.0491	6	0.706	0.0905	6	0.794	0.138									
7	0.592	0.0499	7	0.708	0.0914	7	0.795	0.139									
8	0.594	0.0506	8	0.710	0.0923	8	0.797	0.140	0.270	0.888	0.218						
9	0.597	0.0514	9	0.712	0.0932	9	0.798	0.141	5	0.893	0.224						

Table A1.1 (continued)

R	A	R	L	A	R	L	A
0.100	0.0521	0.150	0.714	0.0941	0.200	0.800	0.142
1	0.0529	1	0.716	0.0950	1	0.802	0.143
2	0.0537	2	0.718	0.0959	2	0.803	0.144
3	0.0545	3	0.720	0.0969	3	0.805	0.145
4	0.0553	4	0.722	0.0978	4	0.806	0.146
5	0.0561	5	0.724	0.0987	5	0.808	0.148
6	0.0568	6	0.726	0.0996	6	0.809	0.149
7	0.0576	7	0.728	0.1005	7	0.810	0.150
8	0.0584	8	0.729	0.1015	8	0.812	0.151
9	0.0592	9	0.731	0.102	9	0.813	0.152
0.110	0.0600	0.160	0.733	0.103	0.210	0.814	0.153
1	0.0608	1	0.735	0.104	1	0.816	0.154
2	0.0616	2	0.737	0.105	2	0.817	0.155
3	0.0624	3	0.738	0.106	3	0.819	0.156
4	0.0632	4	0.740	0.107	4	0.820	0.157
5	0.0640	5	0.742	0.108	5	0.822	0.158
6	0.0648	6	0.744	0.109	6	0.823	0.159
7	0.0656	7	0.746	0.110	7	0.824	0.160
8	0.0664	8	0.747	0.111	8	0.826	0.161
9	0.0672	9	0.749	0.112	9	0.827	0.162

$$R = \frac{\text{Downcomer Rise}}{\text{Diameter}} \qquad \frac{r}{\text{Dia.}} \qquad L = \frac{\text{Weir length}}{\text{Diameter}} = \frac{l_o}{\text{Dia.}} \qquad A = \frac{\text{Downcomer area}}{\text{Tower area}} = \frac{A_D}{A_S}$$

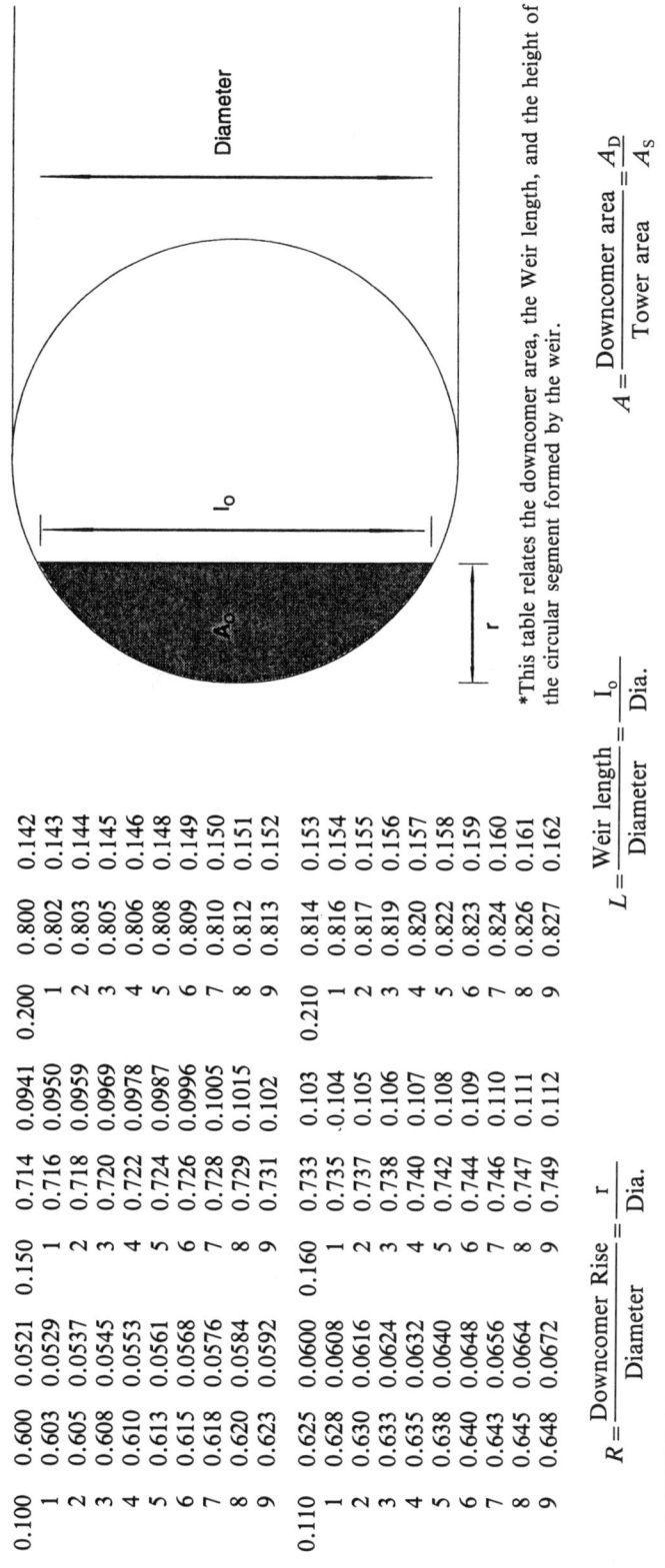

*This table relates the downcomer area, the Weir length, and the height of the circular segment formed by the weir.

APPENDIX 2: A TYPICAL CRUDE ASSAY (KUWAIT CRUDE)

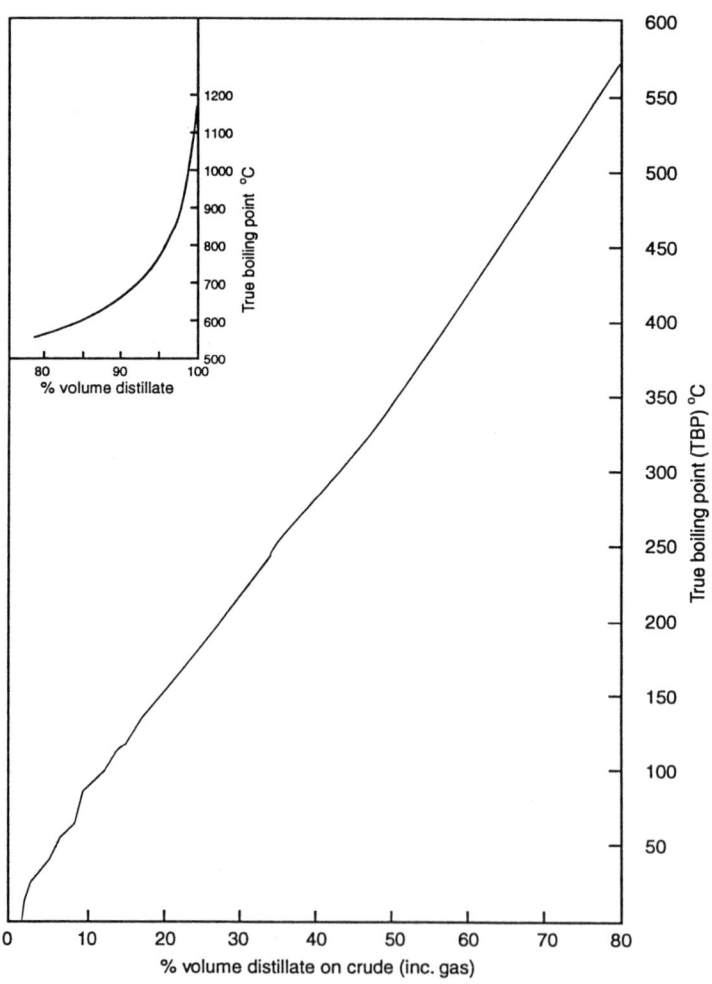

Figure A2.1 True boiling point curve vs percent volume distilled. (Kuwait crude SG. 0.8685)

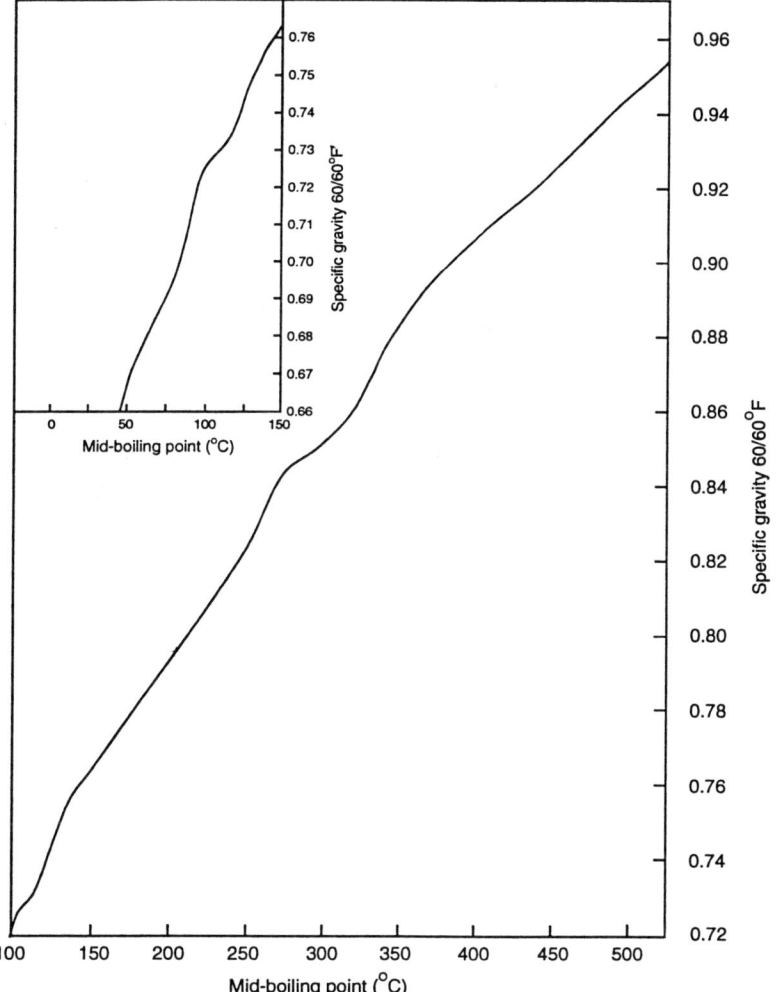

Figure A2.2. Specific gravity vs mid boiling point of fractions

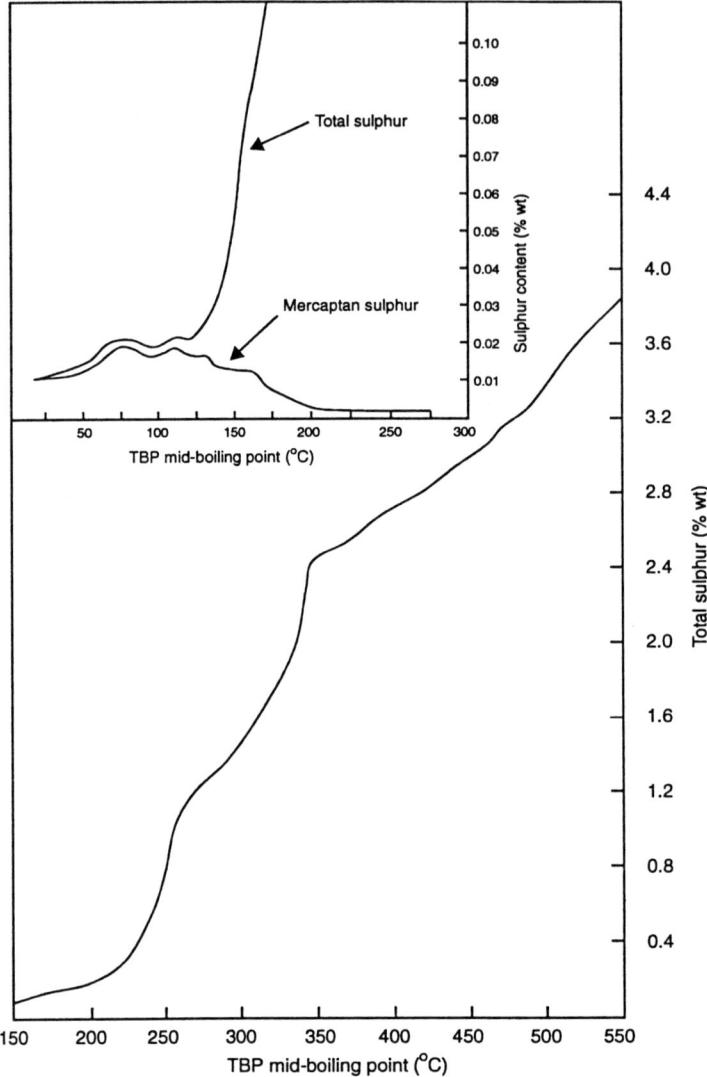

Figure A2.3 Total sulphur % weight vs mid boiling point of fractions

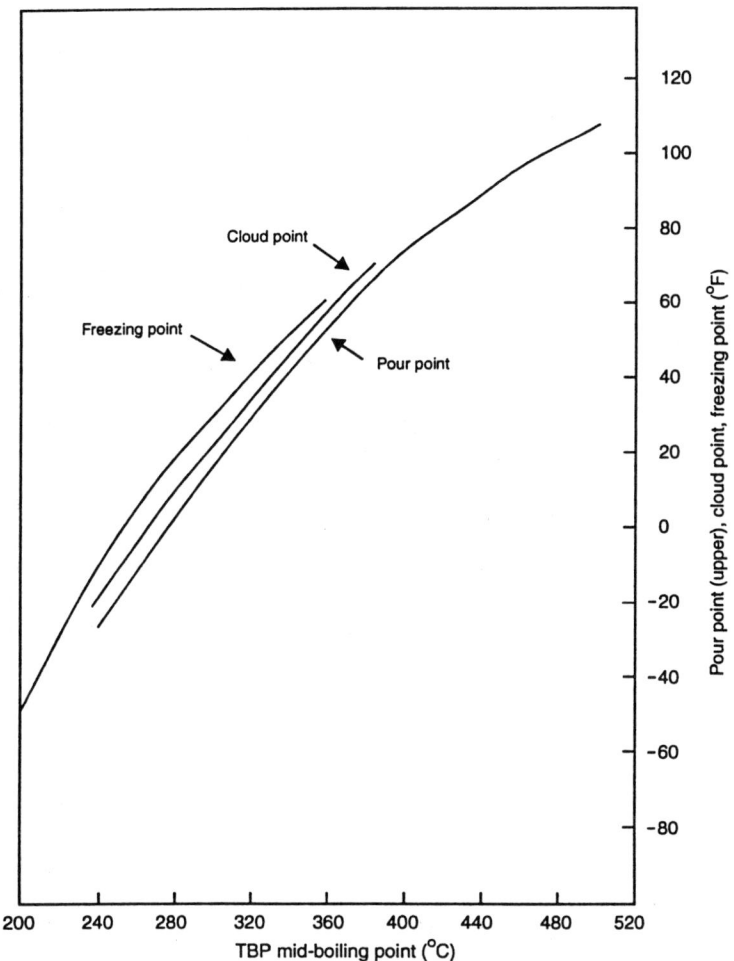

Figure A2.4 Pour point, cloud point and freezing point vs mid boiling point of fractions

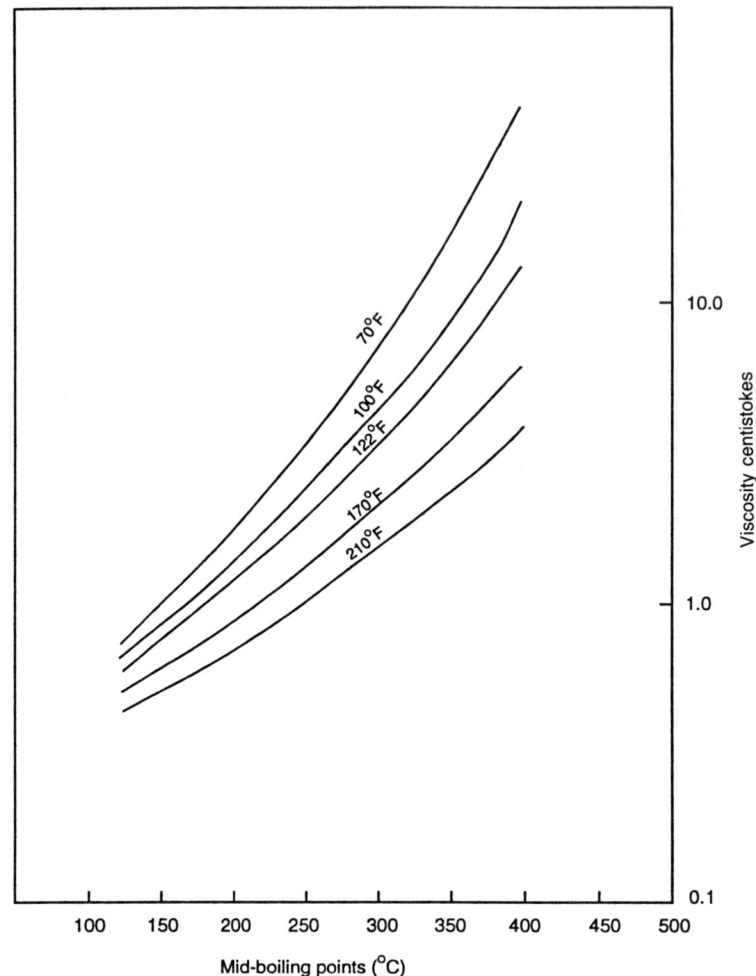

Figure A2.5 Viscosity at various temperatures vs mid boiling point of fractions

APPENDIX 3: TYPICAL PRODUCT QUALITY SPECIFICATIONS

Specifications for Liquified Petroleum Gas

Propane (C_3) LPG product:

Vapour pressure at 100 °F	210 psig (max.)
Residue (by NGAA Test)	2 vol. % (max.)
Volatile sulphur	15 grains/100 Scf (max.)

Butane (C_4) LPG product:

Vapour pressure at 100 °F	70 psig (max.)
95% boiling point	34 °F (max.)
Volatile sulphur	15 grains/100Scf (max.)

Specifications for Motor Gasolines

	Regular	Premium
ASTM distillation		
Minimum of 10% rec'd at °F	158	167
Minimum of 50% rec'd at °F	257	284
Minimum of 90% rec'd at °F	356	392
RVP at 100 °F psig (max.)	10	10
Octane number (Res.) clear	83–92	Above 92
Octane number (Motor) clear	76–85	Above 85
Copper strip (max.)	No. 1	No.1
Gum (max.) mg/100 ml	5	5

Specifications for Commercial Kerosene

Flash point—Abel	115 °F (min.)
ASTM end point	572 °F (max.)
Sulphur (total)	0.13% wt (max.)
Colour Saybolt	+21
Burning test	16 h (min.)
Cloud point	5 °F (max.)

Specifications for Automotive Diesel

Based on ASTM D 975-53T.

Flash point—Pensky–Martens	125 °F (min.)
Pour point	10 °F (max.)
Carbon residue (on 10% residue)	0.35%wt (max.)
ASTM distillation—90% point	675 °F (max.)
Sulphur	0.7%wt (max.)
Viscosity at 100 °F	1.8 cSt (min.)
Viscosity at 122 °F	5.8 (max.)

Specifications for Fuel Oil

Based on ASTM D 396.

	No. 5 fuel	No. 6 fuel
Flash point (PM closed) °F	130 (min.)	150 (min.)
Viscosity at 100 °F cSt	32 (min.)	—
Viscosity at 122 °F cSt	82 (max.)	91 (min.)
		650 (max.)

Specifications for Marine Diesel

Flash point (PM closed) °F	150 (min.)
Carbon residue	1.5 (max.)
Sulphur % wt	1.8 (max.)
Cetane number	30–58
Viscosity at 100 °F cSt	1.7–11.5
ASTM distillation 90% rec'd at °F	600–725

REFERENCES

1 Good, Connel, et al. *Oil and Gas Journal* 30th Dec 1944.
2 Thrift. *Oil and Gas Journal* 4th Sept 1961.
3 Maxwell. *Data Book on Hydrocarbons*.
4 GPSA 9th Edition. *Engineering Data Book*.
5 Packie, JW. *Am. Inst. Chem. Eng.* December 1940.
6 Edmister, W. *Applied Thermodynamics*.
7 Fenske. *Ind. Eng. Chem.* 24 1932.
8 Underwood. *Chem Eng. Progress* 44 603 1948.
9 Gilliland. *Ind. Eng. Chem.* 32 1940
10 "Volume of Aromatics in Reformates". *Chem. Eng. Progress* Vol. 55 No. 6 June 1959.
11 T. A. Cooper and W. P. Ballard *Advances in Petroleum Chemistry and Refining*.
12 Nelson. *Petroleum Refining Third Edition*.

INDEX